TOPOLOGICAL DATA ANALYSIS FOR GENOMICS AND EVOLUTION
Topology in Biology

Biology has entered the age of Big Data. A technical revolution has transformed the field, and extracting meaningful information from large biological data sets is now a central methodological challenge. Algebraic topology is a well-established branch of pure mathematics that studies qualitative descriptors of the shape of geometric objects. It aims to reduce comparisons of shape to a comparison of algebraic invariants, such as numbers, which are typically easier to work with. Topological data analysis is a rapidly developing subfield that leverages the tools of algebraic topology to provide robust multiscale analysis of data sets. This book introduces the central ideas and techniques of topological data analysis and its specific applications to biology, including the evolution of viruses, bacteria and humans, genomics of cancer, and single cell characterization of developmental processes. Bridging two disciplines, the book is for researchers and graduate students in genomics and evolutionary biology as well as mathematicians interested in applied topology.

RAÚL RABADÁN is a Professor at Columbia University, New York. He is Director of the Program for Mathematical Genomics at Columbia University, New York, and the NCI Physics and Oncology Center for Topology of Cancer Evolution and Heterogeneity. Dr Rabadán received his Ph.D. in Theoretical Physics in 2001 and went on to conduct research at the European Laboratory for Particle Physics (CERN) in Switzerland, and at the Institute for Advanced Study (IAS) in Princeton, New Jersey. At Columbia University, he leads a highly interdisciplinary laboratory with researchers from the fields of mathematics, physics, computer science, engineering, and medicine, with the common goal of solving biomedical problems through quantitative computational models.

ANDREW J. BLUMBERG is a Professor in the Department of Mathematics at the University of Texas, Austin. He completed his Ph.D. at the University of Chicago under the supervision of Peter May and Michael Mandell, and was later a National Science Foundation postdoctoral fellow at Stanford. He also spent a year as a member at the Institute for Advanced Study (IAS) in Princeton, New Jersey. His pure mathematics research focuses primarily on homotopy theory and algebraic topology and his applied research focuses on the development of topological and geometric techniques for studying genomic data.

TOPOLOGICAL DATA ANALYSIS
FOR GENOMICS AND EVOLUTION

Topology in Biology

RAÚL RABADÁN

Columbia University, New York

ANDREW J. BLUMBERG

University of Texas, Austin

CAMBRIDGE
UNIVERSITY PRESS

CAMBRIDGE
UNIVERSITY PRESS

University Printing House, Cambridge CB2 8BS, United Kingdom

One Liberty Plaza, 20th Floor, New York, NY 10006, USA

477 Williamstown Road, Port Melbourne, VIC 3207, Australia

314–321, 3rd Floor, Plot 3, Splendor Forum, Jasola District Centre, New Delhi – 110025, India

79 Anson Road, #06–04/06, Singapore 079906

Cambridge University Press is part of the University of Cambridge.

It furthers the University's mission by disseminating knowledge in the pursuit of
education, learning, and research at the highest international levels of excellence.

www.cambridge.org
Information on this title: www.cambridge.org/9781107159549
DOI: 10.1017/9781316671665

First published 2020

Printed in the United Kingdom by TJ International Ltd. Padstow Cornwall

A catalog record for this publication is available from the British Library.

Library of Congress Cataloging-in-Publication Data
Names: Rabadán, Raúl, author. | Blumberg, Andrew J., author.
Title: Topological data analysis for genomics and evolution : topology in
biology / Raúl Rabadán, Columbia University, New York, Andrew J.
Blumberg, University of Texas, Austin.
Description: Cambridge, United Kingdom ; New York, NY : Cambridge University
Press, 2019. | Includes bibliographical references and index.
Identifiers: LCCN 2019002342 | ISBN 9781107159549 (hardback : alk. paper)
Subjects: LCSH: Bioinformatics – Mathematical models. | Computational biology.
| Mathematical analysis.
Classification: LCC QH324.2 .R33 2019 | DDC 570.285–dc23
LC record available at https://lccn.loc.gov/2019002342

ISBN 978-1-107-15954-9 Hardback

This book is dedicated to our families, for their persistent support.
To Jean-Michel, Emma, and Alex.
To Olena, Miriam, and Becky.

Contents

Contributors

Andrew J. Blumberg
University of Texas, Austin

Pablo G. Cámara
University of Pennsylvania, Philadelphia

Joseph Chan
Memorial Sloan-Kettering Cancer Center, New York

Kevin Emmett
Columbia University, New York

M. Riley Meth
University of Texas, Austin

Raúl Rabadán
Columbia University, New York

Daniel Rosenbloom
Columbia University, New York

Preface

Modern biology is awash in data. This situation, the result of a technical revolution in high-throughput genomics, promises rapid scientific advances. However, analyzing the data poses unique challenges. Unlike in physics, there is usually no quantitative biological model that can guide investigation and generate precise predictions; often, we do not even know what the relevant quantities are that could capture the essential behavior of the biological system.

In response to the flood of data, the use of clustering algorithms and dimensionality reduction procedures is now ubiquitous. These families of techniques can be regarded as efforts to describe the *shape* of the data set. Although there have been noted successes, such methods provide only crude descriptions of this shape. The power of these tools, as well as their evident limitations, makes it clear that there would be substantial scientific benefit from richer and more robust methods for understanding geometric structure in data.

Algebraic topology is a well-established branch of pure mathematics that studies qualitative descriptors of the shape of geometric objects. Roughly speaking, the goal of algebraic topology is to reduce questions about comparing shapes to questions about comparing algebraic invariants (e.g., numbers), which are typically easier to solve. Moreover, algebraic topology has had a long tradition of employing combinatorial models of geometric objects, *simplicial complexes*, that are well suited to algorithmic computation.

Topological data analysis is a rapidly developing subfield that leverages the tools and outlook of algebraic topology to provide a methodology for analyzing the shape of data sets. The basic strategy is to assign a family of simplicial complexes to a data set; invariants of the complexes integrate information about the shape of the data across different feature scales.

Our aim in this book is to provide a concise introduction to the central ideas and techniques of topological data analysis and to explain in detail a number of specific applications to biology. We imagine as our idealized readers a modern quantitative biologist or a graduate student in mathematics with a background in topology or geometry and an interest in applied problems. We have three central goals:

1. to equip the modern quantitative biologist with techniques from topological data analysis,
2. to direct mathematicians with training in geometry and topology towards problems of interest to biologists, and
3. to make it easier for mathematicians and biologists to communicate and collaborate.

These goals pose an expositional challenge, as we expect two quite different audiences with different backgrounds. To address this, we have attempted as much as possible to provide a self-contained introduction to the relevant topics along with abundant and detailed references. We assume that the reader has some familiarity with calculus, linear algebra, elementary probability, and basic statistics.

The first part of this book presents the mathematical background necessary to understand topological data analysis and then provides an overview of techniques in the area. These chapters are intended to be read in order, as each one builds on the previous chapters. The second part of this book consists of a collection of distinct biological applications; each chapter can be read independently.

Acknowledgements

This work grew out of the efforts of many people. We would like to thank Arnold Levine, for his vision, his scientific insights, and his enthusiasm. He created an exceptional creative interdisciplinary environment at the Institute for Advanced Study in Princeton, providing the seeds of many of the ideas discussed in this book. Pablo G. Cámara, Joseph Chan, Kevin Emmett, and Daniel Rosenbloom contributed to several sections in the initial draft of the book. M. Riley Meth made many invaluable corrections and contributions to the second draft of the book. Juan Patino Galindo provided feedback on using genomic data for studying evolutionary processes, and, in particular, helped to write an introduction on different methods to study recombination. We are particularly thankful to Timothy Chu, Oliver Elliott, and M. Riley Meth for using their artistic talents to design the illustrations that enliven the book. William Blumberg and Michael Walfish provided careful readings and helpful comments on previous drafts. Jacqueline Aw, Kyle Bolo, Andrew Chen, Ioan Filip, Chioma Madubata, Patrick Van Nieuwenhuizen, Samuel J. Resnick, Richard T. Wolff, and Sakellarios Zairis proofread different sections of the book. Michael Lesnick and Jun-Hou Fung gave the entire book a very careful reading and made numerous helpful comments correcting errors and improving the exposition. The authors gratefully acknowledge many interesting discussions with Nils Baas, Gunnar Carlsson, Ben Greenbaum, Gillian Grindstaff, Hossein Khiabanian, Michael Lesnick, Arnold Levine, Michael Mandell, M. Riley Meth, Bud Mishra, Anthea Monod, Sayan Mukherjee, Vladimir Trifonov, Stephen Walker, and Jiguang Wang. In addition, Raúl Rabadán would like to acknowledge many of his collaborators in biology for the time shared, their patience, and the fun solving many problems together: Uttiya Basu, Riccardo Dalla Favera, Adolfo Ferrando, Antonio Iavarone, Anna Lasorella, Tom Maniatis, Do-Hyun Nam, Gustavo Palacios, Teresa Palomero, Laura Pasqualucci, Abbas Rizvi, and Sagi Shapira among many others.

This book was possible in part due to the funding from the Center for Topology and Evolution of Cancer at Columbia University through the National Cancer Institute (U54 CA193313). The Center brings together mathematicians and cancer biologists to solve some interesting problems in cancer. This book was born from many interesting interactions between mathematicians, computational biologists and cancer researchers, where with more or less success, but always with enthusiasm, we have tried to cross the interdisciplinary borders that separate our disciplines. In addition, Raúl Rabadán would like to acknowledge the National Institute of Health grants, R01 CA179044, R01 GM109018, R01 CA185486 and U54 CA209997, and the Convergence program of Stand Up to Cancer together with National Science Foundation. Both authors acknowledge the National Institute of Health grant R01 GM117591. Andrew Blumberg would also like to acknowledge AFOSR research grant FA9550-15-1-0302.

Introduction

In the long history of humankind (and animal kind, too) those who learned to collaborate and improvise most effectively have prevailed.

Charles Darwin

Knowing is not enough; we must apply. Willing is not enough; we must do.

Johann Wolfgang von Goethe

This book is about the application of algebraic topology to the problem of organizing and describing biological data. The problems this book studies are of recent origin. For much of its history biology was a predominantly descriptive science with comparatively little interaction with mathematics. Explanations of mechanism took place at the level of entire organisms or cells. But over the last century, the development of molecular biology has transformed the field so that it is now data intensive and marked by increasing reliance on mathematics.

This shift began with the discovery of the elemental constituents and rules that govern biological systems at the molecular level. Early highlights included the determination of the structure of DNA, RNA, and proteins, the deduction of some of the processes of information transmission within the cell, and the identification of specific molecular mechanisms underlying particular biological processes.

For a long time, small amounts of data were hard-won in the laboratory; for example, many researchers were focused on elucidating the biological mechanisms of individual genes, the sequences of DNA that are translated into RNA and produce a functional product such as a protein. However, towards the end of the twentieth century, the rate of data production accelerated very rapidly and it became possible to study all the genes of a cell (the entire genome) at once. The publication of the first draft of the human genome [513] in 2001 was a milestone in this revolution, heralding transformation in almost every realm of biology.

An incomplete sampling of the subsequent progress on fundamental problems includes the enumeration of genomic variations in thousands of individuals [122], detailed molecular characterization of thousands of cancers [343], single cell characterization of tumors [401], study of developmental processes [504], and the elucidation of the three dimensional structure of DNA in the nucleus of cells [138, 330].

Mathematics has played a key role in the development of modern molecular biology. The amazing progress in data collection depended in part on the development of mathematical algorithms that supported the assembly of raw DNA sequencing information and enabled the search for genes in the sequences. The development of and continued research on these algorithms is a fascinating and deep story, but it is not the focus of our inquiry here. Rather, we will study mathematical tools for determining the structure of biological processes and mechanisms from the data.

Analyzing biological data is a difficult problem. There is a large amount of data, and it is particularly challenging to work with: high-throughput genomic and transcriptomic data typically resides in very high-dimensional spaces (e.g., on the order of the number of genes in the organism, which can be in the tens of thousands), is frequently extremely noisy, and often reflects poorly understood systematic errors. For example, genetically similar organisms or cells can display different molecular profiles (e.g., present different epigenetic states, express different genes) leading to markedly different experimental measurements.

In short, modern biology has become a data rich discipline, dependent on sophisticated mathematical techniques for both the production of experimental data and its interpretation. In this way, it exhibits kinship with modern physics. But in contrast to the situation in physics, the mathematical models we have to understand genomic processes are in general less descriptive and provide fewer conceptual benefits than the models of physics. One problem is that the immense complexity of fundamental biological systems means that we simply lack good theoretical frameworks to describe them. For example, the enormously complicated cycle controlling gene expression is still not completely understood. Even our knowledge of the basic objects of study is incomplete; we hear almost daily that a new noncoding gene has been identified or that a novel viral species has been associated to a newly reported disease.

The point of departure for this book is a concrete manifestation of this lack of models: to date there has been no real analogue in biology for the role of geometry in physics. Geometry is at the heart of modern physics. This is no surprise; in a sense, modern geometry was invented to describe physical systems. Calculus was developed in order to describe the acceleration of moving bodies. Einstein's theory of general relativity can be succinctly summarized as the contention that gravity curves spacetime, which can be precisely and concisely expressed in the language of differential geometry. In stark contrast, biological data does not naturally appear to have the same kind of rich geometric structure. Typically, all one has is a collection of data points and various choices of a way of measuring the distance between them. Even if there was geometry present, it might be hard to see through the noise.

Our central dogma in this book is that although biological data might not possess the rigid geometric structure that arises in physics, it nonetheless has meaningful coarser geometry; we will broadly refer to this as *shape*. In some sense, this hypothesis is implicit in the standard approach for analyzing genomic data, namely dimensionality reduction and clustering. We can access the geometry of the situation through a distance function that takes as input a pair of data points and outputs a number (larger than 0) that reflects the distance between them. (Here distance is an abstract notion, not a measure of physical distance.) Dimensionality reduction refers to the process of using the distances to embed the data points (which might lie in a 10000-dimensional space) into a low-dimensional space (like the standard two dimensional Euclidean space $\mathbb{R}^2$) in such a way that distances are preserved as much as possible. Clustering refers to the process of grouping the data points into "clusters" such that points within a cluster are much closer to each other than to points in distinct clusters. Often these techniques are combined; clustering algorithms are applied to the results of dimensionality reduction, and we will sometimes refer to the combination as "clustering analysis."

Clustering genomic data has been a very successful way to detect genomic relationships with clinical consequences. In Figure 0.1, there is a representative example of a clustering analysis of mRNA expression data from pancreatic tumors. The data, obtained from samples from 147 patients, consists of vectors of numbers representing the expression levels for various genes. The distance between these expression vectors is roughly speaking a measure of similarity; tumor samples with similar expression profiles are close together. Then the data naturally breaks up into three clusters of points, as indicated in the plot on the left side of Figure 0.1. Each column represents the expression vector of a particular tumor sample; each row represents a particular gene. A point in the square thus encodes the level of expression of a gene in a sample – red means highly expressed, blue suppressed.

It is clear from the picture that points within a cluster have similar expression profiles, but more importantly, these clusters are clinically significant – which cluster a tumor sample is in predicts survival rates. Figure 0.2 graphs the survival curves for the different clusters; squamous pancreatic adenocarnicomas (cluster 2) have noticeably worse survival trajectories. That is, understanding the shape of the expression data as captured by clustering allows us to predict the likely progression of the cancer.

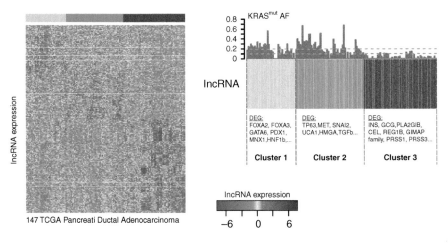

Figure 0.1 Using mRNA expression from long non-coding RNA from 147 patients with pancreatic adenocarcinomas, one can observe three different clusters. Cluster 2 is associated to squamous pancreatic adenocarcinomas. Different clusters reveal molecular mechanisms common to a set of patients. Source: [17]. Reproduced from *Gut*, Luis Arnes, Zhaoqi Liu, Jiguang Wang et al., Published Online First: 10 February 2018. © 2018. With permission from BMJ Publishing Group Ltd.

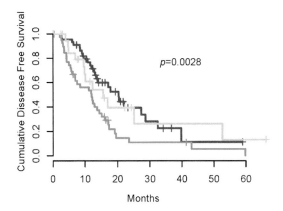

Figure 0.2 Different clusters of pancreatic adenocarcinomas have very different survival profiles. The *y*-axis represents the fraction of patients as a function of time. The colors represent the different clusters. Ideally, we would like to assess the prognosis of a patient based on molecular characteristics, and clustering patients constitutes a simple way of doing that. In addition to the clinical correlates, different clusters could reflect different molecular mechanisms that lead to the disease. Source: [17]. Reproduced from *Gut*, Luis Arnes, Zhaoqi Liu, Jiguang Wang et al., Published Online First: 10 February 2018. © 2018. With permission from BMJ Publishing Group Ltd.

More generally, dimensionality reduction and clustering methods such as PCA, MDS, spectral clustering, non-negative matrix factorization, and so forth are ubiquitous tools for analysis of genomic data. However, despite their successes, clustering algorithms capture only a very limited amount of information about shape – they are sensitive only to how many disconnected pieces a data set should be separated into. And this is often not enough – for example, there are many data sets of tumor samples where the points do not naturally separate into clusters which correlate with clinical outcomes.

Moreover, there are many other questions one can pose about the shape of a data set. For example, a natural question that arises when studying evolutionary phenomena is whether or not genomic data (for example, sequencing information from different flu viruses) can be represented by a tree structure, where the lengths of the branches correspond to distances between data points. To answer this question, an obvious approach is to attempt to determine if the points are better represented not by a tree but by a graph with loops. Such shape information cannot be extracted from clustering, and traditional dimensionality reduction algorithms tend to introduce distortions that obliterate this kind of shape.

Our aim in this book is to make the case that robust algorithms for capturing high-dimensional shape can be effective in situations where clustering fails. Specifically, we want to explain particular mathematical tools from algebraic topology that generalize clustering algorithms, giving rise to a methodology for extracting scientifically meaningful high-dimensional shape information from genomic data.

0.1 Why Algebraic Topology?

Modern algebraic topology was invented by Poincaré to provide tools for describing global properties of differential equations on surfaces. His basic insight was that the qualitative behavior of differential equations depended on the shape of the underlying surface. Algebraic topology studies qualitative and often global properties of geometric objects by constructing *algebraic invariants* of such objects.

By **geometric object**, we mean what we will refer to as a "space"; for the purposes of current discussion this means a subset of Euclidean space (e.g., the surface of a rubber band, or a sheet of paper, or a soda can). By **algebraic**, we mean something like a number or a vector space. By **invariant**, we mean something which is not changed by **smooth deformation**; stretching is allowed, but not tearing, as if we were studying things made out of soft clay. By **global**, we mean something that cannot be figured out by looking at a little piece of the object – one has to inspect the entire thing.

Let us begin with a very simple example. Suppose we want to answer the geometric question of distinguishing between two collections of non-overlapping solid blobs; given a collection of n solid blobs (referred to as the "disjoint union") and the disjoint union of k solid blobs in the plane $\mathbb{R}^2$, we want to decide if these pictures are the same or different. (See Figure 0.3 for a representative example.) An easy way to do this is to count the number of *path components* – the number of distinct pieces that cannot be connected by a path, i.e., a line drawn without removing the pencil from the page. For example, in Figure 0.3, the left hand shape has three path components and the right hand shape has five.

This count is a simple example of an algebraic invariant; it is a number, and it is not affected by smoothly deforming the blobs. Moreover, it is clearly a global quantity – just looking at a little piece of one of the blobs or even any finite subset of the blobs will not suffice to compute it. And using this count allows us to distinguish between geometric objects simply by the algebraic operation of comparing two numbers. Notice that counting path components feels very reminiscent of clustering! And as we will explain, there is a precise relationship between these procedures.

The count of path components is a fairly crude invariant of a space. But there are many more sophisticated invariants which can detect more interesting properties of the shape of a geometric object. Figure 0.4 shows a more difficult version of the question about blobs: how can we distinguish between a circle and a figure-eight?

Figure 0.3 On the left, there are three path components, on the right, five path components.

Figure 0.4 A circle (or annulus) has a single hole; a figure-eight (or union of two annuli) has two holes.

Counting path components cannot distinguish these spaces; there is a single path component in each case. However, if we count the number of "holes" or closed loops, we see that the figure-eight has two holes whereas the circle has one hole. This is another global invariant, the first Betti number, which counts the number of "holes" enclosed by circles in a geometric object. Once again, notice that smoothly stretching the circle and the figure-eight will not change the Betti number.

These examples suggest that algebraic topology provides a powerful methodology for capturing robust global properties of the shape of geometric objects and turning them into algebra. But it is a priori not clear how to use these tools to study real data! The questions we have been discussing above have used spaces that are defined as infinite sets of points, most concisely specified by equations. This observation raises two important issues. First, one might worry that describing spaces in this way does not seem to be algorithmically tractable. Second, the data sets of biology are likely to be finite sets of isolated points – how can we associate a continuous space to a finite set?

0.2 Combinatorial Algebraic Topology

Conveniently, there is a long tradition in algebraic topology of studying *combinatorial models* of geometric objects. By *combinatorial*, we mean a description of a space using only discrete data. Such models are well suited to algorithmic computation. The most important kind of combinatorial model for the approach discussed in this book is the *simplicial complex*. We will give a precise definition of a simplicial complex in Section 1.8, but roughly speaking a simplicial complex should be thought of as a geometric object specified by gluing together a collection of points, line segments, triangles, and higher dimensional analogues called simplices. Simplicial complexes represent spaces up to continuous deformation; they are a satisfactory representation for computing topological invariants.

Figure 0.5 presents examples of the standard pieces (called simplices) that are glued together to form the space represented by a simplicial complex. A k-dimensional simplex has *faces* which are $(k-1)$-dimensional simplices; a simplicial

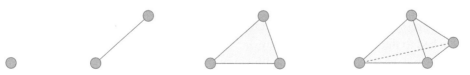

Figure 0.5 Simplicial complexes model spaces made by gluing together standard triangular pieces, called simplices; here we illustrate the 0-, 1-, 2-, and 3-dimensional simplices.

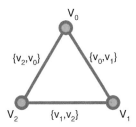

Figure 0.6 A simplicial model of the circle is given by gluing three 1-simplices together at their endpoints.

complex is made by gluing together standard simplices along their faces. For example, the faces of a 1-simplex are the two endpoints. The faces of a 2-simplex are the three edges of the triangle.

To describe a simplicial complex, one simply specifies the number of 0-simplices, 1-simplices, etc. as well as instructions for gluing them together. For example, a simplicial complex consisting only of 0-simplices and 1-simplices is specified by a collection of edges and instructions for attaching them at their endpoints – this is precisely the data of a graph, with 0-simplices the vertices and 1-simplices the edges. That is, a simplicial complex is precisely a higher dimensional generalization of a graph.

For example, consider the complex in Figure 0.6. We have 0-simplices $\{v_0, v_1, v_2\}$ and the 1-simplices $\{\{v_0, v_1\}, \{v_1, v_2\}, \{v_2, v_0\}\}$ where, for example, the 1-simplex $\{v_0, v_1\}$ has faces v_0 and v_1, and is thought of as a line segment connecting the vertices. We think of this complex as *representing* a circle; we are working with spaces up to continuous deformation, and a triangle can be stretched out into a circle.

Notice that there are other natural ways to represent a circle: one could use the "square" specified by the 0-simplices $\{v_0, v_1, v_2, v_3\}$ and the 1-simplices $\{\{v_0, v_1\}, \{v_1, v_2\}, \{v_2, v_3\}, \{v_3, v_0\}\}$.

Figure 0.7 shows how to produce a simplicial model of a solid disk (the circle plus its interior): we could take our first model of the circle above and add the 2-simplex with faces the 1-simplices; we can uniquely specify this 2-simplex as $\{v_0, v_1, v_2\}$. Here the 0-simplices and 1-simplices of the circle specify the *boundary* of the disk; the 2-simplex describes how the *interior* is glued to the boundary. An example of a more complicated simplicial complex is shown below in Figure 0.8; this represents a hollow ball with a circle attached to it (at the vertex v_4) and a line attached to the circle (at the vertex v_2).

The real payoff from working with simplicial complexes is that algebraic invariants of the spaces they represent can be algorithmically computed directly from the combinatorial description. The prototypical example of an algebraic invariant associated to a simplicial complex is the *Euler characteristic*. Suppose that we

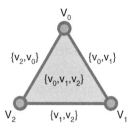

Figure 0.7 A simplicial model of the solid disk is given by gluing three 1-simplices together at their endpoints and gluing a 2-simplex to them at its faces.

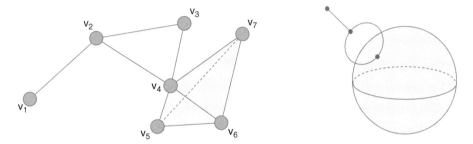

Figure 0.8 The simplicial complex on the left is made by gluing together the standard simplices. Combinatorially, we would write this as having 0-simplices $\{v_1, v_2, v_3, v_4, v_5, v_6, v_7\}$, 1-simplices $\{\{v_1, v_2\}, \{v_2, v_3\}, \{v_2, v_4\}, \{v_3, v_4\}, \{v_4, v_5\}, \{v_4, v_6\}, \{v_4, v_7\}, \{v_5, v_6\}, \{v_5, v_7\}, \{v_6, v_7\}\}$, and 2-simplices $\{\{v_4, v_5, v_6\}, \{v_5, v_6, v_7\}, \{v_4, v_6, v_7\}, \{v_4, v_5, v_7\}\}$. On the right is a space represented by this complex.

have a simplicial complex with V vertices, E 1-simplices, and F 2-simplices and no higher simplices. Then the Euler characteristic is $V - E + F$. In general, the Euler characteristic of a simplicial complex is the alternating sum of the numbers of k-simplices.

For example, the Euler characteristic of a point is clearly 1. The Euler characteristic of a simplicial complex consisting of a single 1-simplex and its two endpoints is $2-1 = 1$. Next, consider the simplicial complex modeling the circle from the discussion above – this is a loop formed by the three vertices and three line segments. This complex has Euler characteristic $3 - 3 = 0$. If we take the model of the circle given by the "square," this also has Euler characteristic $4 - 4 = 0$, and in general any such model of a circle will have n vertices and n 1-simplices and hence Euler characteristic 0.

On the other hand, the Euler characteristic of the disk given by filling in the triangle with a 2-simplex is $3 - 3 + 1 = 1$. Notice that the Euler characteristic of the filled triangle is the same as the Euler characteristic of a point; the Euler characteristic is a topological invariant, as it is insensitive to smoothly crushing the triangle to the central point. Comparing the results for the triangle and the filled triangle, we observe that the Euler characteristic is detecting a topological property, namely that the loop has a hole in the middle.

We can also compute the path components of a space represented by a simplicial complex directly from the complex; this turns out to reduce to a standard problem in graph theory. More generally, one can compute many algebraic invariants directly from simplicial complexes – for instance, the problem of counting holes (i.e., the Betti numbers) can be transformed into an elementary problem in linear algebra, as we shall see in Section 1.10. In summary, provided that we can represent our data using an appropriate simplicial complex, we can apply the computational tools of algebraic topology.

0.3 Topological Data Analysis (TDA)

The kind of biological data we will work with is typically presented as a finite set of points equipped with some kind of distance or dissimilarity measure between the points; a mathematical model of this situation is a *finite metric space*, which is a set X of points equipped with a distance function ∂_X satisfying a few simple axioms. The central question is: given data presented as a finite metric space, how can we robustly produce a simplicial complex such that the algebraic invariants of the simplicial complex reflect the shape of the data? Often, we hypothesize that these points are samples from a probability distribution on some geometric object; Figure 0.9 gives an idealized picture of this situation.

Consideration of clustering guides us to an answer. To explain, we need to make the connection between clustering and components precise, via *single-linkage clustering*, which works as follows.

1. Fix a scale parameter ϵ.
1. Assign two points x and y to the same cluster if they are connected by a path of points (for some k)

$$x = x_0, x_1, x_2, \ldots x_{k-1}, x_k = y$$

such that each point x_i is within a distance ϵ of x_{i+1}.

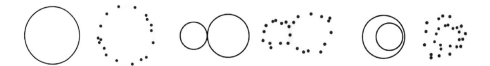

Figure 0.9 On the left, the underlying geometric "ground truth." On the right, finite samples from which we seek to recover the invariants of the circle, figure-eight, and nested circles.

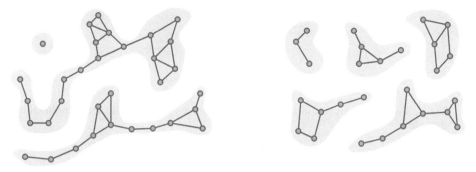

Figure 0.10 The clusters are the path components of the simplicial complexes.

We can interpret single-linkage clustering in terms of simplicial complexes: form the simplicial complex whose vertices are the data points and whose 1-simplices connect points x and y if they are less than a distance ϵ apart. Now the single-linkage clusters are precisely the path components of this simplicial complex; see Figure 0.10 for an illustration. But we can go even further – namely, we can add higher dimensional simplices when groups of points are close in some way. For instance, we could add a 2-simplex for every triple of points $\{x, y, z\}$ such that each pair $\{x, y\}$, $\{x, z\}$, and $\{y, z\}$ has distance less than ϵ. We then hope that the topological invariants of the resulting simplicial complex are capturing qualitative information about the shape of the data set.

This procedure has some attractive properties. Sufficiently small perturbations of the data typically result in small perturbations of the resulting simplicial complex that do not change algebraic invariants. Moreover, the simplicial complex constructed in this fashion reflects the sensible hypothesis that small measured distances between data points are likely to be accurate, but large distances are probably not accurate and should instead be estimated in terms of small distances. So intuitively speaking, it seems plausible that such topological invariants will be robust against certain kinds of noise and corruption, and will reflect real geometric structure of the data.

However, choosing ϵ correctly is difficult; this requires some knowledge of the feature scale of the data. It is illuminating to reflect on what happens to these simplicial complexes as ϵ increases; see Figure 0.11. When ϵ is very small, there are just discrete points (panel A). When ϵ is larger, the resulting simplicial complex has interesting geometric structure (panels C and D). And when ϵ is very large, everything is connected and there is no information recovered at all (panel E).

In the example above, it is not clear what the "correct" value of ϵ is, as the underlying topology is not evident. The best we can say is that there is a wide range of values for ϵ in which there is non-trivial topology. In simple cases, however, we

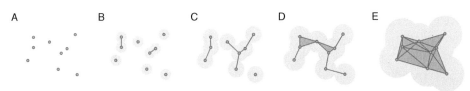

Figure 0.11 As ϵ grows, more and more simplices are added to the simplicial complex.

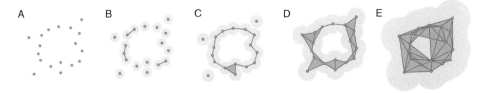

Figure 0.12 As ϵ grows, topological features appear. In panels C and D, the circle can be detected.

might hope to extract more precise topological hypotheses; we illustrate how this might work in Figure 0.12.

When ϵ is small, again the result is just discrete points (panel A). As ϵ grows, adjacent points begin to link up (panel B). But there is a wide range in which ϵ results in adjacent points along the circle being connected without connecting points across the circle (panels C and D); this is an illustration of the importance of privileging "short" distances over "long" ones. One way of looking at the situation is to observe that for these ϵ, distances between points that are less than ϵ accurately reflect distances along the circle. When ϵ is large enough, connections across the circle "short-circuit" the complex (panel E), and we eventually again obtain a completely connected complex.

As both of the preceding examples make clear, it is a priori very difficult to guess what the correct feature scale should be. There might be multiple scales at which we expect to see meaningful topological features, or it might even be the case that no single scale correctly encodes the salient features. A basic philosophy underlying topological data analysis is that scale issues should be handled simply by encoding the complexes for all ϵ simultaneously and keeping track of how they change as ϵ changes. This leads to a series of new algebraic invariants, which reflect the *persistence* of topological features across scales. By using these invariants, topological data analysis provides tools for robustly describing multiscale shape information of data.

In recent years, there has been an explosion of work in this area; however, many interesting problems remain to be solved. For instance, there are still many questions about the relationship between statistical practice and the invariants of topological data analysis. Nonetheless, part of our motivation for writing this book is that already the field is sufficiently mature for there to be many interesting applications to biological data. With this in mind, we now explain why topological data analysis is a potentially very useful tool to analyze biological data. We begin by explaining the kinds of biological problems that we will focus on in this book.

0.4 Genetics and Genomics

We will focus on biological questions arising from the perspectives of modern genetics (the study of genes, the fundamental units of heredity) and genomics (the study of genomes, the collections of all genes in an organism). These questions have been chosen to illustrate how topological data analysis can be used to address biological problems. Genetics and genomics are particularly amenable to the application of topological methods: there is a great need for mathematical tools to study the shape of large amounts of large scale experimental data, and the standard methods in use are comparatively crude.

At a high level, most of the problems in genetics can be posed in a simple fashion. There are two "spaces" of interest, the space of genotypes (the set of possible genomes) and the space of phenotypes (the set of observable characteristics of an organism that could occur in a particular environment); scientific questions are typically about describing and understanding a function that maps genotype into phenotype (e.g., see Figure 0.13). Such a function specifies which genetic alterations lead to a particular phenotype. Conversely, the function also determines the

$$\mathcal{G} \quad \text{x} \quad \mathcal{E} \quad \xrightarrow{f} \quad \mathcal{P}$$
$$f(\texttt{genotype, environment}) \longmapsto \texttt{phenotype}$$

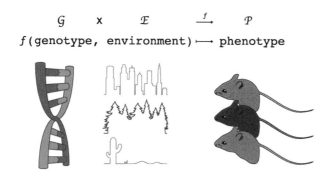

Figure 0.13 Many problems in genetics can be posed as the study of functions from the space of genes and genomes (the genotype) and the environment to the observable characteristics (the phenotype). Variation in genes and genomes between different organisms causes changes in observable characteristics, such as protein structure, protein function, disease survival, and many other potential phenotypes.

most interesting phenotypes to look at when studying variations in a particular gene. For example, cancer genetics studies the impact of mutations in cancer cells (genotypes) on clinical manifestations of cancer, notably on tumor growth, tissue invasion, and metastasis (phenotypes). To understand this relationship, one studies the molecular mechanisms of this association: how a mutation changes a protein, how this change affects the cell, and how such changes lead to the observed phenotypes.

At a very high level, the genome can be understood as a long word whose letters are the four nucleotide bases, denoted (A, C, G, and T or U, in the case of RNA). The length of this word varies dramatically across different organisms. The shortest, called viroids, are a few hundred bases. Humans have roughly three billion bases. And plant genomes can be two orders of magnitude larger (e.g., the genome of *Paris japonica*, a rare and beautiful plant from alpine regions in Japan, has a genome of 150 billion bases). The situation is further complicated by the fact that in multicellular organisms, such as humans, different cells will have similar but not necessarily identical genomes. Mathematically, different organisms can be regarded as producing distinct points in the genotype space, and so can different cells from a single organism.

However, the most interesting sources of variation come from the fact that genomes are not stable objects; they change over time. Specifically, errors occur when the genome of an organism is copied to produce offspring. The simplest types of mistakes are *point mutations*, where at a particular place in the genome one base is replaced with another one. However, more drastic changes can occur; sections of the genome can be lost or duplicated, or there can be more wholesale scrambling. In Figure 0.14 we show the typical order of magnitude of the size of the genome of different organisms along with the mutation rates (i.e., the probability of a point mutation at a particular spot per replication). Genome sizes and mutation rates vary by orders of magnitude; organisms with shorter genomes tend to be prone to mutations. A pervasive and more complicated phenomenon is that different organisms can exchange genomic information, resulting in new genomes which shuffle together the original genetic information. These processes produce clouds of points in the genome space; the problem is then to understand the relationship between these point clouds and resulting phenotypic changes.

However, the phenotype space is harder to specify and often more complex than the genotype space. Examples of phenotypic characteristics include the expression of different mRNAs, the expression of proteins, the shapes of these proteins, the shape of the cell, the ability to grow and replicate, the susceptibility to different stimuli, the ability to respond to an infection, and the size and weight of a multicellular organism. Obviously we do not expect an exhaustive enumeration of scientifically important observable characteristics. Instead, different areas of biology focus on specific choices of salient phenotype; for example, in evolutionary

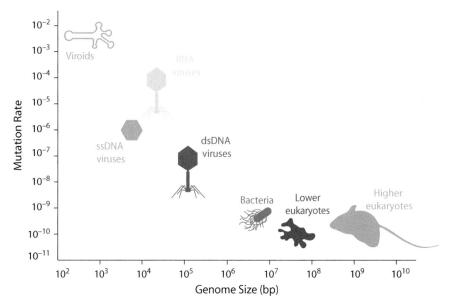

Figure 0.14 The size of genomes varies by many orders of magnitude. Viroids are small (a few hundred bases) sequences of free RNA that can infect plants. The genomes of RNA viruses (like influenza) are usually around 10,000 bases, the genomes of DNA viruses are typically 100,000 bases, and the genomes of bacteria can be millions of bases. Some plant genomes can reach 100 billion bases (*Paris japonica*). There is a fascinating relationship between the size of the genome and the number of mistakes per replication (mutation rate), represented here in the *y*-axis.

biology we might be interested in fitness or the ability to proliferate in some particular environment. In the context of tumors, proliferation, invasion of new tissues, and survival rate are all interesting phenotypes to study.

Finally, the environment is also an important factor determining when genetic variation will cause changes in the phenotype. Many genes are only expressed in certain circumstances, and beneficial alterations in one environment could be detrimental in another one. As a first approximation to reduce the complexity of the problem, it is common to fix or reduce the number of environmental factors to a few conditions that are suspected to be germane to the phenotype under study.

0.5 Why Is Topological Data Analysis Useful in Genomics?

Our contention is that topological data analysis provides novel and effective tools to attack the problem of inferring the relationship between genotypic events and changes in phenotype. For example, understanding the shape of the data in genome space reveals the way that certain phenotypical changes arise. To support our claim, in this book we describe a number of biological problems where the methods of topological data analysis reveal new and interesting phenomena. In each case, biological data is presented as a finite collection of points equipped with a distance, i.e., a finite metric space. Topological invariants of associated simplicial complexes then turn out to encode biologically relevant quantities. Here, we describe three illustrative examples (see Figure 0.15).

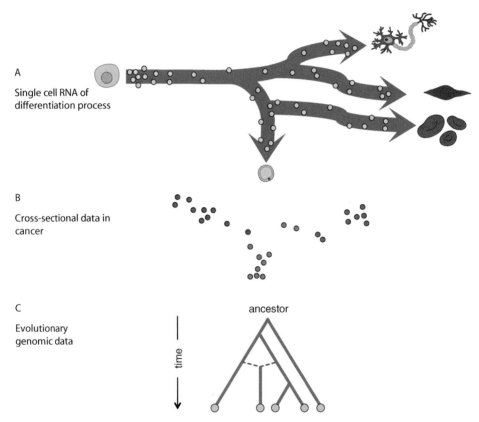

Figure 0.15 Examples of biological point cloud data: (A) Starting from stem cells, different cell types arise from a process of differentiation over time. Single cell approaches provide information about differentiation, where each point corresponds to a particular cell. Important questions include characterizing distinct subpopulations/expression programs/specific surface markers, and determining how cells decide their fate. (B) Each point represents the tumor of a patient. Questions in this space concern the classification of patients according to their molecular profile, association of location in this space with survival, determination of mechanisms of drug resistance, and the identification of specific pathways implicated in tumor progression. (C) Each dot (tree leaf) represents a genome. Traditionally, we expect evolutionary processes to be described as trees. But there are many examples of phenomena (e.g., recombinations) that do not fit into this framework. This raises the question of how to describe the relationship between genomes.

Our first example concerns the process of differentiation (panel A). A baby animal begins from a single cell that divides and differentiates to generate an incredibly complicated collection of organs, tissues, and cell types. Differentiation is usually represented as a branching process, with a root, the stem cell, giving rise to descendant cells of many different types. All of our cells share a common genome: a particular cell type is characterized by the *expression* of specific genes, i.e., by the amounts of messenger RNA (mRNA) or protein generated from each gene. Transcription, the generation of RNA copies from DNA, follows a carefully orchestrated program in which certain genes are turned on in consonance with other genes. This transcription program is regulated by proteins that control the expression of multiple genes; the regulation ensures that the right amounts of RNA are produced at the right times. Cells of similar type have related transcriptional programs and thus similar gene expression profiles – comparing the expression of genes of individual cells sampled along the process of differentiation can reveal the specific mechanisms that determine what type of cell will arise.

Until recently, studying the process of differentiation was complicated by the fact that experimental techniques commingled genomic information from many cells, each potentially in a different stage of differentiation. However, single cell expression technologies now allow the measurement of the transcriptional state of single cells throughout the differentiation process. The transcriptional state of a cell can be described by a vector $(e_1, e_2, \ldots, e_G)$, where e_i is a measure of the amount of mRNA produced from gene i and G is the total number of genes measured. Given this data, we wish to characterize different cell states and types and infer the trajectory of the differentiation process. This problem can be formulated as reconstructing low-dimensional geometric structure from a sample of points (cells sequenced) in a high-dimensional ambient space (with dimension given by the number of genes). Both the Euclidean distance and the correlation between expression vectors can be used to provide metrics on transcription vectors, and clustering using these metrics has been applied to great effect. Framed in this fashion, the problem of analyzing differentiation using single cell data is clearly a potential application area for the tools of topological data analysis.

Our second example focuses on cancer (panel B). A cancerous tumor is the result of the accumulation of mutations that lead to uncontrolled cell growth; for example, mutations that alter the cell division cycle by reducing apoptosis (cell death), enhancing blood supply, and increasing generation or responsiveness to growth-promoting signals. Tissue samples from tumors in patients can be sequenced to identify these mutations and to try to determine how expression differs between tumors and normal tissue. While each individual tumor is the result of specific genomic alterations, almost all currently available therapies are generic. Moreover, it is often unclear why some tumors are cured by treatment whereas others progress. The goal of *precision medicine* is to provide doctors with guidance enabling the deployment of therapies tailored to a particular patient's tumors.

Given sequencing data from a variety of tumors, one can study the relationships between these tumors in the hope of finding improved tumor classifications, discovering specific molecular mechanisms of progression and drug resistance, and eventually providing specific therapeutic options based on the tumor's molecular profile. Sequencing data taken from the same tumor at different times gives insight into tumor progression and development. Traditional computational approaches cluster the data and use the clusters to try to characterize the spectrum of alterations, pathways, and the clinical characteristics (e.g., survival). However, often the structure of the data does not support unambiguous division into clusters; in this case, the task of grouping patients is very difficult, as the number of clusters becomes a matter of opinion and many of the samples remain unclassified. Better tools for understanding and characterizing the shape of the tumor data have the potential to provide valuable information about clinical relationships. As we explain, sophisticated geometric models of the space of tumors as well as simplicial complexes associated to the metric space of sequencing data reveal biologically meaningful structure invisible to clustering algorithms.

Our final example has to do with evolution (panel C). Darwin first proposed the phylogenetic tree as a means to represent the evolution of phenotypic attributes. Since then, methods in molecular phylogenetics have been developed to characterize evolutionary relationships between species. These approaches generally assume that genomic information is solely passed from parents to children.

However, it has long been known that more complex modes of genetic exchange can occur, including lateral gene transfer in bacteria, recombination and reassortment in viruses, viral integration in eukaryotes, and fusion of genomes of symbiotic species. These "horizontal inheritance" phenomena can cause serious concerns about the reliability of inference of evolutionary relationships.

For example, traditional phylogenetic classifications of microorganisms have relied on evolutionary relationships inferred from 16S ribosomal RNA, a highly conserved genomic region between bacteria and archea species. However, as this region accounts for under 1% of the complete genome in most species, the vast majority of genetic information is ignored. Since horizontal inheritance is pervasive, the remaining 99% of genes might tell a very different evolutionary story. This problem becomes acute in viruses that lack 16S or other universal genes. Such challenges underscore the need for approaches to describe the shape of evolutionary processes in a more general way, free from the constraints of the tree representation. As a first step, it would be very useful to have criteria to determine when tree representations are inadequate. The rapidly growing number of sequenced microbial genomes provides fertile ground for developing and testing new approaches to quantifying both vertical and horizontal evolutionary processes.

While recent developments in phylogenetic networks provide ways to identify instances of non-tree-like events, the field does not have a widely accepted framework to visualize and quantify the frequency, scale, and significance of horizontal evolution. Although phylogenetic trees can be visually complex, from a topological standpoint they are very simple mathematical objects: a tree is a simplicial complex with only 0-simplices and 1-simplices that contains no loops. In contrast, simple kinds of horizontal evolution can be represented by the presence of loops. Thus, computing Betti numbers of the simplicial complexes produced from sequencing data can detect horizontal evolution.

0.6 What Is in This Book?

This book is aimed at two distinct audiences: quantitative biologists interested in applying new mathematical tools to the study of genomics, and mathematicians and computer scientists interested in understanding geometric problems that arise in modern genetics and genomics. As a consequence, we have written neither a traditional mathematics textbook nor a standard biology textbook.

In the first part of the book, we begin by giving a rapid but comprehensive review of the mathematical background for topological data analysis (TDA). We state definitions and theorems, and provide many examples, but do not give proofs; our goal is to provide context for understanding the TDA framework and also to provide detailed references for the reader interested in achieving a deeper understanding. We assume that the reader has some familiarity with calculus, linear algebra, elementary probability, and basic statistics.

In Chapter 1, we give a brief introduction to the basic ideas of algebraic topology, including discussion of algebraic background (linear algebra and abstract algebra), basic point-set topology, simplicial complexes, and the construction of homology groups. In Chapter 2, we give an overview of topological data analysis, focused on the theory surrounding persistent homology. We review the machinery for understanding topological invariants of data sets in terms of associated simplicial complexes, explain persistent homology and the basic structural theorems, and describe the Mapper algorithm. In Chapter 3, we describe the emerging and active area of research integrating topological data analysis with the methods of statistics; this is a necessity for the use of these tools to analyze scientific data and perform inference. In Chapter 4, we give a brief overview of the area of manifold learning, which is closely related to topological data analysis, and review mathematical models of spaces of phylogenetic trees.

In the second part of the book, we explore some biological applications. In Chapter 5, we study the topology of point clouds in genomic space using persistent homology and the geometry of phylogenetic spaces. Specific examples include viruses (influenza and HIV), bacteria, and humans. Chapter 6 provides a concise introduction to cancer genomics; among the applications, we use topological data analysis to study the evolution of tumors in collections of patients, to describe the stratification of patients, and to capture the association between genomic data and sensitivity to diverse therapeutic agents. Next, in Chapter 7, we turn to a new type of data that is particularly well suited to TDA tools: expression profiles of large collections of single cells. In Chapter 8 we study the three dimensional structure of DNA using persistent homology, with examples from bacteria and human cells. Finally in Chapter 9 we use a mapping of time-series data into finite metric spaces to extract periodic features. Each of these chapters contains background information on the relevant biological problem and can be read independently.

Part I
Topological Data Analysis

1

Basic Notions of Algebraic Topology

> ... geometry is the art of reasoning well from badly drawn figures; however, these figures, if they are not to deceive us, must satisfy certain conditions; the proportions may be grossly altered, but the relative positions of the different parts must not be upset ...
>
> *Henri Poincaré*

Modern algebraic topology arose in order to provide quantitative tools for studying the "shape" of geometric objects without using distances. It assigns algebraic invariants (e.g., numbers) to geometric objects in a way that depends only on the relative, not absolute, positions of points. In this chapter, we motivate and introduce the basic ideas of algebraic topology. This material provides a conceptual framework for understanding the tools of topological data analysis and their application to real data. We do not provide a complete treatment, and in particular we omit proofs of the theorems. At the beginning of each section, we provide a reference to a comprehensive source for the material.

Although algebraic topology is not yet a standard tool in genomics, the study of shape is already ubiquitous – clustering techniques are widely used to analyze data in all domains of molecular biology. For example, we can represent the expression profile of genes in cancer patients as points in a high-dimensional Euclidean space. Patients that have similar expression profiles will have points that are close together. A clustering algorithm can then be employed to classify expression profiles of cancer patients and thereby illuminate some of the distinct molecular mechanisms underlying the disease.

Recall that a clustering algorithm assigns to a finite collection of points X equipped with a distance function ∂_X a partition of the points of X, i.e., a collection of subsets $C_i \subseteq X$ such that

1. the $\{C_i\}$ do not overlap, so $C_i \cap C_j = \emptyset$ for all $i \neq j$, and
2. together the $\{C_i\}$ cover all of X, so that $\bigcup_i C_i = X$.

These subsets C_i are the clusters. Typically, clustering algorithms seek to generate partitions so that points within a given cluster are closer together than points in distinct clusters.

A representative clustering algorithm, single-linkage clustering in Euclidean space, takes as input a set of points $X \subseteq \mathbb{R}^n$ and a fixed $\epsilon > 0$, and assigns points x and y to the same cluster if there is a path of points

$$x = x_0, x_1, x_2, \ldots x_{k-1}, x_k = y$$

such that $\|x_i - x_{i-1}\| < \epsilon$ for $1 \le i \le k$. (Here for $x, y \in \mathbb{R}^n$, $\|x - y\|$ denotes the Euclidean distance between the points x and y, see Example 1.3.6.) In other words, we connect points if they are closer than ϵ; clusters are groups of connected points.

The methodology of clustering is motivated by the same focus on relative information as in algebraic topology. Specifically, clustering is a useful technique for analyzing data in circumstances in which the data is very noisy, so relative information is more reliable than absolute information. In fact, the connection between clustering and algebraic topology is very close: as we shall see in Section 1.3.2, single-linkage clustering has an interpretation in terms of a standard topological invariant.

In contrast to clustering techniques, which typically work on a collection of separated points (referred to as a "point cloud"), algebraic topology has traditionally concerned itself with continuous objects with infinitely many points which can be arbitrarily close together, e.g., a sphere. A first question we might ask is "what is the continuous analogue of the clustering algorithm described above?" Roughly speaking, the answer to this question will be as follows: a "cluster" should consist of all points which can be connected by a smooth path.

In order to make sense of this, we need a precise definition of a geometric object and of a smooth path through a geometric object. In the continuous setting, this is done using the notion of a *topological space*. The study of basic properties of topological spaces is typically referred to as *point-set topology*. We begin by giving a little background about sets and then reviewing the concept of a metric space, which provides a rich source of examples of topological spaces.

Guide for the Reader

Our expositional choices in this chapter (and in this part of the book more broadly) are motivated by our belief that in order to safely use mathematical tools, it is important to understand where they come from and how they fit into a broader ideological context. As a consequence, we have not adopted the maximally streamlined approach (which might start directly with simplicial complexes) to mathematical background. Instead, we have endeavored to "start from the beginning," and give a rapid but thorough introduction to the ideas of algebraic topology.

On the other hand, we are aware that the volume of material below might pose challenges to the energy of readers who have less math background. For someone interested in a minimal path through this section, we might recommend skipping to Section 1.8 and reading prior material as necessary to proceed. Strictly speaking, only Sections 1.8 through 1.12 are required for the rest of the book. Nonetheless, we hope that there are some readers from biology who find the broader introductory material useful.

1.1 Sets

All of the mathematical objects we will study herein are built on top of sets. Although the construction of rigorous axiomatizations of set theory is subtle and complicated, we can get by with a fairly naive view of the foundations. An excellent textbook that covers the material we use (and more) is Halmos' *Naive Set Theory* [224].

We will regard a *set* as simply an unordered collection of objects, referred to as *members* or *elements*. We require that the elements of a set be unique. A *finite* set has finitely many elements; otherwise, the set is *infinite*.

Example 1.1.1.

1. The empty set, denoted $\emptyset$, is the set with no elements.
2. The integers $\mathbb{Z}$ is the set $\{\ldots, -5, -4, -3, -2, -1, 0, 1, 2, 3, 4, 5, \ldots\}$; an element of $\mathbb{Z}$ is a number. Similarly, the natural numbers $\mathbb{N} = \{0, 1, 2, \ldots\}$, the rational numbers $\mathbb{Q}$ consisting of the fractions $\{\frac{p}{q}\}$ where p and q are relatively prime, and the real numbers $\mathbb{R}$ are sets. Note that rigorously constructing the real numbers as a set is complicated; although informally we are used to working with them as decimals, the construction requires some machinery we will discuss below.
3. The Euclidean vector spaces $\mathbb{R}^n$ are sets; the elements are the vectors $(x_1, x_2, \ldots, x_n)$, where each $x_i \in \mathbb{R}$.
4. The collection of possible bases in a DNA strand, $\{A, G, C, T\}$, is a set.
5. The expression vectors from a collection of samples from a cancerous tumor form a set, e.g., a set of vectors $\{(30, 50, 10, \ldots), (10, 16, 29, \ldots), \ldots\}$, where each element is a vector and each entry in an element of the set is an expression value at a particular position on a gene.
6. In general, a finite set can be specified as a list of elements, e.g., $\{a, b, 4\}$, which has elements a, b, and 4. These elements could be specified by a condition, e.g., the set of people named "Harold" in New York.

7. In contrast, "tall people in Boston" does not describe a set; the term "tall" is not an adequately specific description by itself. On the other hand "living people over six feet tall in Boston" is a well-defined characterization of a set.

There are several familiar constructions of new sets from old that will be of particular relevance for our work. First, given a set X, we can form new sets by taking only certain elements from X; we have seen examples of this above.

Definition 1.1.2. A *subset Y* of a set X is a set Y such that every element $y \in Y$ is an element of X. We write $Y \subseteq X$ to denote a subset of X.

Second, given a finite set of sets $\{X_i\} = \{X_1, X_2, \dots, X_k\}$, we can form the set of tuples.

Definition 1.1.3. Let $\{X_i\}$ be a finite set of sets. The *Cartesian product* is the set specified as

$$\prod_i X_i = \{(x_1, x_2, \dots, x_k) \mid x_i \in X_i\}.$$

Example 1.1.4.

1. Almost by definition, the standard xy-plane $\mathbb{R}^2$ can be identified with the product $\mathbb{R} \times \mathbb{R}$,
2. and more generally

$$\mathbb{R}^n \cong \prod_{i=1}^n \mathbb{R}.$$

Given two sets X and Y, we can form the *union*

$$X \cup Y = \{z \mid z \in X \text{ or } z \in Y\}$$

and *intersection*

$$X \cap Y = \{z \mid z \in X \text{ and } z \in Y\}.$$

More generally, for a collection $\{X_i\}$ of sets we can form the union $\cup_i X_i$ or intersection $\cap_i X_i$ of all of them.

If S_1 and S_2 are sets, a function $f\colon S_1 \to S_2$ is a rule that produces an element of S_2 for each element of S_1. We often refer to functions between sets as *maps* or *maps of sets*. Given two maps $f\colon X \to Y$ and $g\colon Y \to Z$, the composite $g \circ f$ takes $x \in X$ to $g(f(x)) \in Z$.

Definition 1.1.5. A map of sets $f\colon X \to Y$ is defined as follows.

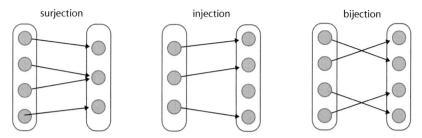

Figure 1.1 A surjective map hits everything. Injective maps take distinct elements to distinct places. Bijective maps are both injective and surjective.

- *Surjective* or *onto* if for every point $y \in Y$, there is at least one $x \in X$ such that $f(x) = y$; that is, f hits all the points of Y.
- *Injective* or *one-to-one* if for any two points $x, y \in X$ that are not the same, $f(x) \neq f(y)$; that is, no point of Y is hit more than once.
- *Bijective* if it is injective and surjective.

See Figure 1.1 for an illustration of these three properties of a map of sets.

Example 1.1.6.

1. The map $f \colon \mathbb{R} \to \mathbb{R}$ specified by $f(x) = x^2$ is not injective, since -2 and 2 both go to 4, and it is not surjective, since no negative numbers are hit.
2. The map $\{a, b, c\} \to \{d\}$ that takes every element to d is not injective since a and b both go to d, but it is surjective.

It is extremely useful to develop a criterion for considering sets "the same" that is weaker than requiring that they be identical. For this, we introduce the notion of the inverse of a function. Recall that the identity map $\mathrm{id}_X \colon X \to X$ is simply the function defined to be $f(x) = x$.

Definition 1.1.7. The function $f \colon X \to Y$ has *inverse* $g \colon Y \to X$ if the composite $g \circ f \colon X \to Y \to X$ is id_X, the identity on X, and the composite $f \circ g \colon Y \to X \to Y$ is id_Y, the identity on Y.

Notice that any bijective map $f \colon X \to Y$ has an inverse $f^{-1} \colon Y \to X$ where $f^{-1}(y)$ is defined to be the unique element of X that has image y.

Definition 1.1.8. Two sets X and Y are *isomorphic* if there exists a bijection $f \colon X \to Y$. We often write $X \cong Y$ and leave the functions f and g implicit.

We will refer to f as an *isomorphism*.

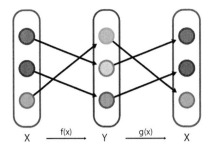 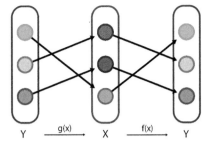

Figure 1.2 An isomorphism between two finite sets can be described in terms of a permutation.

Example 1.1.9.

1. Two finite sets X and Y are isomorphic if and only if they have the same number of elements. See Figure 1.2 for an example of an isomorphism between two finite sets.
2. The map $\mathbb{R}^2 \to \mathbb{R}^2$ that takes (x, y) to $(-x, -y)$ is an isomorphism; its inverse is itself.
3. The map $\mathbb{N} \to \mathbb{Z}$ that takes 0 to 0, 1 to 1, 2 to -1, and in general is specified by the formula

$$f(x) = \begin{cases} 0, & x = 0 \\ \frac{x+1}{2}, & x \text{ odd} \\ -\frac{x}{2}, & x \text{ even} \end{cases}$$

is an isomorphism.

Elaborating on the observation that finite sets are isomorphic if and only if they have the same number of elements, we can use isomorphisms to talk about the size of infinite sets.

Definition 1.1.10. A set S is *countable* if there exists a bijection $f \colon \mathbb{N} \to S$, where $\mathbb{N}$ denotes the natural numbers $\{0, 1, 2, \ldots\}$.

Countable sets are the smallest kind of infinite sets.

Example 1.1.11.

1. Clearly the set of natural numbers $\mathbb{N}$ is countable. The set of integers $\mathbb{Z}$ is also countable, by the bijection given above in Example 1.1.9.
2. A little bit of work shows that the set of rational numbers $\mathbb{Q}$ is countable.

3. Famously, Cantor showed that $\mathbb{R}$ is uncountable, which means that it is *bigger* in a precise sense than any countable set. More generally, $\mathbb{R}^n$ is uncountable for any $n > 0$.

Two sets can be isomorphic in many different ways; for example, there are many isomorphisms between any two finite sets of the same size. In general, composing an isomorphism between two different sets X and Y with an isomorphism from Y to itself will produce a new isomorphism from X to Y.

We will often want to work with sets "up to isomorphism." Formally, we do this using the fact that isomorphism of sets is an *equivalence relation*.

Definition 1.1.12. Let S be a set and let $\sim$ be a relation on S, i.e., a collection of tuples (x, y) with $x, y \in S$. Given such a tuple (x, y), we write $x \sim y$. Then $\sim$ is an *equivalence relation* when the following holds.

1. For all $x, y \in S$, if $x \sim y$ then $y \sim x$.
2. For all $x \in S$ we have $x \sim x$.
3. For all $x, y, z \in S$, if $x \sim y$ and $y \sim z$, then $x \sim z$.

Isomorphism of sets clearly satisfies these properties. The collection of all sets isomorphic to X is called the *isomorphism class* of X; often we will be interested in a set only up to its isomorphism class. However, we have to be a little bit careful when formalizing the idea of an isomorphism class; the isomorphism class of a set is usually not itself a set! Instead, it is a larger object, referred to as a class. The issue is that Russell's paradox shows that the "set of all sets" cannot exist: the set of all sets would have to contain in particular the set that does not contain itself as an element, and this is a contradiction. The paradox rules out certain appealing but naive axioms about which sets can exist: in particular, certain constructions that intuitively seem like they should produce sets in fact do not, but rather produce larger objects.

1.2 Metric Spaces

It is very common to represent experimental data as a set of measurements, together with a distance between every pair of measurements. For example, genomic expression data is often presented as a collection of arrays of the form $\{x_1, x_2, \ldots, x_k\}$, where $x_i \in \mathbb{R}$ is a number representing the expression of the ith measured gene. The distance between two expression vectors could be the standard Euclidean distance or it could be a correlation function, depending on the specific situation. Mathematically, this kind of setup is captured by the notion of a *metric space*. There are many good treatments of metric spaces; Kaplansky's *Set Theory and Metric Spaces* is a particularly accessible elementary treatment [284].

A metric space is a set X equipped with a distance function, referred to as the metric, that satisfies a few simple axioms encapsulating the salient features of the usual Euclidean distance in $\mathbb{R}^n$. Specifically, we have the following definition.

Definition 1.2.1. A *metric space* is specified by a pair (X, ∂_X) where X is a set and ∂_X is a function

$$\partial_X : X \times X \to \mathbb{R}$$

that assigns a non-negative real number to each pair of points of X such that the following holds.

1. The metric ∂_X detects whether two points are the same, in the sense that

$$\partial_X(x_1, x_2) = 0 \iff x_1 = x_2.$$

2. The metric ∂_X is *symmetric* in that

$$\forall x, y \in X, \qquad \partial_X(x, y) = \partial_X(y, x).$$

3. The metric ∂_X satisfies the *triangle inequality*:

$$\forall x, y, z \in X, \qquad \partial_X(x, z) \leq \partial_X(x, y) + \partial_X(y, z).$$

The most interesting of these axioms is the triangle inequality. See Figure 1.3 for pictures of triangles on the surface of a cylinder and a sphere; the triangle inequality is evident. (Here the metric on these surfaces is computed by the length of shortest path.)

Remark 1.2.2. Particularly in biological applications, we sometimes encounter *dissimilarity measures* which are not quite metrics. For example, the Kullback-Leibler divergence (see Remark 3.2.32) is not symmetric, the Gromov-Hausdorff distance (see Definition 2.4.4) on the set of metric spaces can be zero for metric

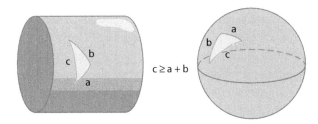

Figure 1.3 No matter how curved or distorted triangles in a metric space are, the length of any side must always be shorter than the sum of the other two sides.

spaces that are not identical, and many common dissimilarity measures (e.g., the Bray-Curtis dissimilarity measure [70]) do not satisfy the triangle inequality. In the first two kinds of examples, it is easy to construct a metric that captures the salient properties of the dissimilarity function – for instance, by symmetrizing (making a new metric $\partial'_X = \min(\partial_X(x, y), \partial_X(y, x)))$ or identifying points such that $\partial_X(x, y) = 0$ when $x \neq y$. Fixing triangle inequality violations is more subtle (e.g., see [196] for interesting recent progress).

Example 1.2.3. The most familiar and important examples of metric spaces are the Euclidean spaces $\mathbb{R}^n$; these are defined as the n-tuples $\{(x_1, x_2, \ldots, x_n) \mid x_i \in \mathbb{R}\}$ equipped with the standard distance metric

$$\partial_{\mathbb{R}^n}((x_1, x_2, \ldots, x_n), (y_1, y_2, \ldots, y_n)) = \sqrt{(x_1 - y_1)^2 + (x_2 - y_2)^2 + \ldots + (x_n - y_n)^2}.$$

A natural family of examples of metric spaces come from metrics induced by weighted graphs. Particularly interesting examples of graph metrics come from trees with weighted edges; this kind of metric space will be important in work on modeling evolutionary phenomena using phylogenetic trees, as we will see in Section 5.2.

Example 1.2.4. A *graph* is specified by a set of vertices and a set of edges connecting the vertices. A weighted graph has weights (nonnegative numbers) attached to the edges. More precisely, a weighted graph is a tuple $G = (V, E, W)$ with vertex set V, edge set $E \subset V \times V$, and weights $W \colon E \to \mathbb{R}^{\geq 0}$.

Regarding this graph as undirected and stipulating that there are no edges with non-zero weight from any vertex v to itself, the graph metric on a weighted graph is a metric on the set of vertices of the graph. The metric is defined so that the distance between vertices v and w is the minimal length of a path connecting v and w:

$$\partial_G(v, w) = \min_{v, z_0, z_1, \ldots, z_k, w \mid z_i \in V} \left(W(v, z_0) + \sum_{i=0}^{k-1} W(z_i, z_{i+1}) + W(z_k, w) \right).$$

(See Figure 1.4.)

The metrics we have described so far are continuous, in the sense that distances can in principle be any real number. But many interesting metrics are discrete. For example, the *Hamming distance*, which is a metric on strings that counts the number of differences, takes values in the natural numbers. The Hamming distance is a basic concept in information and coding theory.

Example 1.2.5. Fix an alphabet Σ, i.e., a set of symbols we will call letters. Let x and y be words of length n with letters in Σ. Then the *Hamming distance* between x and y is defined to be the number of positions at which the letters of x and y differ:

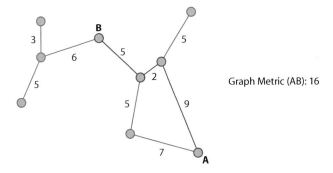

Figure 1.4 The length of the shortest path between A and B in the weighted graph gives the distance between them.

$$\partial_H(x, y) = \#\{i \mid x_i \neq y_i\}.$$

For example, if $\Sigma = \{A, C, G, T\}$, then

$$\partial_H(ACGT, ACAA) = 2.$$

An important point to emphasize is that there can be many distinct metrics on the same underlying set. For instance, in genomic data considered as words in $\{A, G, C, T\}$ there are, in addition to the Hamming distance, other well-motivated biologically relevant distances (see Section 5.2). As another example, a common distance metric used for gene expression data represented as points in $\mathbb{R}^n$ is the Pearson correlation distance.

Example 1.2.6. For $x, y \in \mathbb{R}^n$, define the *Pearson correlation distance* between x and y to be

$$\partial_{\text{cor}}(x, y) = 1 - \frac{\sum_{i=1}^n (x_i - \bar{x})(y_i - \bar{y})}{\sqrt{\sum_{i=1}^n (x_i - \bar{x})^2 (y_i - \bar{y})^2}},$$

where $\bar{x} = \frac{1}{n} \sum_{i=1}^n x_i$ and $\bar{y} = \frac{1}{n} \sum_{i=1}^n y_i$.

The existence of a distance function allows us to define many familiar notions from calculus; we review these now, as this is the prototype for the definitions of elementary topology. For instance, for each point x in a metric space and $\epsilon > 0$, we can specify the ϵ-neighborhoods of x to describe points that are close to x. Specifically, we have the *open balls* and *closed balls*

$$B_\epsilon(x) = \{z \in X \mid \partial_X(z, x) < \epsilon\} \quad \text{and} \quad \bar{B}_\epsilon(x) = \{z \in X \mid \partial_X(z, x) \leq \epsilon\}.$$

We can always *separate* two distinct points x and y by taking a ball B_1 around x and a ball B_2 around y such that $B_1 \cap B_2 = \emptyset$; if $\partial_X(x, y) = \epsilon$, we can set $B_1 = B_{\frac{\epsilon}{4}}(x)$ and $B_2 = B_{\frac{\epsilon}{4}}(y)$, for example. (See Figure 1.5.)

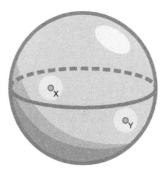

Figure 1.5 Any pair of distinct points in a metric space can be separated by open balls around them.

Elaborating on this, the existence of a metric allows us to talk about convergence of sequences. A sequence of points in X will be a function $\mathbb{N} \to X$, i.e., a sequence

$$\{x_i\} = x_0, x_1, x_2, x_3, \ldots$$

for $x_i \in X$.

Definition 1.2.7. For a metric space (X, ∂_X), an infinite sequence of points $\{x_i\}$ *converges* to a point $x \in X$ if for any $\epsilon > 0$, there exists a positive integer N such that $\partial_X(x_k, x) < \epsilon$ for all $k > N$.

Informally speaking, the definition of convergence simply means that if we go out far enough in the sequence, all the points are arbitrarily close to x. (See Figure 1.6 for a picture of what this means.)

Example 1.2.8. Consider the sequence

$$\left\{\frac{1}{n}\right\} = 1, \frac{1}{2}, \frac{1}{3}, \ldots, \frac{1}{100}, \ldots$$

This sequence converges to 0; for any ϵ, it is clear that we can find an N such that for $n > N$,

$$\left|\frac{1}{n} - 0\right| = \frac{1}{n} < \epsilon.$$

Specifically, take N to be the smallest integer larger than $\frac{1}{\epsilon}$.

A more subtle notion is that of a *Cauchy sequence*; this is a sequence of points that *ought* to converge somewhere, in the following sense.

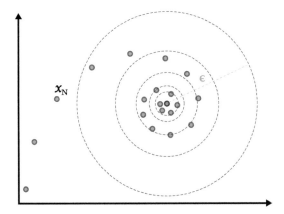

Figure 1.6 For any ball around the point of convergence, all but finitely many points of the convergent sequence are within that ball. (Note that in the picture there are only finitely many points, due to limits of resolution.)

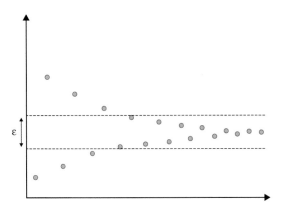

Figure 1.7 The points in a Cauchy sequence get closer and closer together but need not converge.

Definition 1.2.9. For a metric space (X, ∂_X), a *Cauchy sequence* is a sequence of points $\{x_i\}$ such that for all $\epsilon > 0$, there exists an N such that $\partial_X(x_j, x_k) < \epsilon$ for $j, k > N$.

Although the points in a Cauchy sequence get closer and closer together (see Figure 1.7), it is not necessarily the case that all Cauchy sequences converge to a point $x \in X$.

Example 1.2.10. Consider the set of rational numbers $\mathbb{Q}$ equipped with the standard metric, i.e., the distance between x and y is $\partial(x, y) = |x - y|$. Then the sequence

$$\{3, 3.1, 3.14, 3.141, 3.1415\ldots\}$$

(where each new number in the sequence has an additional digit of π) is a Cauchy sequence and "wants" to converge to π, but π is not in $\mathbb{Q}$!

This possible failure of Cauchy sequences to coincide with convergent sequences motivates the following definition.

Definition 1.2.11. A metric space (X, ∂_X) is *complete* if every Cauchy sequence converges to a point $x \in X$.

Example 1.2.12. The Euclidean spaces $\mathbb{R}^n$ are all complete; $\mathbb{R}$ can in fact be constructed by formally adding to $\mathbb{Q}$ points for each Cauchy sequence to converge to.

As Example 1.2.12 indicates, there is a tension between the size of a metric space and whether it is complete; $\mathbb{Q}$ is countable but not complete. Adding points to $\mathbb{Q}$ to make it complete yields $\mathbb{R}$, which is uncountable. Although metric spaces of interest are often not countable, there is frequently a countable subset $X' \subset X$ that is dense, in the following sense.

Definition 1.2.13. A subset $X' \subset X$ is *dense* if for all $x \in X$ and $\epsilon > 0$ there exists a point $z \in X'$ such that $\partial_X(x, z) < \epsilon$. That is, for any point X, there exists an arbitrarily close approximation in X'.

For example, $\mathbb{Q}$ is dense in $\mathbb{R}$; any real number can be approximated to any precision by a finite-length decimal.

Definition 1.2.14. A metric space (X, ∂_X) is *separable* if there exists a countable subset $X' \subset X$ that is dense in X.

Example 1.2.15. All of the Euclidean spaces $\mathbb{R}^n$ are separable; any point can be approximated by a point with rational coordinates.

A closely related notion is the idea of an ϵ-net (Figure 1.8).

Definition 1.2.16. Let (X, ∂_X) be a metric space. A subset $X' \subset X$ is ϵ-*dense* if for every $x \in X$ there exists $z \in X'$ such that $\partial_X(x, z') < \epsilon$. (So a dense set is ϵ-dense for every ϵ.) An ϵ-*net* is a subset $X' \subset X$ that is ϵ-dense.

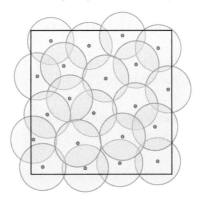

Figure 1.8 Any point in the square is within ϵ of the blue points at the centers of the circles.

In order to understand when ϵ-nets exist, we need to have ways to talk about the size of a metric space. In order to define the size, we first need to review the notion of inf and sup.

Definition 1.2.17. Given a subset $A \subset \mathbb{R}$, a *lower bound* for A is an element $x \in \mathbb{R}$ such that for all $a \in A$, $x \leq a$. Then the *infimum* $\inf(A)$ is the greatest lower bound, if one exists. Similarly, an *upper bound* for A is an element $y \in \mathbb{R}$ such that for all $a \in A$, $a \leq y$. Then the *supremum* $\sup(A)$ is the least upper bound, if one exists.

The sup and inf are distinct from the max and min, respectively, because they might not lie in A itself.

Definition 1.2.18. Let (X, ∂_X) be a metric space. The *diameter* of a subset $A \subset X$ is the supremum

$$\sup_{x,y \in X} \partial_X(x, y).$$

We must write sup rather than max because there might not be any pair of points which realizes the bound. (Note also that the diameter can be ∞, when there is no upper bound!)

Another way to talk about this is to observe that a subset $A \subset X$ has finite diameter when there exists $a \in A$ such that $A \subset B_\kappa(a)$ for some κ; more generally, such a set will be referred to as bounded. (See Figure 1.9 for an example of this.)

An even stronger notion is that of being *totally bounded*.

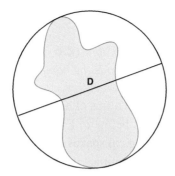

Figure 1.9 The diameter of a subset of a metric space can be approximated by taking a ball that completely encloses the subset.

Definition 1.2.19. Let (X, ∂_X) be a metric space. Then X is *totally bounded* if for every $\epsilon > 0$, there exists a finite cover of X by balls of radius ϵ, i.e., a collection of balls $\{B_\epsilon(x_i)\}$ whose union is X.

In $\mathbb{R}^n$ a subset is bounded if and only if it is totally bounded, but in general, a bounded space need not be totally bounded. For example, a metric space with infinitely many points such that all interpoint distances are 1 is bounded but not totally bounded.

Lemma 1.2.20. *Let (X, ∂_X) be a totally bounded metric space. Then for any ϵ we can find a finite ϵ-net in X.*

An important theme in modern mathematics is that the structure of mathematical objects (e.g., sets or metric spaces) can be completely understood in terms of functions between them. We describe a framework that allows us to be precise about this in Section 1.7 below (where we introduce basic concepts of category theory). From this perspective, an essential next step is to define a function between metric spaces.

At a minimum, a map between metric spaces (X, ∂_X) and (Y, ∂_Y) should involve a function of sets $f\colon X \to Y$. But we would like to require that the function also respect the metric structures on X and Y, in some sense. There are different ways to do this; we now discuss the familiar notion of a *continuous map*.

Definition 1.2.21. Let (X, ∂_X) and (Y, ∂_Y) be metric spaces. A map $f\colon X \to Y$ is *continuous* if for every sequence $\{x_i\}$ in X converging to x the sequence $\{f(x_i)\}$ converges in Y to $f(x)$.

An important property of continuous maps is that they compose.

Lemma 1.2.22.   *Let (X, ∂_X), (Y, ∂_Y), and (Z, ∂_Z) be metric spaces. If $f : X \to Y$ and $g : Y \to Z$ are continuous, then so is the composition $g \circ f : X \to Z$.*

Continuity can also be defined in terms of a traditional ϵ-δ definition; this is easy to show directly. We explain this below in Example 1.3.20, where we generalize the notion of continuity to topological spaces.

For metric spaces, it is also sometimes useful to consider a stronger notion of continuous where the "expansion" of the map is bounded.

Definition 1.2.23.   A map $f : X \to Y$ between metric spaces (X, ∂_X) and (Y, ∂_Y) is *Lipschitz* with constant κ if for all $x_1, x_2 \in X$ the inequality

$$\partial_Y(f(x_1), f(x_2)) \leq \kappa \partial_X(x_1, x_2)$$

holds.

Any Lipschitz map is continuous, but the converse does not hold in general.

1.3 Topological Spaces

The motivating idea of point-set topology is to relax the requirement of a distance and define a weaker and more flexible notion of *closeness* that still allows us to formalize the notions that lead to calculus (i.e., continuity and convergence). This is the basis for elementary analysis, which studies the foundations of calculus. A classic textbook for point-set topology is Munkres [369]; there are many excellent analysis books, of which Rudin [440] is a canonical example.

The basic observation that leads to the development of point-set topology is that most of the concepts we defined for metric spaces in Section 1.2 were or could be phrased in terms of the metric balls $B_\epsilon(x)$. A topological space can be thought of as simply a set with a well-behaved collection of subsets that act like metric balls. This abstraction is extremely useful, for a number of reasons: many metrics can lead to the same topology, some important topological spaces (notably those arising in algebraic geometry) do not come from a metric, and many basic constructions (e.g., gluing) are much more complicated to express in the context of a metric.

Definition 1.3.1.   A *topological space* is a pair $(X, \mathcal{U})$, where X is a set and $\mathcal{U}$ is a collection of subsets of X, which we refer to as *open sets*. The open sets satisfy the following conditions.

1. Both the empty subset $\emptyset$ and X are elements of $\mathcal{U}$.
2. Any union of elements of $\mathcal{U}$ is an element of $\mathcal{U}$.
3. The intersection of a finite collection of elements of $\mathcal{U}$ is an element of $\mathcal{U}$.

A subset $Z \subseteq X$ is *closed* if the complement of Z in X is open.

Any metric space gives rise to a topological space.

Example 1.3.2. Let (X, ∂_X) be a metric space. Then we say that a subset $A \subset X$ is *open* if for every $z \in A$, there exists ϵ such that $B_\epsilon(z) \subseteq A$. The open sets make X into a topological space.

But the definition of a topological space is sufficiently flexible so as to allow a variety of strange examples. For instance, any set has two trivial topologies.

Example 1.3.3. Any set X can be given the following two topologies.

1. The *discrete topology*, in which any subset $Y \subset X$ is an open set. In particular, the points themselves are open sets. As the name suggests, in this topology the points should be thought of as maximally separated from one another.
2. The *indiscrete topology*, in which the only open sets are the entire set X and $\emptyset$. In this topology, the points should be thought of as being arbitrarily close to each other.

However, the most frequently occurring examples are very familiar. In order to specify a topological space, one typically gives a *base* for the topology.

Definition 1.3.4. A *base* for a topological space $(X, \mathcal{U})$ is a collection of open sets $\{U_\alpha\}$ such that any open set is a union of elements of the base. Given simply a set X, a collection of sets $\{B_\alpha\}$ is a base if every $x \in X$ is in some B_α and given $x \in B_\alpha \cap B_\beta$, there exists $B_\gamma \subseteq B_\alpha \cap B_\beta$ such that $x \in B_\gamma$.

The importance of the intrinsic definition is that we can define a topology on X given a base.

Lemma 1.3.5. *Given a set X and a base $\{B_\alpha\}$, we can define a topology on X where a set U is open if for each $x \in U$ there exists B_α such that $x \in B_\alpha \subseteq X$. (And we will often refer to this as the topology generated by a base.)*

As Lemma 1.3.5 makes clear, the base of a topological space is modeled on the open balls of a metric space.

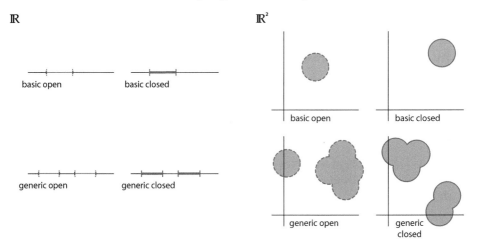

Figure 1.10 Basic open and closed sets in Euclidean space are *balls*; more general open and closed sets are generated by union and intersection.

Example 1.3.6. Euclidean space with the topology generated by the open balls $B_\epsilon(x) = \{y \in \mathbb{R}^n \mid \|x - y\| < \epsilon\}$, for $x \in \mathbb{R}^n$ and $\epsilon > 0$. See Figure 1.10 for some examples of open and closed sets in this topology.

Example 1.3.7. In fact, we can conveniently describe the topology of Example 1.3.2 on a metric space (X, ∂_X) as generated by the base of the open balls $B_\epsilon(x) = \{y \in X \mid \partial_X(y, x) < \epsilon\}$, for $x \in X$ and $\epsilon > 0$.

An important class of topological spaces are those with a countable base; these are called *second countable*. Example 1.3.6 is a second countable topological space; we can take the base using only the balls with rational radii.

The example of the topology induced by a metric has a particularly important property that we now highlight. Specifically, recall that in a metric space we can separate points in the sense that given two distinct points $x, y \in X$, we can choose balls $B_{\epsilon_1}(x)$ and $B_{\epsilon_2}(y)$ such that $B_{\epsilon_1}(x) \cap B_{\epsilon_2}(y) = \emptyset$; we simply take $\epsilon_1, \epsilon_2 < \frac{\partial_X(x,y)}{2}$. It turns out to be very useful to consider topological spaces that have this property, even if the topology is not generated by a metric.

Definition 1.3.8. A topological space $(X, \mathcal{U})$ is *Hausdorff* if for any pair of distinct points $x, y \in X$ there exist open sets U_x and U_y such that $x \in U_x$, $y \in U_y$, and $U_x \cap U_y = \emptyset$.

If $(X, \mathcal{U})$ is a topological space, any subset $Y \subset X$ can be given the structure of a topological space in a natural fashion induced from the topology on X. This

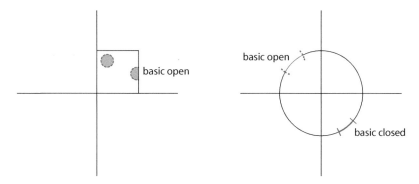

Figure 1.11 Left: Basic open sets in the subspace topology on the unit square. Right: Basic open sets in the subspace topology on the unit circle.

is referred to as the *subspace topology* on Y, and is a very important source of examples of topological spaces.

Definition 1.3.9. Let $(X, \mathcal{U})$ be a topological space and $Y \subset X$ a subset. Then the *subspace topology* on Y is defined by taking the open sets to be $\{Y \cap U \mid U \in \mathcal{U}\}$.

Example 1.3.10. The subspace topology on the unit square $[0, 1] \times [0, 1] \subset \mathbb{R}^2$ has basic open sets that are either balls (when the ball is completely contained within the square) or the intersection of balls with the square; see Figure 1.11 for examples.

Example 1.3.11. Let S^1 denote the standard unit circle in $\mathbb{R}^2$; that is, $S^1 = \{(x, y) \subset \mathbb{R}^2 \mid x^2 + y^2 = 1\}$. We topologize S^1 using the subspace topology as a subset of $\mathbb{R}^2$; see Figure 1.11 for examples.

Just as with sets, another standard way to produce new topological spaces from old is via the Cartesian product (recall Definition 1.1.3).

Definition 1.3.12. Let X and Y be topological spaces, with the topologies specified by open sets $\{U_\alpha\}$ and $\{V_\beta\}$ respectively. Then the product

$$X \times Y = \{(x, y) \mid x \in X, y \in Y\}$$

is a topological space with a base for the topology given by the open sets $\{U_\alpha \times V_\beta\}$; we refer to this as the *product topology*.

A topological space is designed to be a minimal structure in which we can talk about "closeness," in a precise sense.

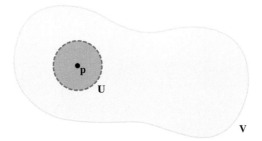

Figure 1.12 *V* is a neighborhood of *p*.

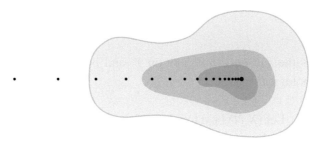

Figure 1.13 Smaller and smaller open sets around the point of convergence still contain all but finitely many points in the approaching sequence.

Definition 1.3.13.   Given a point $x \in X$, we define a *neighborhood* of x to be a set $V \subseteq X$ such that there is an open set $U \subseteq V$ and $x \in U$. (See Figure 1.12.)

Immediately, we can use this definition to specify the notion of convergence of a sequence (Figure 1.13).

Definition 1.3.14.   A sequence of points $\{x_i\}$ *converges* to p if for any neighborhood V of p there exists an N such that $x_n \in V$ for $n \geq N$.

Considering Example 1.3.6, we see that in Euclidean space this means that for any ϵ, there exists an n such that $x_n \in B_\epsilon(x)$, i.e., $\|x_n - x\| < \epsilon$. In particular, when restricted to $\mathbb{R}$, the definition recovers the usual notion from elementary calculus of convergence of a sequence. More generally, Definition 1.3.14 coincides with Definition 1.2.7 in a metric space given the metric topology.

Topological spaces also admit an extremely useful notion of size. This takes a little bit more work to specify without explicit reference to a distance function. In order to define this, we need the notion of a *cover*.

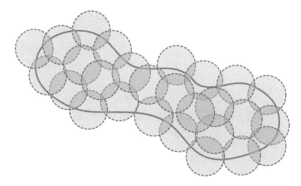

Figure 1.14 An open cover of U is a collection of open sets whose union contains U.

Definition 1.3.15. An *open cover* of a set U in a topological space X is a collection of open sets $\{U_\alpha\}$, with each $U_\alpha \subset X$, such that $U \subseteq \bigcup U_\alpha$.

For example, the collection of all balls $B_\epsilon(x)$ as x varies over the points of $\mathbb{R}^n$ is an open cover of $\mathbb{R}^n$. (See Figure 1.14.) A subcover of a cover is a subset whose union still contains U.

Definition 1.3.16. A topological space X is *compact* if any open cover of X has a finite subcover.

Example 1.3.17.

1. Every finite set is compact.
2. The sphere $\{x, y, z \mid x^2 + y^2 + z^2 = 1\}$ with the subspace topology is compact.
3. No Euclidean space $\mathbb{R}^n$ is compact for $n > 0$.

Compact sets are "small" in a basic sense. The notion of compactness is a way of formalizing the properties of the closed and bounded subsets of $\mathbb{R}^n$.

Theorem 1.3.18. *A subset $X \subseteq \mathbb{R}^n$ regarded as a metric space is compact if and only if it is closed and bounded.*

1.3.1 Maps between Topological Spaces

We now turn to consider the correct notion of a map between topological spaces. We want to restrict ourselves to maps $f : X \to Y$ which satisfy certain properties expressing compatibility with the topologies on X and Y. Roughly speaking, we want continuous maps to have the property that "nearby" points in X are taken to "nearby" points in Y.

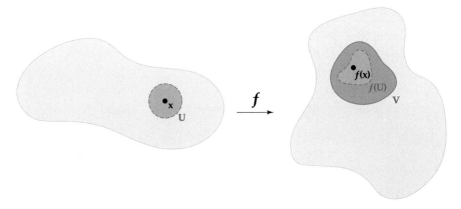

Figure 1.15 A function is continuous if for every neighborhood V around $f(x)$, we can find a neighborhood U of x whose image $f(U)$ sits inside it.

Definition 1.3.19. Let $(X, \mathcal{U}_X)$ and $(Y, \mathcal{U}_Y)$ be topological spaces. A map $f \colon X \to Y$ is *continuous at a point* x if for every neighborhood V of $f(x)$, there exists a neighborhood U of x such that $f(U) \subseteq V$ (Figure 1.15). The map f is *continuous* if it is continuous at every point $x \in X$.

It is instructive to work out exactly what this means in the case of the standard metric topology on $\mathbb{R}$.

Example 1.3.20. A map $f \colon \mathbb{R} \to \mathbb{R}$ is continuous at a point $x \in \mathbb{R}$ if for every open ball $B_\epsilon(f(x))$, there exists an open ball $B_\delta(x)$ such that $f(B_\delta(x)) \subseteq B_\epsilon(f(x))$. Put another way, for every $\epsilon > 0$, there exists $\delta > 0$ such that $|x - y| < \delta$ implies that $|f(x) - f(y)| < \epsilon$. That is, we have recovered precisely the usual ϵ-δ notion of continuity.

More generally, in any metric space, maps are continuous in the sense of Definition 1.2.21 if and only if they are continuous in the sense of Definition 1.3.19.

Generalizing Lemma 1.2.22, the composition of continuous maps is continuous.

Lemma 1.3.21. *Let $(X, \mathcal{U}_X)$, $(Y, \mathcal{U}_Y)$, and $(Z, \mathcal{U}_Z)$ be topological spaces and suppose we have continuous maps $f \colon X \to Y$ and $g \colon Y \to Z$. Then the composite $g \circ f \colon X \to Z$ is continuous.*

Continuous maps out of simple "test spaces" that are well understood play an important role in algebraic topology; for example, we can now define a path in terms of maps out of the unit interval.

Definition 1.3.22. A *path* from x to y in a topological space $(X, \mathcal{U}_X)$ is a continuous function $\gamma \colon [0, 1] \to X$ such that $\gamma(0) = x$ and $\gamma(1) = y$. Here $[0, 1]$ is given the subspace topology it inherits as a subset of $\mathbb{R}$.

The notion of a path captures many familiar examples, but the price of the generality is that strange examples are also permitted.

Example 1.3.23.

1. A path γ in $\mathbb{R}^n$ is just a curve that could be drawn without lifting up the pen (see Figure 1.16). Note that these can be surprisingly complicated: there are famous examples of "space-filling" curves, which are precisely paths that touch every point of $\mathbb{R}^2$.
2. A path γ in S^2 is a smooth curve on the surface of the sphere.
3. A path in a space given the discrete topology must be a constant map.

We now return to considering the continuous analogue of clustering; in light of Definition 1.3.22, this is straightforward – we replace the discrete paths by continuous ones.

Definition 1.3.24. Let $(X, \mathcal{U}_X)$ be a topological space. Two points $p, q \in X$ are *path-connected* if there exists a continuous path $\gamma \colon [0, 1] \to X$ such that $\gamma(0) = p$ and $\gamma(1) = q$.

It is clear that the relation of being path-connected is an equivalence relation (reparametrizing paths to obtain transitivity), and so the following definition makes sense.

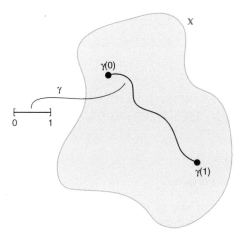

Figure 1.16 A path γ is a continuous map $\gamma \colon [0, 1] \to X$.

Definition 1.3.25. We define the *path components* of a topological space $(X, \mathcal{U}_X)$ to be the collection of subsets of X such that x, y are in the same subset if and only if there is a path joining them.

We can think of the path components of X as giving a *continuous clustering* of the points of X; roughly speaking, two points are in distinct path components when they are separated by a "gap" in space. An important property of path components is that continuous maps of spaces give rise to maps of path components; this fact, referred to as *functoriality*, is essential for calculations (see Figure 1.17).

Lemma 1.3.26. *Let X and Y be topological spaces. Given a continuous map $f: X \to Y$, there is an induced map of sets between the path components of X and the path components of Y.*

1.3.2 Homeomorphisms

The construction of the set of path components is an example of a *topological invariant*; for two topological spaces that are "the same" in a suitable sense, the sets of path components should be isomorphic. To be precise about this, we need to describe when we will consider two topological spaces to be the same.

Definition 1.3.27. Topological spaces $(X, \mathcal{U}_X)$ and $(Y, \mathcal{U}_Y)$ are *homeomorphic* if there exists a bijection $f: X \to Y$ such that both f and f^{-1} are continuous maps.

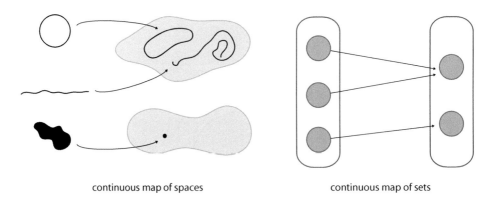

continuous map of spaces   continuous map of sets

Figure 1.17 A continuous map of spaces induces a map of sets of path components. Here, the black space is the union of three components and the blue space the union of two.

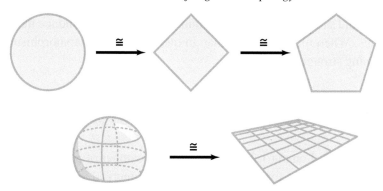

Figure 1.18 Two spaces are homeomorphic if there is a continuous bijection between them with continuous inverse. On the top, the circle is deformed into a pentagon. On the bottom, a sphere with the bottom cut off can be stretched onto a plane.

In this situation, we refer to f as a *homeomorphism*. Intuitively, two spaces are *homeomorphic* when they are related by a *continuous deformation*; roughly speaking, this means they are related by stretching and bending without introducing tears or gluing things together. See Figure 1.18 for a few examples of homeomorphic spaces.

Example 1.3.28.

1. The xy-plane $\mathbb{R}^2$ and a punctured two-dimensional sphere (i.e., a sphere where we have removed a point at one of the poles) are homeomorphic; there is a homeomorphism that "unwraps" the sphere. This homeomorphism is very familiar; this is a stereographic projection, used for example to make maps.
2. A square, a circle, and an octagon are all homeomorphic – we can define a homeomorphism by smoothing out the corners of the square and octagon, or alternatively adding kinks to the circle.
3. Famously, a coffee cup and a solid torus (a doughnut) are homeomorphic.

We can write $X \cong Y$ when two spaces X and Y are homeomorphic. The relation of homeomorphism is an equivalence relation on spaces:

1. it is reflexive (clearly $X \cong X$ via the identity map),
2. symmetric ($X \cong Y$ implies that $Y \cong X$), and
3. transitive (if $X \cong Y$ and $Y \cong Z$, composing the homeomorphisms shows that $X \cong Z$).

Recall from Lemma 1.3.26 that continuous maps of spaces induce maps of path components. When the continuous map in question is a homeomorphism, we can say something stronger.

Lemma 1.3.29. *Let $f: X \to Y$ be a homeomorphism. Then f induces a bijection between the set of path components of X and the set of path components of Y.*

We can interpret Lemma 1.3.29 to say that the number of path components is a topological invariant of a topological space. This numerical invariant is interesting, insofar as it allows us to distinguish spaces very easily.

Corollary 1.3.30. *Let X and Y be topological spaces. If X and Y have different numbers of path components, then X and Y are not homeomorphic. (Of course, two spaces with the same number of path components need not be homeomorphic!)*

We can directly relate the notion of path components to the problem of clustering discrete data, in a precise sense. First consider the case in which (M, ∂_M) is a finite metric subspace of $\mathbb{R}^n$. Fix a scale parameter $\epsilon \geq 0$. Then the topological space formed as the union

$$\mathcal{N} = \bigcup_{x \in M} \bar{B}_{\frac{\epsilon}{2}}(x)$$

has the property that the path components of $\mathcal{N}$ recover the clusters obtained via single-linkage clustering with parameter ϵ. However, a general finite metric space will not come with an embedding into $\mathbb{R}^n$; for this reason, it is useful to recast the clustering problem using a discretized topological model that encodes the same basic data.

To this end, we consider a construction which associates a graph to (M, ∂_M).

Definition 1.3.31. Let (M, ∂_M) be a finite metric space and fix $\epsilon \geq 0$. Define the associated *neighborhood graph* $G_\epsilon(M)$ to have *vertices* given by the points of M, and an *edge* (v_i, v_j) connecting v_i and v_j if and only if $\partial_M(v_i, v_j) \leq \epsilon$.

Regarding the graph as a topological space, we can give a graph-theoretic description of the path components.

Lemma 1.3.32. *Two vertices v_i and v_j in a graph G are in the same path component if there exists a collection of edges $(v_i, v_{k_1}), (v_{k_1}, v_{k_2}), \ldots, (v_{k_m}, v_j)$ where each pair of adjacent edges shares a vertex.*

It is now evident that the components of the graph associated to (M, ∂_M) correspond to the clusters given by single-linkage clustering with parameter ϵ.

1.4 Continuous Deformations and Homotopy Invariants

We have now arrived at the beginnings of homotopy theory; two excellent modern textbooks are by May [342] and Hatcher [235]. We observed in Lemma 1.3.29 in the previous section that the set of path components of a space X is a topological invariant, in the sense that if $f: X \rightarrow Y$ is a homeomorphism then the induced map on path components is an isomorphism. However, counting path components is much weaker than deciding whether two spaces are homeomorphic.

1. A circle and a point $\{x\}$ are not homeomorphic but have the same number of path components.
2. As an even simpler example, a disk $\{x \mid x \in \mathbb{R}^2, \|x\| \leq 1\}$ and a point $\{x\}$ have the same number of path components. However, they are clearly not homeomorphic (there is no map from $D^n \rightarrow \{x\}$ that is a bijection).

These examples motivate a search for a notion of equivalence that is weaker than homeomorphism and closer to comparing counts of path components. In particular, it seems reasonable to want a weaker kind of equivalence for which a point and a disk look the same but a point and a circle look different.

In order to introduce such a notion of equivalence, we will introduce the idea of a homotopy. A homotopy specifies a relationship between continuous maps from $X \rightarrow Y$; we will subsequently use this to define a kind of "approximate" homeomorphism.

Definition 1.4.1. Let X and Y be topological spaces. Then two continuous maps $f, g: X \rightarrow Y$ are *homotopic* if there exists a continuous map (called a *homotopy*) $h: X \times [0, 1] \rightarrow Y$ such that

$$\begin{cases} h(x, 0) = f(x) \\ h(x, 1) = g(x). \end{cases}$$

We write $f \simeq g$ when f and g are homotopic.

We think of $t \in [0, 1]$ as parametrizing a family of maps interpolating between f and g; for each t, h induces a continuous map $h(-, t): X \rightarrow Y$. The continuity condition on h means that these maps vary "smoothly" as the parameter changes. In fact, for maps to Euclidean space, this description can be made precise as follows. (See also Figure 1.19.)

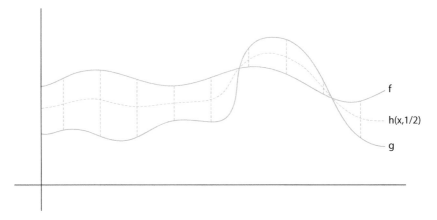

Figure 1.19 Two maps $\mathbb{R} \to \mathbb{R}$ are homotopic via linear interpolation. We can think of this as if we represented the graphs of f and g as rubber bands and dragged one to the other.

Example 1.4.2. Any two continuous maps $f, g \colon \mathbb{R}^m \to \mathbb{R}^n$ are homotopic; the homotopy is specified by interpolation as

$$h(x, t) = (1 - t)f(x) + t(g(x)).$$

The relation of being homotopic is an equivalence relation on the set $\mathrm{Map}(X, Y)$ of continuous maps between topological spaces X and Y. As in the previous examples, only transitivity is non-trivial to check. Assume that for $f, g, h \colon X \to Y$, we have $f \simeq g$ via the homotopy H_1 and $g \simeq h$ via the homotopy H_2. Then a homotopy H_3 defined as

$$\begin{cases} H_3(t, x) = H_1(2t, x) & 0 \leq t \leq \frac{1}{2} \\ H_3(t, x) = H_2(2t - 1, x) & \frac{1}{2} < t \leq 1 \end{cases}$$

shows that $f \circ h$.

The notion of a homotopy now allows us to weaken the definition of homeomorphism; we will consider continuous maps $f \colon X \to Y$ that admit continuous inverses *up to homotopy*. Specifically, we have the following definition.

Definition 1.4.3. Let X and Y be topological spaces. Then X and Y are *homotopy equivalent* if there exist continuous maps

$$f \colon X \to Y \qquad \text{and} \qquad g \colon Y \to X$$

such that

$$f \circ g \simeq \mathrm{id}_Y \qquad \text{and} \qquad g \circ f \simeq \mathrm{id}_X.$$

(Here id_X and id_Y denote the identity maps on X and Y.) In this case, we write $X \simeq Y$ and we refer to f and g as *homotopy equivalences*.

Example 1.4.4.

1. Any spaces X and Y which are homeomorphic (via maps f and g) are also homotopy equivalent; the required homotopies are

$$h_1 : X \to X \qquad h_1(x, t) = x$$
$$h_2 : Y \to Y \qquad h_2(y, t) = y$$

   since $f \circ g = \mathrm{id}_X$ and $g \circ f = \mathrm{id}_Y$.
2. For a disk $B_\epsilon(x) \subset \mathbb{R}^2$, the inclusion $i \colon \{x\} \to B_\epsilon(x)$ and the constant map $p \colon B_\epsilon(x) \to \{x\}$ induces a homotopy equivalence. The composite $p \circ i$ is equal to the identity, and for the composite $i \circ p$, we use the "radial contraction"

$$h((r, \theta), t) = (tr, \theta),$$

   where here we are representing the disk using polar coordinates. See the left panel of Figure 1.20 below for a picture of this process.
3. Recall (from Example 1.3.11) that S^1 denotes the standard unit circle. A cylinder $[0, 1] \times S^1$ is homotopy equivalent to the circle; the maps are the inclusion $S^1 \to [0, 1] \times S^1$ that takes $(x, y) \mapsto (0, (x, y))$ and the collapse that takes $(t, (x, y)) \mapsto (x, y)$. Once again, the composite of the inclusion and the collapse is the identity and the other composite is homotopic to the identity via the homotopy

$$h(t, (s, x, y)) = (ts, x, y).$$

   See the right panel of Figure 1.20 for a picture of this process.

Homotopy equivalence is an equivalence relation on spaces:

1. it is reflexive (clearly $X \simeq X$ via the identity homotopy),
2. symmetric ($X \simeq Y$ implies that $Y \simeq X$, using the same homotopy in the opposite direction), and
3. transitive; this is the only property that is not immediate. The key idea is that given homotopy equivalences $f_1 \colon X \to Y$ and $f_2 \colon Y \to Z$ (with inverses g_1 and g_2), we can build a homotopy from $(f_2 \circ f_1) \circ (g_1 \circ g_2)$ to the identity of Z by using the homotopy from $f_1 \circ g_1$ to the identity of Y on the interval $[0, \frac{1}{2}]$ and the homotopy from $f_2 \circ g_2$ to the identity of Z on the interval $[\frac{1}{2}, 1]$.

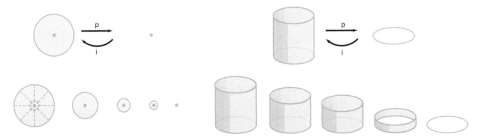

Figure 1.20 Radially shrinking a disk realizes the homotopy equivalence between a point and a disk. A cylinder shrinks along its length to a circle.

Definition 1.4.5.   We will refer to the equivalence class of a space under the relation of homotopy equivalence as its *homotopy type*.

To understand homotopy equivalence, it is useful to consider the notion of a *deformation retraction*.

Definition 1.4.6.   Let $A \subset X$ be a subspace. Then A is a *deformation retraction* of X if there exists a homotopy $H\colon X \times I \to X$ such that $H(x, 0) = x$, $H(x, 1) \in A$, and $H(a, 1) = a$.

A deformation retraction specifies a homotopy equivalence between A and X. Not all homotopy equivalences are deformation retractions, but one can show that two spaces X and Y are homotopy equivalent if and only if there is a space Z such that X and Y are each deformation retractions of Z.

Lemma 1.3.29 showed that counting path components of a space was a homeomorphism invariant. In fact, it is an invariant of homotopy equivalence.

Lemma 1.4.7.   *Let X and Y be topological spaces such that there is a homotopy equivalence $f\colon X \to Y$. Then f induces a bijection between the set of path components of X and the set of path components of Y.*

In order to study homotopy equivalences, it turns out to be useful to consider the set obtained by taking *homotopy classes* of maps; two continuous maps are in the same homotopy class if they are homotopic.

Definition 1.4.8.   Let X and Y be topological spaces. The set of *homotopy classes* of maps from X to Y, denoted $\{X, Y\}$, is the set of equivalence classes in $\text{Map}(X, Y)$ under the equivalence relation given by homotopy.

1.4.1 Homotopy Groups

An essential insight from early in the development of algebraic topology is the idea that homotopy classes of maps from certain "test spaces" capture the homotopy type of a topological space. The test spaces we need are the standard spheres.

Definition 1.4.9. Let D^n denote the n-dimensional unit disk in $\mathbb{R}^n$ defined as

$$D^n = \left\{ (x_1, \ldots, x_n) \in \mathbb{R}^n \mid \sum_{i=1}^{n} x_i^2 \leq 1 \right\}$$

and let S^{n-1} denote the $(n-1)$-dimensional unit sphere in $\mathbb{R}^n$ defined as

$$S^{n-1} = \left\{ (x_1, \ldots, x_n) \in \mathbb{R}^n \mid \sum_{i=1}^{n} x_i^2 = 1 \right\}.$$

Observe that there is a natural inclusion $S^{n-1} \to D^n$ as the boundary.

Notice that $D^1 = [-1, 1] \subseteq \mathbb{R}^1$, $S^0 = \{-1, 1\} \subseteq \mathbb{R}^1$, and so forth. We regard D^n and S^{n-1} as topologized using the subspace topology, with regard to the standard topology on $\mathbb{R}^n$.

Now we define the *homotopy groups*. These will be sets with some additional algebraic structure, which we will describe informally below and then more precisely in Section 1.6.4. For this definition, we use the notion of a *based homotopy*, which is simply a homotopy $H: X \times I \to Y$ that has the property that for specified basepoints $x \in X$ and $y \in Y$, $H(x, t) = y$ for all t.

Definition 1.4.10. Let X be a topological space and $x \in X$ a point. Choose a point $p \in S^n$. Then for $n \geq 0$, as a set, the nth *homotopy group* $\pi_n(X, x)$ is the set of based homotopy classes $\{S^n, X\}$ where the point p is sent to x.

Up to isomorphism, the homotopy groups are independent of the choice of basepoint in the spheres S^n, but might change depending on the chosen basepoint in the target space X. For example, if X has many path components, then $\pi_n(X, x)$ will depend on which path component x lies in.

Example 1.4.11.

1. When $n = 0$, $\pi_0(X, x)$ is the set of path components of X.
2. When $n = 1$, $\pi_1(X, x)$ is called the *fundamental group*, the set of homotopy classes of loops in X that start and end at x. (See Figure 1.21.)
3. The fundamental group $\pi_1(S^1, x)$, where x is any point of the circle, has elements in bijection with $\mathbb{Z}$; each homotopy class of maps from $S^1 \to S^1$ can

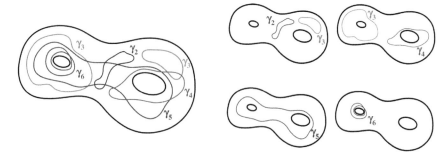

Figure 1.21 The fundamental group of a space X is the set of homotopy classes of loops.

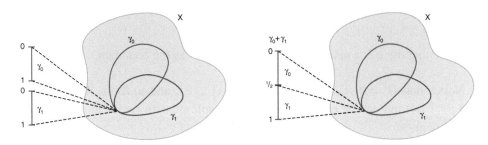

Figure 1.22 A loop $S^1 \to X$ is represented by a map $[0, 1] \to X$ with the same value on 0 and 1. Two loops γ_0 and γ_1 are added by reparameterizing, doing γ_0 on $[0, \frac{1}{2}]$ and γ_1 on $(\frac{1}{2}, 1]$.

be characterized by how many times it wraps around, and in which direction it goes.

The fundamental group of X records information about "holes" in X; a loop is homotopic to the constant map at a point unless it goes around a hole in X. (Of course, the loop might go around many times or it might wind around multiple holes; the intricacies of the geometry are reflected in the additional algebraic structure.)

When $n \geq 1$, π_n has additional algebraic structure; given two basepoint preserving maps from $S^1 \to X$, we can "add" them to get a new loop by doing first one, then the other. (See Figure 1.22.)

More generally, given two pointed maps $f_1, f_2 \colon S^n \to X$, we can make a new one by "pinching" a radial belt of the sphere to a point, forming two copies of the sphere, and then considering the new map that does f_1 on one bulb and f_2 on the other. (See Figure 1.23.)

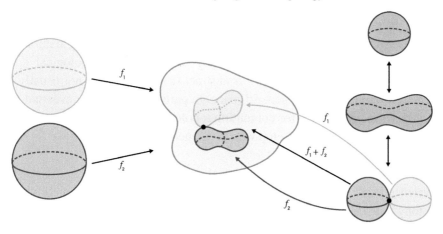

Figure 1.23 Two maps $f_1, f_2 \colon S^n \to X$ are added by taking a sphere, pinching it around the radius to produce two spheres joined at a point, and then doing f_1 on one "bulb" and f_2 on the other.

In fact, not only can we add, we can subtract as well. In Section 1.6, we quickly review the abstract framework for this kind of algebraic structure; in Section 1.6.4, we return to discuss the homotopy groups in more detail.

Another important property of the homotopy groups is that they behave nicely in the presence of continuous maps. Specifically, restating Lemma 1.4.7 in this language, we have the following result.

Lemma 1.4.12. *Let X and Y be topological spaces and $f \colon X \to Y$ a homotopy equivalence. Then for any $x \in X$ there is an isomorphism of sets $\pi_0(X, x) \cong \pi_0(Y, f(x))$.*

More generally, we have the following result.

Proposition 1.4.13. *Let X and Y be topological spaces and $f \colon X \to Y$ a homotopy equivalence. Then for any $x \in X$, there is an isomorphism of sets $\pi_n(X, x) \cong \pi_n(Y, f(x))$.*

The most pressing question about the homotopy groups is now to what degree there is a converse to Proposition 1.4.13. An answer to this question and a justification of the use of spheres as test objects is provided by the theory of CW complexes.

1.5 Gluing and CW Complexes

When contemplating practical work with topological spaces, a very natural question arises: how do we concretely specify the data of a topological space? Definition 1.3.1 is very well suited for abstract reasoning, but is not usually convenient as a way to present a generic space. In particular, since our eventual goals involve devising algorithms for computing topological invariants that are tractable on computers, we want to develop means of encoding topological spaces that are discrete.

If we restrict attention to the question of working with spaces up to homotopy equivalence, then we obtain additional flexibility. The idea is now to model a given homotopy type by particularly nice spaces; in a precise sense, it turns out that we can always replace an arbitrary topological space by one which has a very regular topological structure. This approach is based on an inductive description of a topological space in terms of building blocks that are easily understood, namely disks and spheres.

In order to describe the topology on spaces built up in this way, we begin by describing the *quotient topology*. To motivate this construction, consider the interval $[0, 1]$, topologized with the subspace topology from $\mathbb{R}$. Gluing together the two endpoints $\{0\} \subset [0, 1]$ and $\{1\} \subset [0, 1]$ should produce a circle. The quotient topology is a way to make this precise.

Proposition 1.5.1. *Let X be a topological space and Y a set. Let $p \colon X \to Y$ be a surjective map. Then we can make Y a topological space by specifying that a subset $U \subset Y$ is open when $p^{-1}(U)$ is an open set in X. We call the topology on Y the quotient topology.*

Equivalently, given a continuous surjection of topological spaces $p \colon X \to Y$, we can identify a criterion for when the topology on Y is the quotient topology.

Proposition 1.5.2. *Given a surjective map of topological spaces $p \colon X \to Y$, we say that p is a quotient map provided that $U \subseteq Y$ is an open set in Y if and only if $p^{-1}(U) \subseteq X$ is an open set in X. In this case, the topology on Y is the quotient topology.*

We can now identify the usual topology on the unit circle S^1 as the quotient topology.

Example 1.5.3. Let $p \colon [0, 1] \to S^1$ be the map specified by $x \mapsto (\cos(2\pi x), \sin(2\pi x))$. Then p is a quotient map. (See figure 1.24.)

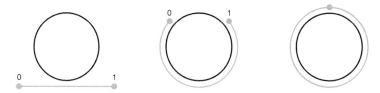

Figure 1.24 The unit interval wraps around the circle, joined at the endpoints.

Given a topological space, it is often useful to have a more intrinsic way of producing a surjective map $f: X \to Y$, where Y is a set; by intrinsic, we mean defined in terms of some sort of "gluing" data on X. For this, we need the notion of a partition.

Definition 1.5.4. Given a topological space X, we let a *partition* of X be a decomposition

$$X = \bigcup X_i, \quad \text{where} \quad X_i \cap X_j = \emptyset, i \neq j.$$

A partition specifies an equivalence relation on the points of X, where x and y are equivalent when $x, y \in X_i$.

The basic idea is that all of the points in each X_i are going to be glued together.

Definition 1.5.5. Given a partition $\{X_i\}$ of X, the *quotient space* of the partition is a topological space with points the set of partitions. The topology is induced by the surjective map $X \to \{X_i\}$ which takes $x \in X$ such that $x \in X_i$ to X_i. Put another way, we are topologizing the set of equivalence classes determined by the partition.

For instance, if we take the partition of $[0, 1]$ specified by $\{0, 1\}$ and the points $\{x\}$ in the open interval $(0, 1)$, we generate the usual topology on S^1 as in Example 1.5.3. A rich source of partitions comes from circumstances in which we want to glue a space X to a space Y along a map from $Z \subset X$ to Y.

Definition 1.5.6. Let X and Y be topological spaces, $Z \subseteq X$ a subspace of X, and $f: Z \to Y$ a continuous map. Define a partition on the disjoint union $X \coprod Y$ with sets

$$\begin{cases} \{x\} & \forall x \in X - Z, \\ \{y\} & \forall y \in Y - f(Z), \\ \{z, f(z)\} & \forall z \in Z. \end{cases}$$

The *gluing* $X \cup_f Y$ is the quotient space associated to this partition.

Figure 1.25 Cell attachment involves gluing on a disk along its boundary; here, the boundary circle of the blue disk is glued to the red loop on the surface.

For example, Definition 1.5.6 allows us to regard S^1 as obtained by gluing two copies of $[0, 1]$ along the map that identifies the endpoints. More generally, Definition 1.5.6 allows us to regard S^n as built by gluing two copies of D^n along the boundary $S^{n-1} = \partial D^n \subset D^n$.

Now, we will describe an inductive process for constructing a topological space by repeatedly gluing on disks along their boundaries, as follows (see Figure 1.25).

1. Let X_0 be a set of points, given the discrete topology. These are the *zero cells*.
2. Form X_1 by *attaching* copies of D^1 to X_0 by gluing them along their boundaries – that is, we are given the data of continuous maps

$$f_\alpha \colon \partial D_1 = S^0 \to X_0$$

(referred to as *attaching maps*), and for each one we look at the quotient $D_1 \cup_{f_\alpha} X_0$ of the disjoint union $X_0 \coprod D^1$ where we identify the points $z \in S^0 \subseteq D^1$ and $f_\alpha(z) \in X_0$. The intervals glued in during this stage are referred to as 1-cells.
3. Then we repeat, attaching copies of D^2 to X_1 by gluing them along their boundaries – in this case, the data of the attaching maps is given by continuous maps $f_\beta \colon S^1 \to X_1$, and we form the corresponding union $D^2 \coprod_f X_0$. The disks glued in during this stage are referred to as 2-cells.
4. And so on ...

Formalizing this, we have the following definition.

Definition 1.5.7. A *finite CW complex* is a topological space obtained as a finite union $\bigcup_i X_i$ in which each stage X_i is obtained from X_{i-1} by gluing on copies of D^i as above. (The topology is the natural quotient topology induced by the gluing, and is independent of the order in which cells are attached.)

The subspace $X_n \subset X$ is referred to as the *n-skeleton*, and consists of *k*-cells for $k \le n$; if there are no cells of dimension larger than m, then the CW complex A is referred to as *m-dimensional*. Notice that the essential data of the CW complex is contained in the number of cells and the attaching maps, and the *n*-skeleton encodes all of the attaching data for objects of dimension less than n.

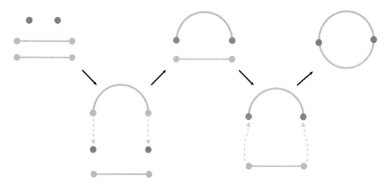

Figure 1.26 The circle S^1 can be formed by gluing two intervals along their boundaries.

Remark 1.5.8. It is also possible to consider an infinite attachment process of this kind, but the construction of the topology on the infinite union requires some care.

Example 1.5.9.

1. Any graph can be realized as a CW complex with one 0-cell for each vertex and a 1-cell for each edge (glued to the relevant vertices).
2. The circle S^1 can be given the structure of a CW complex in which $X_0 = \{0\}$ and X_1 is obtained by the map that attaches $[-1, 1]$ to 0 via the map from $\{-1, 1\}$ that takes both points to 0.
3. The circle can also be given many CW structures, as follows: take n 0-cells (points), where $n \geq 2$. Label these points as $\{x_1, \ldots, x_n\}$. Then take n 1-cells (intervals) and attach them sequentially to connect x_1, x_2, then x_2, x_3, then x_i, x_{i+1}, and finally x_n, x_1. (See Figure 1.26.)
4. In general, a sphere can be given a CW structure by taking a single 0-cell and a single n-cell and gluing the n-cell to the 0-cell along the map that sends the entire boundary to the point.
5. A torus (the surface of a doughnut) can be given the structure of a CW complex by taking a single 0-cell, two 1-cells, and a 2-cell. The two 1-cells are glued to the 0-cell to form a figure-eight, and then the 2-cell is glued to the figure-eight to make the torus. (See Figure 1.27.)

We now describe two ways to construct new CW complexes out of old that cover many interesting examples.

Definition 1.5.10. Let X and Y be CW complexes. Then $X \times Y$ has the structure of a CW complex where the cells are the products of the cells of X and Y.

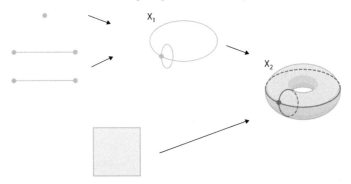

Figure 1.27 The torus can be built up by gluing together two intervals and then a
two-cell to the resulting figure-eight.

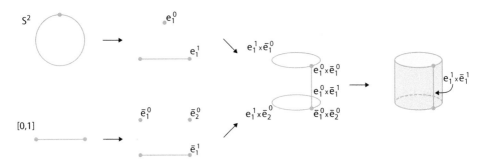

Figure 1.28 The cylinder is the product of a circle and an interval.

To be more explicit, given a cell D^n attached to X along $f: S^{n-1} \to X$ and D^m
attached to Y along $g: S^{m-1} \to Y$, we can attach a cell $D^{n+m} \cong D^n \times D^m$ to $X \times Y$
along the map $S^{n+m-1} \to X \times Y$ determined by the homeomorphism

$$S^{n+m+1} \cong (D^n \times S^{m-1}) \cup (S^{n-1} \times D^m),$$

the maps f and g, and the inclusions $D^n \to X$ and $D^m \to Y$.

Example 1.5.11.

1. The standard cylinder $S^1 \times [0, 1]$ can be given a CW complex structure as the product
   of the CW complex S^1 and the CW complex $[0, 1]$. (See Figure 1.28.)
2. The torus can be given a CW complex structure as the product of the CW complexes
   $S^1 \times S^1$.

A *subcomplex* of a CW complex is just a closed subspace determined by taking
only some of the cells.

Figure 1.29 Collapsing one of the copies of S^1 inside the cylinder $S^1 \times [0, 1]$ to a point results in a cone.

Definition 1.5.12. Let X be a CW complex and A a subcomplex; then the *quotient* X/A has a CW complex structure consisting of the cells of X that are not contained in A, along with a new 0-cell representing A. (An attaching map $\gamma \colon S^n \to X$ gives rise to an attaching map $\gamma' \colon S^n \to X \to X/A$.)

Taking the cylinder from Example 1.5.11 and taking the quotient $S^1 \times [0, 1]/S^1 \times \{0\}$ gives rise to a model for the CW complex structure on a cone; see Figure 1.29.

There are three essential results about CW complexes that justify focus on these combinatorial models of spaces.

1. Replacing an attaching map in a CW complex by a homotopic map does not change the homotopy type.
2. A homotopy equivalence $X \to Y$ of CW complexes can be detected algebraically in terms of the homotopy groups π_n.
3. Any reasonable topological space can be approximated up to homotopy equivalence by a CW complex, and for an arbitrary topological space there is an approximation up to a weak kind of equivalence. (See Definition 1.6.32 below.)

The first observation tells us that the data of a CW complex is entirely contained in the homotopy classes of the attaching maps. The second and third observations imply that if we are working up to homotopy equivalence, CW complexes are a good model for general spaces and that homotopy equivalence classes can be studied algebraically. That is, CW complexes provide a class of spaces which are constructed according to a recipe from basic building blocks and are well suited to work up to homotopy equivalence. To make the last two observations precise (notably in Theorem 1.6.31), we need to develop some algebraic background.

The next section, which briefly reviews abstract algebra (notably group theory and ring theory), may be particularly difficult for readers new to the subject. On a first reading, a quick perusal of Section 1.6.6 for a refresher on linear algebra might suffice; such readers could then skip to Section 1.7, which introduces ideas from category theory.

1.6 Algebra

A central goal of algebraic topology is to produce suitable *algebraic* invariants of topological spaces to allow us to determine whether two spaces are homeomorphic or homotopy equivalent. For example, the function which takes a topological space to the number of path components is an example of such an invariant; by Lemma 1.4.12, this invariant can serve to distinguish certain spaces with different homotopy types.

Early on in the development of the subject, it was recognized that more discriminatory power could be obtained by considering more structured algebraic objects than numbers as repositories for topological invariants. For example, the set of path components is a richer invariant than simply its size. The point is that there are no maps between numbers, but there are maps of sets – and we have seen in Lemma 1.3.26 that a continuous map of spaces induces a map of sets of path components.

It turns out that keeping even more algebraic structure leads to invariants that are computable and very informative. For example, consider the problem of distinguishing the circle from the figure-eight. Looking at homotopy classes of maps from S^1, both of these have an infinite number. But in the circle, the homotopy classes are all "multiples" of the basic one which wraps around once, and in the figure-eight all of the homotopy classes are built from combinations of the classes which wrap around one circle or the other. Algebraic invariants provide a way to make precise the intuitive notion of being "built from" or "generated by" these basic loops, and therefore let us tell these spaces apart.

In order to describe these algebraic invariants, we now turn to a quick review of the background from abstract algebra that we need. Again, our treatment is very terse and selective; we refer the reader to one of the many excellent treatments of abstract algebra, for example Artin's *Algebra* [22] or Lang's *Undergraduate Algebra* [314]. We begin by reviewing the theory of groups.

1.6.1 Groups

A group is a set with the additional structure of an "addition" operation.

Definition 1.6.1. A set G is equipped with the structure of a group if there is a distinguished element $e \in G$ and functions

$$G \times G \to G \qquad (g_1, g_2) \mapsto g_1 +_G g_2$$

and

$$G \to G \qquad g \mapsto -g$$

such that

1.

$$\forall x \in G, \qquad e +_G x = x = x +_G e,$$

2.

$$\forall x \in G, \qquad x +_G (-x) = e = (-x) +_G x,$$

3. and

$$\forall x, y, z \in G, \qquad x +_G (y +_G z) = (x +_G y) +_G z.$$

We will often write $g_1 + g_2$ rather than $g_1 +_G g_2$ and usually write 0 for e, in analogy with the notation. We sometimes use "multiplicative" notation and write $g_1 g_2$ rather than $g_1 +_G g_2$, 1 for e, and g^{-1} for the inverse of g.

Put another way, a group is a set equipped with an "addition" operation that is associative, has a unit element, and such that every element $x \in G$ has an inverse. The definition of a group is an abstraction of familiar objects from arithmetic.

Example 1.6.2.

1. The integers $\mathbb{Z}$ under the standard addition operation form a group; $x +_\mathbb{Z} y = x + y$ for $x, y \in \mathbb{Z}$. The unit is $0 \in \mathbb{Z}$, and the inverse of x is $-x$.
2. The real numbers $\mathbb{R}$ under the standard addition operation form a group; $x +_\mathbb{R} y = x + y \in \mathbb{R}$. The unit is $0 \in \mathbb{R}$ and the inverse of x is $-x$.
3. The non-zero real numbers $\mathbb{R} - \{0\}$ under multiplication form a group; the operation is $(x, y) \mapsto xy$ for $x, y \in \mathbb{R} - \{0\}$. The unit is $1 \in \mathbb{R}$ and the inverse of x is $\frac{1}{x}$. (It is the existence of inverses that requires us to restrict to non-zero reals!)
4. The set of all polynomials in $\mathbb{R}$ of degree k in a single variable t,

$$\mathcal{P}_k = \{a_0 + a_1 t + \ldots + a_k t^k \mid a_0, a_1, \ldots, a_k \in \mathbb{R}\},$$

is a group under addition of polynomials, i.e.,

$$(a_0 + a_1 t + \ldots + a_k t^k) + (b_0 + b_1 t + \ldots + b_k t^k) = (a_0 + b_0) + (a_1 + b_1)t + \ldots + (a_k + b_k)t^k.$$

The identity element is 0 and the inverse of $p(x)$ is $-p(x)$.
5. The set $C(\mathbb{R})$ of all continuous functions $f : \mathbb{R} \to \mathbb{R}$ is a group under pointwise addition, i.e.,

$$f +_{C(\mathbb{R})} g = (f + g)(x) = f(x) + g(x).$$

The identity element is the zero function $f(x) = 0$ and the inverse of a function f is $-f$.
6. The set of $n \times n$ matrices with real elements $M_n(\mathbb{R})$ is a group under matrix addition. The unit element is the zero matrix and the inverse of A is the matrix $-A$.

7. The set of invertible $n \times n$ matrices $GL_n(\mathbb{R})$ is a group under matrix multiplication where the unit element is the identity matrix and the inverse of A is the inverse matrix A^{-1}.

The example of $GL_n(\mathbb{R})$ is particularly interesting, since this group has the property that the operation is not commutative, i.e., $AB \neq BA$ in general.

Definition 1.6.3. A group G is *abelian* if for all $x, y \in G$, we have $x +_G y = y +_G x$.

Example 1.6.4. All of the examples above in Example 1.6.2 are abelian except for $GL_n(\mathbb{R})$.

The examples of groups we have discussed above are "numerical." But historically, groups arose from symmetries and rigid transformations of physical objects; for example, the set of rotations of an object in space forms a group. More abstractly, the symmetries of a finite set form a group.

Example 1.6.5.

1. The set of symmetries of a square is the group generated by two elements r and f; r is the counterclockwise rotation and f is the flip across a diagonal. These are subject to certain relations, as indicated in Figure 1.30; the group has 8 elements. In general, the *dihedral groups* D_n describe the symmetries of a regular n-gon in the plane, and have $2n$ elements.
2. The set of rotations of the unit cube $[-1, 1] \times [-1, 1] \times [-1, 1] \subset \mathbb{R}^3$ about the z-axis is the circle group S^1; we can parametrize the elements as $e^{i\theta}$, with group operation $e^{i\theta_1} e^{i\theta_2} = e^{i(\theta_1 + \theta_2)}$. (See Figure 1.31.)

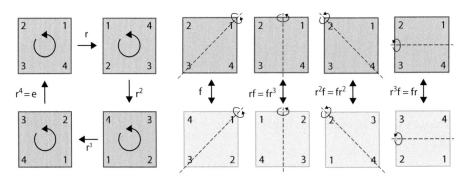

Figure 1.30 The rotation and the flip across a diagonal specify two basic symmetries of a square. Together these generate a group of order 8, the dihedral group D_4.

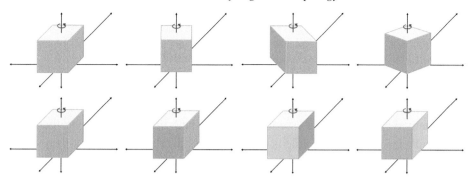

Figure 1.31 There is a natural action of the circle on a cube that rotates the cube around the z-axis. On the top row, we see some snapshots of this rotation. There is a natural subgroup isomorphic to $\mathbb{Z}/4$ inside of S^1 determined by rotations by $90°$; the action of this group on the cube is shown on the bottom row.

3. The set of rotations of $\mathbb{R}^3$ about the origin forms a group, the *special orthogonal group* $SO(3)$. This can be described as the set of orthogonal 3×3 matrices (i.e., matrices A such that $A^{-1} = A^T$) with determinant 1. The group operation is matrix multiplication. The identity is the identity map (i.e., the rotation that leaves everything fixed) and the inverse of a rotation is the "opposite" rotation.

4. Let S be an ordered set with n elements. The set of permutations of S (i.e., bijective maps $S \to S$) forms a group. The identity is the permutation that leaves every element of S in place, the group operation is given by composition of permutations, and the inverse of a permutation is the permutation that "undoes" it.

Another important arithmetic example comes from *modular arithmetic*.

Definition 1.6.6. For x and y in $\mathbb{Z}$, define $x = y \mod n$ if $x - y = kn$, for some $k \in \mathbb{Z}$. The *congruence class* of x modulo n is a subset of the form

$$\{x + kn \mid k \in \mathbb{Z}\}.$$

The classical long division algorithm implies that a congruence class has a unique smallest nonnegative representative, the remainder r when we write $x = qn + r$ via long division.

Example 1.6.7. The set of congruence classes modulo n, which we can represent as $\{0, 1, 2, \ldots, n-1\}$, forms a group that we denote by $\mathbb{Z}/n$. The identity element is 0, addition is given by letting the sum of x and y be $x + y \mod n$, and the inverse of x is $n - x \mod n$.

The preceding example has a special structure; it is a *cyclic group*, in the sense that every element other than the identity is generated by sums of a distinguished generator, for example 1. The integers $\mathbb{Z}$ are an *infinite cyclic group*, with generator 1. But not all groups are cyclic; for example, $SO(3)$ is very far from being cyclic.

1.6.2 Homomorphisms

A fundamental tenet of modern mathematics is that to understand a collection of mathematical objects it is essential to understand the maps between them. An important aspect of this principle is that invariants should "take maps to maps." We have already seen this at work in the context of topological spaces and continuous maps: a continuous map of spaces induces a map between sets of path components. In Section 1.7, we will describe an abstract framework for formalizing this insight.

In the meantime, we want to describe the correct notion of a map between groups. Recall that we singled out the class of continuous maps when describing functions between topological spaces; these were the functions that were suitably compatible with the topologies of the domain and range. Correspondingly, we are primarily interested in functions between groups which respect the group structure, in the sense of the following definition.

Definition 1.6.8. A map $f: G_1 \to G_2$ is a *group homomorphism* if

$$f(0) = 0 \quad \text{and} \quad f(x +_{G_1} y) = f(x) +_{G_2} f(y) \quad \forall x, y \in G_1.$$

Example 1.6.9.

1. The natural inclusion $\mathbb{Z} \to \mathbb{R}$ is a group homomorphism.
2. The projection $\mathbb{Z} \to \mathbb{Z}/m$ specified by the formula

$$x \mapsto x \mod m$$

   is a group homomorphism.
3. The derivative

$$\frac{d}{dt} : \mathcal{P}_k \to \mathcal{P}_{k-1}$$
$$a_0 + a_1 t + a_2 t^2 + \ldots + a_k t^k \mapsto a_1 + 2a_2 t + \ldots + k a_k t^{k-1}$$

   is a group homomorphism.
4. The trace of a square matrix with real entries (the sum of the diagonal elements) specifies a group homomorphism

$$\mathrm{Tr}: M_n(\mathbb{R}) \to \mathbb{R}.$$

Associated to a homomorphism $f: G_1 \to G_2$ are certain distinguished subsets of G_1 and G_2.

Definition 1.6.10. Let $f: G_1 \to G_2$ be a group homomorphism (Figure 1.32).

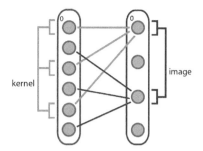

Figure 1.32 The kernel of a homomorphism f is the set of points that go to 0; the image is the set of points that f hits.

- The *kernel* of f, ker $f \subseteq G_1$, is the set of elements x such that $f(x) = 0$.
- The *image* of f, im $f \subseteq G_2$, is the set of elements y such that $y = f(x)$ for some x.

Generalizing the notion of an isomorphism of sets from Definition 1.1.8, we have the following version in the context of groups and group homomorphisms.

Definition 1.6.11. A group homomorphism $f\colon G_1 \to G_2$ is an *isomorphism* if there exists an inverse group homomorphism $g\colon G_2 \to G_1$ such that f and g demonstrate an isomorphism of sets between G_1 and G_2.

Equivalently, we have the following characterization.

Lemma 1.6.12. *Let G_1 and G_2 be groups. A group homomorphism $f\colon G_1 \to G_2$ is an isomorphism if and only if it is a bijection. As a consequence, f is an isomorphism if and only if $\ker f = \{0\}$ and $\operatorname{im} f = G_2$.*

Both ker f and im f are themselves groups, with operations inherited from G_1 and G_2 respectively. These are *subgroups* of G_1 and G_2, as we now explain.

1.6.3 New Groups from Old

Many groups of interest arise via constructions that start from an existing group. The simplest is to consider subsets of a group that inherit the structure of a group themselves.

Definition 1.6.13. A *subgroup* of a group G is a subset $H \subseteq G$ such that H is a group in its own right with the operation and unit inherited from G. That is,

1. the identity element $0 \in G$ is an element of H,
2. for any $h \in H$, $-h$ is in H, and

3. for all $h_1, h_2 \in H$, the sum $h_1 + h_2$ is in H.

We have already seen some examples of subgroups.

Example 1.6.14.

1. The special orthogonal group $SO(3)$ is a subgroup of $GL_3(\mathbb{R})$.
2. The set $\{x \in \mathbb{Z} \mid x \text{ even}\}$ is a subgroup of $\mathbb{Z}$ under addition.
3. The set $\mathcal{P}_k$ of degree at most k polynomials is a subgroup of $\mathcal{P}_{k+1}$.
4. The set $\mathcal{P}_k$ of degree at most k polynomials is a subgroup of $C(\mathbb{R})$.

The following lemma provides many other examples of subgroups.

Lemma 1.6.15. *Let $f: G_1 \to G_2$ be a group homomorphism. Then $\ker f \subseteq G_1$ is a subgroup of G_1 and $\operatorname{im} f \subseteq G_2$ is a subgroup of G_2.*

The preceding lemma is a simple exercise in the properties of group homomorphisms; for the first part, if $f(g_1) = 0$ and $f(g_2) = 0$, then

$$f(g_1 + g_2) = f(g_1) + f(g_2) = 0 + 0 = 0.$$

Given a suitable subgroup $H \subset G$, we can "collapse it out" by forming the quotient group G/H of G by a subgroup H, which is akin to the quotient topology discussed above in Proposition 1.5.1. The idea is to specify that in G/H all elements of H are identified. We will define the quotient in the setting of an abelian group G; when G is not abelian, only certain subgroups permit the construction of the quotient group.

Definition 1.6.16. Let G be an abelian group and $H \subset G$ a subgroup. Then the *quotient group* G/H is given by the set of *cosets* $gH = \{gh \mid h \in H\}$ as g varies, with group operation $(g_1 H)(g_2 H) = (g_1 g_2)H$.

(Note that a small check is required to verify that the definition of the quotient group is independent of choice of coset representative.)

Example 1.6.17.   For $\mathbb{Z}$ and the subgroup $3\mathbb{Z} = \{3k \mid k \in \mathbb{Z}\}$, the quotient $\mathbb{Z}/3\mathbb{Z}$ is isomorphic to the construction of $\mathbb{Z}/3$ described in Example 1.6.7.

A basic structural property of group homomorphisms can be usefully described in terms of the quotient group.

Theorem 1.6.18. *Let G_1 and G_2 be groups and $f\colon G_1 \to G_2$ be a group homomorphism. Then there is an isomorphism*

$$\operatorname{im} f \cong G_1/\ker f.$$

(This is true even if G_1 is not abelian; the kernel of a homomorphism allows the construction of the quotient.)

As an elaboration of this result, we can describe a large class of groups in terms of *generators and relations*.

Definition 1.6.19. A group G is *finitely generated* if there exists a finite set $S \subseteq G$ such that any $g \in G$ can be written as a (finite) sum of elements in S.

For example, any finite group is of course finitely generated. The integers $\mathbb{Z}$ are finitely generated with generator 1. On the other hand, the rationals $\mathbb{Q}$ are not finitely generated. Clearly, a finitely generated group must be countable; therefore, $\mathbb{R}$ is not finitely generated.

Definition 1.6.20. A group is *free* if there exists a collection of elements $\{g_\alpha\}$ (called the *generators*) such that every element $g \in G$ can be uniquely written as a finite sum

$$\sum_i n_i g_{\alpha_i}$$

for $n_i \in \mathbb{Z}$.

Free groups are easy to work with because group homomorphisms $F \to G$, where F is free, can be described simply as set maps from the generators of F to G. That is, to specify such a group homomorphism f, it suffices to give the data of where each generator lands in G,

$$f\left(\sum_i n_i g_{\alpha_i}\right) = \sum_i n_i f(g_{\alpha_i}).$$

Theorem 1.6.21. *Any finitely generated group G is isomorphic to the quotient of a free group by a subgroup described by specifying products of generators that are equal to 1.*

We refer to the generators of the free group as the generators of G and the products describing the subgroup as the relations of G. From an algorithmic perspective, a presentation of a group in terms of generators and relations is essential.

Example 1.6.22.

1. The integers $\mathbb{Z}$ can be represented as having the identity element 0, a single generator 1, and no relations. Here the element -1 must exist and is distinct from 1, and in general we have a description as

$$\mathbb{Z} \cong \{\ldots, -1 + (-1) + (-1), -1 + (-1), -1, 0, 1, 1 + 1, 1 + 1 + 1, \ldots\}.$$

2. The cyclic group $\mathbb{Z}/3$ is the quotient of the free group $\mathbb{Z}$ by the subgroup of relations $\{3k \mid k \in \mathbb{Z}\}$. Another way to express this is that $\mathbb{Z}/3$ can be described as having an identity element, a single generator g, and the single relation $g^3 = 1$. Then explicitly this representation describes $\mathbb{Z}/3$ as the set $\{1, g, g^2\}$ with the usual multiplication of polynomials as the group operation; $g^{-1} = g^2$, since $(g)(g^2) = g^3 = 1$.

Remark 1.6.23. Note that an interesting problem arises in this context, namely, the problem of deciding when two "words" representing group elements are equal. For instance, in the group with generator $\{x\}$ and relation $x^4 = 1$, one might ask whether x^8 and x^{16} are the same. This is known as the *word problem* for a group, and it is an important classical result that this is *undecidable*. That is, there does not exist any algorithm (computer program) to solve this problem in general! This hardness result is the core of many demonstrations that certain mathematical questions are undecidable.

However, for our purposes it will suffice to consider free abelian groups. A free abelian group with one generator is an infinite cyclic group and is isomorphic to $\mathbb{Z}$. In order to describe free groups with more generators, we need the notion of a product.

Definition 1.6.24. Let G_1 and G_2 be groups (not necessarily abelian). Then the Cartesian product $G_1 \times G_2$ denotes the group structure on the Cartesian product of sets with identity element $(0_{G_1}, 0_{G_2})$, operation

$$(g_1, g_2) + (g_1', g_2') = (g_1 + g_1', g_2 + g_2'),$$

and the inverse of (g_1, g_2) is $(-g_1, -g_2)$.

Lemma 1.6.25. *A free abelian group with k generators is isomorphic to a product of k copies of $\mathbb{Z}$: one copy of $\mathbb{Z}$ for each generator.*

In this case, any finitely generated abelian group G is isomorphic to a quotient $\mathbb{Z}^n/H$, for a subgroup $H \in \mathbb{Z}^n$; here n is the size of a set S of generators. More precisely, we have the following fundamental characterization, the *structure theorem for finitely generated abelian groups*.

Theorem 1.6.26. *Let G be a finitely generated abelian group. Then there is an isomorphism*

$$G \cong \underbrace{\mathbb{Z} \times \mathbb{Z} \times \ldots \mathbb{Z}}_{k} \times \mathbb{Z}/p_1^{n_1} \times \mathbb{Z}/p_2^{n_2} \times \ldots \times \mathbb{Z}/p_m^{n_m}.$$

Here the p_i are prime and not necessarily distinct.

The number k of factors of $\mathbb{Z}$ is known as the *rank* of G. The part of G that does not consist of copies of $\mathbb{Z}$ is often referred to as the *torsion*. The rank is unique and the torsion is unique up to rearrangement.

1.6.4 The Group Structure on $\pi_n(X, x)$

We now return to justify referring to the homotopy groups $\pi_n(X, x)$ (from Definition 1.4.10) as groups. Specifically, we explain the following theorem.

Theorem 1.6.27. *When $n > 0$, the set of homotopy classes of maps*

$$\pi_n(X, x) = \{(S^n, *), (X, x)\}$$

can be given the structure of a group, where the identity element is the constant map and the composition is given by "composing" maps.

We begin by considering the case of $\pi_1(X, x)$. Given two loops $\gamma_1, \gamma_2 \colon S^1 \to X$, we can produce a new loop as follows. Regard the maps γ_1 and γ_2 as paths (maps from $[0, 1]$ to X) such that

$$\gamma_1(0) = \gamma_2(0) = \gamma_1(1) = \gamma_2(1) = x.$$

Then define $\gamma_1\gamma_2 \colon [0, 1] \to X$ to be the loop specified by the formula

$$(\gamma_1\gamma_2)(t) = \begin{cases} \gamma_1(2t) & 0 \le t < \frac{1}{2}, \\ \gamma_2(2t - 1) & \frac{1}{2} \le t \le 1. \end{cases}$$

That is, we reparameterize and do γ_1 on the first half of the interval and γ_2 on the second half of the interval (see Figure 1.33). Since $\gamma_1(0) = \gamma_2(1) = x$, this defines a map $S^1 \to X$.

Note that the composition we have just defined is not associative prior to passing to homotopy classes of maps; that is, $(\gamma_1\gamma_2)\gamma_3$ is not the same map as $\gamma_1(\gamma_2\gamma_3)$. Specifically, given $\gamma_1, \gamma_2, \gamma_3 \colon S^1 \to X$, $(\gamma_1\gamma_2)\gamma_3$ does γ_1 on $[0, \frac{1}{4})$, γ_2 on $[\frac{1}{4}, \frac{1}{2})$ and γ_3 on $[\frac{1}{2}, 1]$ whereas $\gamma_1(\gamma_2\gamma_3)$ does γ_1 on $[0, \frac{1}{2})$, γ_2 on $[\frac{1}{2}, \frac{3}{4})$, and γ_3 on $[\frac{3}{4}, 1]$. However, there is a natural straight-line homotopy connecting $(\gamma_1\gamma_2)\gamma_3$ to $\gamma_1(\gamma_2\gamma_3)$; see Figure 1.34.

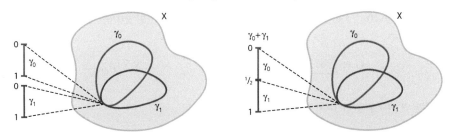

Figure 1.33 Two loops γ_0 and γ_1 are added by reparameterizing, doing γ_0 on $[0, \frac{1}{2})$ and γ_1 on $[\frac{1}{2}, 1]$.

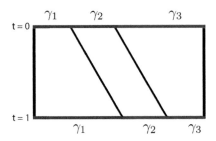

Figure 1.34 A linear homotopy connects the two associativity parameterizations.

Analogously, we define the inverse of $\gamma \colon S^1 \to X$ to be the loop traversed in the opposite direction:

$$\gamma^{-1}(t) = \gamma(1 - t).$$

Once again, note that $\gamma\gamma^{-1}$ is not equal to the constant map until we pass to homotopy classes of maps; there is a homotopy connecting $\gamma\gamma^{-1}$ to the constant map that takes all of S^1 to x.

Generalizing this, we can put a group structure on $\pi_n(X, x)$ for $n > 1$ as follows. We regard maps from $S^n \to X$ as maps from $[0, 1]^n \to X$ which take the boundary of $[0, 1]^n$ to x and again compose by reparametrizing. We have choices about how to reparameterize; fixing an index $1 \le i \le n$, we define

$$\gamma_1\gamma_2(x_1, x_2, \ldots, x_n) = \begin{cases} \gamma_1(x_1, x_2, \ldots, 2x_i, \ldots, x_n) & x_i \in [0, \frac{1}{2}) \\ \gamma_1(x_1, x_2, \ldots, 2x_i - 1, \ldots, x_n) & x_i \in [\frac{1}{2}, 1]. \end{cases}$$

(See Figure 1.35.)

Once again, there is a homotopy that makes this associative. In fact, for $n > 1$, we have the following improvement of Theorem 1.6.27.

Theorem 1.6.28. *For $n > 1$, the homotopy group $\pi_n(X, x)$ is abelian.*

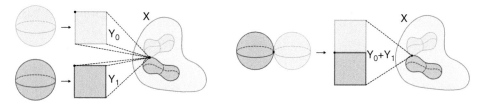

Figure 1.35 Two maps from spheres γ_0 and γ_1 are added by reparameterizing, doing γ_0 on the upper square and γ_1 on the lower square.

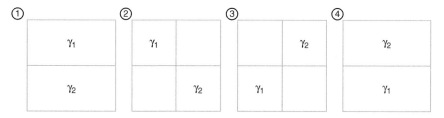

Figure 1.36 The commutativity homotopy involves moving two squares past each other. Here the unlabeled squares are sent to the basepoint.

A picture of the commutativity homotopy that proves Theorem 1.6.28 is shown in Figure 1.36.

Given a continuous map $f: X \to Y$, composition defines a map

$$\pi_n(X, x) \to \pi_n(Y, f(x))$$

via

$$\gamma: S^1 \to X \mapsto (f \circ \gamma): S^1 \to X \to Y.$$

In fact, this map specifies a group homomorphism when $n > 0$.

Lemma 1.6.29. *Let $f: X \to Y$ be a continuous map of spaces. There are induced group homomorphisms for $n > 0$*

$$\pi_n(X, x) \to \pi_n(Y, f(x)).$$

The importance of the homotopy groups as algebraic invariants is provided by the following two theorems. First, homotopy groups are invariants of the homotopy type.

Proposition 1.6.30. *Let $f: X \to Y$ be a homotopy equivalence. Then the induced group homomorphism*

$$\pi_n(X, x) \to \pi_n(Y, f(x))$$

is an isomorphism.

Although the converse to this is not in general true, we have the following basic result.

Theorem 1.6.31 (Whitehead).   *Let* $f: X \to Y$ *be a continuous map of CW complexes such that the induced maps* $\pi_n(X, x) \to \pi_n(Y, f(x))$ *are isomorphisms for every* $n \geq 0$ *and* $x \in X$. *Then* f *is a homotopy equivalence between* X *and* Y.

We say that a map f that induces isomorphisms of homotopy groups as in Theorem 1.6.31 is a *weak homotopy equivalence*.

Definition 1.6.32.   Let $f: X \to Y$ be a continuous map of topological spaces. Then f is a *weak homotopy equivalence* (or *weak equivalence*) if the induced group homomorphisms

$$\pi_n(X, x) \to \pi_n(Y, f(x))$$

are isomorphisms for every $n \geq 0$ and $x \in X$.

This is a central definition in modern algebraic topology. Moreover, it turns out that any topological space X is weakly homotopy equivalent to a CW complex. (Warning: note that not every space is homotopy equivalent to a CW complex. For example, the sequence $\{\frac{1}{n}\}$ along with its limit point 0 is not homotopy equivalent to a CW complex. See also discussion of the "long line", e.g., in [369, §10].)

Weak homotopy equivalence is not an equivalence relation on spaces, but we work with the transitive closure, which is the smallest equivalence relation it generates.

Definition 1.6.33.   We will refer to the equivalence class of a space under the relation of weak homotopy equivalence as its *weak homotopy type*.

We now have a number of different equivalence relations on topological spaces. These relations are progressively weaker – the relationship between them can be summarized as follows.

1. If two spaces X and Y are homeomorphic, then they are homotopy equivalent.
2. If two spaces X and Y are homotopy equivalent, then they are weakly homotopy equivalent.

Theorem 1.6.31 shows that for CW complexes, the latter two equivalence relations coincide. In contrast, determining when a homotopy equivalence is even homotopic to a homeomorphism is quite difficult; a restricted version of this problem is the subject of *surgery theory*.

Although homotopy groups are easy to define and the Whitehead theorem implies that they are complete invariants of the homotopy type of a CW complex (in the presence of a continuous map), the best known algorithms for computing them in general are intractable. As a consequence, we are led to search for algebraic invariants which are rich enough to distinguish a wide class of spaces but can be easily computed.

1.6.5 Rings and Fields

We return to the basic examples of the abelian groups $(\mathbb{Z}, +, 0)$ and $(\mathbb{R}, +, 0)$, the integers and the real numbers with group operation given by addition. These groups have additional structure, namely a second operation – multiplication. Moreover, multiplication interacts nicely with addition, for example, the distributive property tells us that $x(y + z) = xy + xz$.

Definition 1.6.34. A *ring* is a set R that has an abelian group structure (with operation denoted by + and identity by 0) along with a distinguished element $1 \in R$ and an additional operation

$$R \times R \to R \qquad (x, y) \mapsto xy$$

such that

$$\forall x \in G, \quad 1x = x = x1,$$

and

$$x(yz) = x(yz).$$

In addition, we require that the new operation satisfy the distributive law with respect to the abelian group structure:

$$x(y + z) = xy + xz.$$
$$(x + y)z = xz + yz.$$

A ring has both an additive identity element (typically written 0) and a multiplicative identity element (typically written 1). A *multiplicative inverse* for an element $x \in R$ is an element y such that $xy = 1$; typically we write x^{-1} for the multiplicative inverse. An element $x \in R$ that has a multiplicative inverse is called a *unit*. Not all elements of a ring have multiplicative inverses.

Definition 1.6.35. A *field* F is a ring such that for all $x \in R$ such that $x \neq 0$ (where 0 denotes the additive identity), x has a multiplicative inverse x^{-1} such that $xx^{-1} = x^{-1}x = 1$.

Example 1.6.36.

1. The integers $\mathbb{Z}$ with addition and multiplication form a ring, but not a field as there is no multiplicative inverse for any $x \neq \pm 1$.
2. The rational numbers $\mathbb{Q}$ with addition and multiplication form a ring and in fact a field; the inverse of $\frac{p}{q}$ is $\frac{q}{p}$, which is well defined as long as $p \neq 0$.
3. The set of congruence classes $\mathbb{Z}/m$ forms a ring, where multiplication is also computed by taking the remainder of xy when divided by m. When m is prime, this is in fact a field; the inverse can be computed using the long division algorithm. The fields $\mathbb{Z}/p$ are referred to as finite fields of order p.

In addition to $\mathbb{R}$, the most important fields for our purposes are the rational numbers $\mathbb{Q}$ and the finite fields $\mathbb{Z}/p$ (which are often denoted $\mathbb{F}_p$). For any field F, we can consider a vector space with F as the scalars. Although we assume that the reader has some familiarity with linear algebra in the context of the fields $\mathbb{R}$ and $\mathbb{C}$, we quickly review linear algebra from a more abstract perspective.

1.6.6 Vector Spaces and Linear Algebra

Linear algebra studies the geometric structure of solutions to systems of linear equations; these turn out to form lines and (hyper)planes. It is a central example of the power of using algebraic structures to encode geometry. There are an enormous number of textbooks on linear algebra. For an abstract treatment, Axler's book [24] is very clearly written. For applications, Meyer's book is an excellent introduction [349].

The basic object in linear algebra is the vector space, which is an abstraction of some parts of the structure of Euclidean space.

Definition 1.6.37. Let F be a field. An *F-vector space* is an abelian group V with an additional operation called *scalar multiplication*

$$F \times V \to V \qquad (x, v) \mapsto xv$$

that is

1. associative, $x_1(x_2 v) = (x_1 x_2)v$,
2. distributive with respect to addition in F, $(x_1 + x_2)v = x_1 v + x_2 v$,
3. distributive with respect to the group operation in V, $x(v_1 + v_2) = xv_1 + xv_2$, and
4. compatible with the multiplicative unit in F, $1v = v$.

We call the elements of V vectors.

Example 1.6.38.

1. The field F itself gives a first example of a vector space.
2. The set $\{0\}$ is a vector space.
3. When $F = \mathbb{R}$, familiar examples of vector spaces are given by $\mathbb{R}^n$, where $\mathbb{R}$ acts by multiplication in each component.
4. More generally, for any field F, the product $F^n = \prod_{i=1}^{n} F$ of n copies of F is a vector space where F acts by componentwise multiplication.

Other basic examples of vector spaces are given by *subspaces*.

Definition 1.6.39. A *subspace* W of a vector space V is a subgroup such that $kw \in W$ for all $k \in F, w \in W$. (That is, W is closed under addition in V and scalar multiplication.)

Vector spaces can sometimes be decomposed into pieces by subspaces.

Definition 1.6.40. Let U and W be subspaces of the vector space V. If $U \cap W = \{0\}$, the *direct sum* $U \oplus W$ is defined to be the collection

$$U \oplus W = \{u + w \mid u \in U, w \in W\}.$$

More generally, given two vector spaces V_1 and V_2 we can define the external direct sum $V_1 \oplus V_2$ to consist of pairs (v_1, v_2) for $v_1 \in V_1$ and $v_2 \in V_2$, with the operations defined coordinatewise. Then regarding V_1 and V_2 as subspaces of $V_1 \oplus V_2$ (via $\{(v_1, 0)\}$ and $\{(0, v_2)\}$, respectively), $V_1 \oplus V_2$ arises as their direct sum as in Definition 1.6.40.

Although a priori it appears that subspaces could take on many forms, in fact, it turns out that all examples of finite-dimensional vector spaces look like the examples in 1.6.38. For example, the subspaces of $\mathbb{R}^2$ are $\{0\}$, $\mathbb{R}^2$ itself, and lines that pass through the origin. Each such line looks like a copy of $\mathbb{R}$. Similarly, the subspaces of $\mathbb{R}^3$ are $\{0\}$, lines through the origin (which look like $\mathbb{R}$), planes through the origin (which look like $\mathbb{R}^2$), and $\mathbb{R}^3$ itself. To be precise about this fact, we need the notion of a basis, which generalizes the idea of the coordinate axes in Euclidean space.

Definition 1.6.41. Let V be a vector space. For a subset $B = \{b_1, b_2, \ldots, b_n\} \subseteq V$,

1. *B spans V* if any vector $z \in V$ can be written as a sum

$$z = \sum_{i=1}^{n} a_i b_i, \quad a_i \in F,$$

i.e., any vector admits a representation as a weighted sum of basis elements,

2. the set B is *linearly independent* if the only solution to the equation

$$\sum_{i=1}^{n} a_i b_i = 0$$

is $a_i = 0$ for all i, and

3. B is a basis for a vector space V if it *spans* and is *linearly independent*.

Linear independence is a way of saying that a set of vectors has no redundancy, in the following sense.

Lemma 1.6.42. *The set B is linearly independent if and only if when $z \in V$ can be written as a sum*

$$z = \sum_{i=1}^{n} a_i b_i,$$

then this representation is unique, i.e., the values $\{a_i\}$ are unique.

Example 1.6.43.

1. In $\mathbb{R}^2$, the standard unit vectors along the axes $(1, 0)$ and $(0, 1)$ form a basis.
2. In $\mathbb{R}^2$, the vectors $(3, 4)$ and $(-1, 1)$ form a basis. In fact, any two non-collinear vectors form a basis. (See Figure 1.37 and Figure 1.38 for an example.)
3. In $\mathbb{R}^3$, any three vectors that do not all lie in the same plane form a basis.
4. More generally, in $\mathbb{R}^n$, any n vectors that do not all lie in the same *hyperplane* (i.e., subspace of strictly smaller dimension) form a basis.

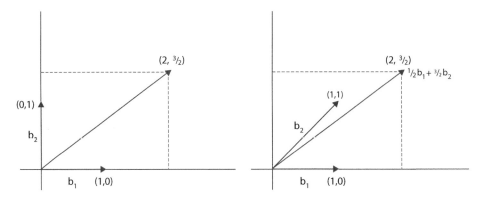

Figure 1.37 Any vector in $\mathbb{R}^2$ can be written uniquely as a linear combination $a_1 v_1 + a_2 v_2$ as long as v_1 and v_2 do not lie on the same line. We illustrate this for the vector $(2, 3/2)$.

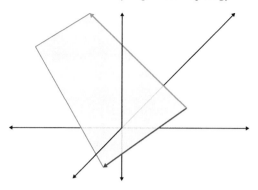

Figure 1.38 Any two-dimensional subspace of $\mathbb{R}^3$ is a plane; two vectors that specify the plane provide a basis.

By providing coordinates for describing points in vector spaces, bases are essential for calculation. They also give rise to the notion of dimension of a vector space.

Proposition 1.6.44. *Any basis for a vector space V has the same size.*

In light of the preceding proposition, the following definition makes sense.

Definition 1.6.45. The *dimension* of a vector space V is the size of a basis.

In fact, the dimension is a complete invariant of finite-dimensional vector spaces. To be precise, we need to define the notion of a map between vector spaces.

Definition 1.6.46. Let V and W be vector spaces. A *linear transformation* $f: V \to W$ is a map of sets such that

$$f(ax + by) = af(x) + bf(y).$$

That is, a linear transformation is a group homomorphism that preserves scalar multiplication.

The kernel and image of a linear transformation $f: V_1 \to V_2$ are subgroups of V_1 and V_2 respectively. In fact, they are vector spaces themselves.

Lemma 1.6.47. *Let $f: V_1 \to V_2$ be a linear transformation. Then $\ker f$ is a subspace of V_1 and $\operatorname{im} f$ is a subspace of V_2.*

One of the appealing things about linear transformations is that they can be expressed in a concise and algorithmically tractable way. Since a vector space is the set of linear combinations of basis elements, a linear transformation can be specified simply in terms of its action on the basis. Put another way, linear transformations can be specified by matrices; the ith column of the matrix describes the effect of the linear transformation applied to the basis vector b_i.

Definition 1.6.48.   A linear transformation $f\colon V \to W$ is an *isomorphism* if it is injective and surjective, or equivalently if there is an inverse transformation $g\colon W \to V$ such that $g \circ f = \mathrm{id}_V$ and $f \circ g = \mathrm{id}_W$.

Theorem 1.6.49.   *Any vector space of dimension n is isomorphic to F^n.*

The homotopy groups $\pi_n(X, x)$ are groups that are very hard to compute. The basic topological invariants that will be our algorithmic focus take values in vector spaces; the fact that a linear transformation can be specified by a matrix will ensure that computation is tractable. Before we introduce these invariants, we will have a brief interlude about category theory, which provides a formal context to describe the invariants.

1.7 Category Theory

The basic topological invariants we study are *functions* that take as input topological spaces (represented by CW complexes or simplicial complexes) and output finitely generated abelian groups or vector spaces:

$$\left\{ \begin{array}{c} \text{finite} \\ \text{simplicial complexes} \end{array} \right\} \to \left\{ \begin{array}{c} \text{abelian} \\ \text{groups} \end{array} \right\}.$$

However, these invariants are better than functions, as they turn out to have an additional essential property: they take continuous maps between spaces to group homomorphisms. We have already seen an example of this in Lemma 1.6.29, which states that a continuous map $f\colon X \to Y$ induces a group homomorphism $\pi_k(X, x) \to \pi_k(Y, f(x))$. Formalizing this property of algebraic invariants was one of the original motivations for the invention of *category theory*.

Category theory provides a language for capturing common phenomena in different domains. For example, the notion of an isomorphism has appeared in a variety of different contexts in this chapter. A motivating idea at the core of the development of category theory is the notion that properties of mathematical objects (e.g., topological spaces) can often be characterized entirely in terms of maps from other objects. We have seen this philosophy at work already in our discussion of homotopy groups. Properties that can be expressed purely in terms of such data are often referred to as *formal*; a common slogan is that category theory is a way to make formal things formal. We give a very brief overview of category theory; the classic text is Mac Lane [337]. Riehl has written two excellent recent books, [428] which is a more elementary introduction and [427] which is an in-depth discussion from the perspective of algebraic topology. Spivak's book [480] strives to provide context for categorical notions in applications.

Definition 1.7.1. A *category* $\mathcal{C}$ is a collection of objects ob($\mathcal{C}$) and for each pair of objects $x, y \in$ ob($\mathcal{C}$) a set of *morphisms* or maps $\mathrm{Hom}_{\mathcal{C}}(x, y)$ satisfying the following conditions.

1. For all objects $w, x, y \in \mathcal{C}$, there is a composition map

$$\mathrm{Hom}_{\mathcal{C}}(x, y) \times \mathrm{Hom}_{\mathcal{C}}(w, x) \to \mathrm{Hom}_{\mathcal{C}}(w, y)$$

   that takes the morphisms $f\colon w \to x$ and $g\colon x \to y$ to the composite morphism $g \circ f\colon w \to y$.
2. There is a distinguished element $\mathrm{id}_x \in \mathrm{Hom}_{\mathcal{C}}(x, x)$, the identity map.
3. Composition is associative and unital. Associativity means that given $f \in \mathrm{Hom}(w, x)$, $g \in \mathrm{Hom}(x, y)$, and $h \in \mathrm{Hom}(y, z)$, we have the equality of composites

$$(h \circ g) \circ f = h \circ (g \circ f).$$

   Unitality means that

$$\mathrm{id}_x \circ f = f = f \circ \mathrm{id}_w.$$

The composition map is written in the "backwards" order above in order to align with the standard notation for composition, i.e. $(g \circ f)(-) = g(f(-))$.

Remark 1.7.2. The sophisticated reader will notice that we are being incautious about set theory and using the somewhat vague term "collection"; as we discussed in Section 1.1, Russell's paradox tells us that there is no "set of all sets," and so there cannot be a set of objects for the category of sets. We refer the reader to the category theory references for more discussion of this point.

We have many familiar examples of categories underlying the notions we have already seen.

Example 1.7.3.

1. The category Set with objects sets and morphisms maps of sets.
2. The category Grp with objects groups and morphisms homomorphisms.
3. The category Vect with objects vector spaces and morphisms linear transformations.
4. The category Top with objects topological spaces and morphisms continuous maps.

5. The category Met of metric spaces and metric maps (i.e., maps $f: X \to Y$ such that $\partial_Y(f(x_1), f(x_2)) \le \partial_X(x_1, x_2)$).
6. The category Ho(Top) with objects topological spaces and morphisms homotopy classes of continuous maps.
7. A partially ordered set forms a category. For example, $\mathbb{N}$ is a category with objects the elements of $\mathbb{N}$ and a morphism between x and y if $x \le y$.

Moreover, for any category we can obtain new categories by taking subsets of the collection of objects and morphisms.

Definition 1.7.4. A category $\mathcal{D}$ is a *subcategory* of a category $\mathcal{C}$ if each object of $\mathcal{D}$ is an object of $\mathcal{C}$ and for every $x, y \in \mathrm{ob}(\mathcal{D})$, we have

$$\mathrm{Hom}_{\mathcal{D}}(x, y) \subseteq \mathrm{Hom}_{\mathcal{C}}(x, y).$$

When we have equality in the previous inclusion, $\mathcal{D}$ is called a *full* subcategory of $\mathcal{C}$.

Example 1.7.5.

1. The category Ab with objects abelian groups and morphisms homomorphisms is a full subcategory of Grp.
2. The category of finite dimensional vector spaces and linear transformations is a full subcategory of Vect.
3. The category of topological spaces and morphisms the homeomorphisms is a subcategory of Top, although it is not full.

In any category, there is an intrinsic notion of two things being "the same" that comes directly from the data of the category.

Definition 1.7.6. Let $\mathcal{C}$ be a category. A map $f \in \mathrm{Hom}_{\mathcal{C}}(x, y)$ is an isomorphism if there exists $g \in \mathrm{Hom}_{\mathcal{C}}(y, x)$ such that

$$f \circ g = \mathrm{id}_y \in \mathrm{Hom}_{\mathcal{C}}(y, y) \qquad \text{and} \qquad g \circ f = \mathrm{id}_x \in \mathrm{Hom}_{\mathcal{C}}(x, x).$$

The notion of a categorical isomorphism encompasses all of the definitions we have seen so far.

Example 1.7.7.

1. In Set, an isomorphism is an isomorphism of sets (as defined in Definition 1.1.8).
2. In Grp, an isomorphism is an isomorphism of groups (as defined in Definition 1.6.11).
3. In Vect, an isomorphism is an isomorphism of vector spaces (as defined in Definition 1.6.48).

4. In Top, an isomorphism is a homeomorphism (as defined in Definition 1.3.27).
5. In Ho(Top), an isomorphism is the equivalence class of a homotopy equivalence (as defined in Definition 1.4.3).

Since the only properties we can express in a category are described in terms of morphisms and the result of composing morphisms, the notion of a *commutative diagram* is of basic importance. A commutative diagram refers to a collection of objects and morphisms such that any morphisms between two objects coincide. For example, in the commutative square

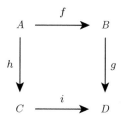

we are expressing the compatibility requirement that $g \circ f = i \circ h$ as a morphism in $\mathrm{Hom}_{\mathcal{C}}(A, D)$.

The structure of the category itself can encode many interesting properties of objects; we now give some examples.

Definition 1.7.8. An *initial object* in a category is an object c such that $\mathrm{Hom}_{\mathcal{C}}(c, z)$ consists of a single point for any z. That is, there is a unique morphism from c to any other object.

Dually, a *terminal object* is an object d such that $\mathrm{Hom}_{\mathcal{C}}(z, d)$ consists of a single point for any z.

These notions are not necessarily unique, although they are unique up to isomorphism, i.e., any two initial or terminal objects are isomorphic.

Example 1.7.9.

1. In Set, the initial object is the empty set $\emptyset$ and any one-point set is a terminal object. We will denote a choice of terminal object by $*$.
2. In Grp the initial object is the trivial group and the terminal object is also the trivial group.
3. In Top the initial object is $\emptyset$ and the one-point space is a terminal object. We will again denote a choice of terminal object by $*$.

The point here (no pun intended) is that the special properties of the one-point set or the one-point space can be expressed in a way which generalizes to any category; the properties can be expressed solely in terms of data about maps to and from other objects.

Moreover, commutative diagrams allow us to succinctly express algebraic properties. For instance, a group is an object G in the category Set along with a morphism $m\colon G \times G \to G$, a morphism $u\colon * \to G$, and a morphism $i\colon G \to G$ such that the following holds.

1. The diagram

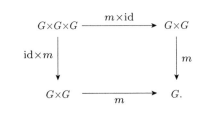

commutes; this expresses associativity.
2. The diagrams

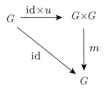

and

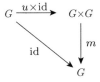

commute; this expresses the property of the identity element.
3. The diagrams

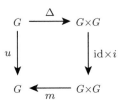

and

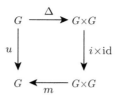

commute, where $\Delta\colon G \to G{\times}G$ is the diagonal map specified by the assignment $x \mapsto (x, x)$ and $u\colon G \to G$ is the composite $G \to * \to G$ specified by the unique map $G \to *$ and the unit map $u\colon * \to G$. These diagrams express the property of the inverse.

We can also describe gluing constructions (e.g., the attaching of cells in Definition 1.5.6) purely in terms of categorical data. Suppose that we have a diagram

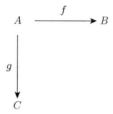

in some category $\mathcal{C}$. Explicitly, this means that

1. A, B, and C are objects in the category $\mathcal{C}$,
2. f is an element of $\mathrm{Hom}_{\mathcal{C}}(A, B)$ and g is an element of $\mathrm{Hom}_{\mathcal{C}}(A, C)$.

We will refer to the data of this diagram as D. We now want to explain how to give a general construction of an object that is produced by "gluing" B to C along A.

To motivate the abstract definition, it is instructive to consider how to describe such a construction. Within category theory, the only way we can express the properties of such a gluing is to talk about morphisms either into or out of it, i.e., to talk about the gluing in terms of its relationship to other objects. Let us consider how to specify a map out of the gluing of B and C along A, to some other object X. Such a map should be determined by maps

$$B \to X \quad \text{and} \quad C \to X$$

that agree on the image of $A \to B$ and $A \to C$. Moreover, we would like the gluing to be the "smallest" such object. We can make all of this precise as follows.

Definition 1.7.10. The *pushout* of D is an object P equipped with morphisms $p_1 \colon B \to P$ and $p_2 \colon C \to P$ such that the square

commutes, and for any pair of morphisms $a \colon B \to X$ and $b \colon C \to X$ such that $a \circ f = b \circ f$ there is a unique morphism $h \colon P \to X$ such that the diagram

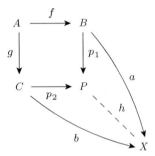

commutes.

The requirement that for *any* maps a and b there is a map $h \colon P \to X$ enforces the condition that P be the smallest candidate, up to isomorphism; if there were another object P′ that satisfied the same property as P, then P would map to P′ and P′ would map to P and by the uniqueness of the induced mappings P and P′ would be isomorphic.

Example 1.7.11.

1. In Set, the pushout of the maps $\emptyset \to \{0, 1, 2\}$ and $\emptyset \to \{7, 8, 9\}$ is the set $\{0, 1, 2, 7, 8, 9\}$.
2. More generally, the pushout in Set of the maps $\emptyset \to B$ and $\emptyset \to C$ is the disjoint union of B and C, i.e., the set consisting of all the elements of B and C.
3. In Set, the pushout of the maps

$$f \colon \{0, 1\} \to \{3, 4, 5\} \qquad f(0) = 3, \ f(1) = 4$$

and

$$g \colon \{0, 1\} \to \{a, b, c\} \qquad g(0) = g(1) = a$$

is the union of $\{3, 4, 5\}$ and $\{a, b, c\}$ with a identified with 3 and 4.

4. More generally, the pushout in Set of maps $A \to B$ and $A \to C$ is the set specified by taking the disjoint union of B and C and identifying $f(a)$ and $g(a)$.

As a set, the pushout in the category of topological spaces is described by the pushout in sets. However, we need to specify the topology on this identification. We have already seen how to perform this kind of construction in our discussion of the quotient topology.

Example 1.7.12. Let $f\colon A \to B$ be a continuous map of topological spaces. The pushout of the diagram

where $A \to *$ is the unique map taking all of A to $*$, is the quotient space generated by the partition of B given by $\{b\}$ for $b \in B - f(A)$ and $f(A)$. That is, the pushout is isomorphic to the quotient $B/f(A)$.

Example 1.7.13. Let B be a cylinder $S^1 \times [0,1]$, C a point $*$, and A be the circle S^1. Take $f\colon A \to B$ to be the inclusion $S^1 \to S^1 \times [0,1]$ specified by $x \mapsto (x,0)$ and $g\colon A \to *$ to be the unique map taking all $x \in S^1$ to the point $*$. Then the pushout

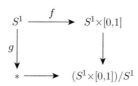

is a cone (see Figure 1.39).

The description of the quotient topology in terms of the pushout gives rise to the following interesting characterization.

Corollary 1.7.14. *Let $f\colon A \to B$ be a continuous map of topological spaces. A map from the quotient space $B/f(A) \to X$ is determined by a map $B \to X$ which takes all of A to a point.*

More generally, we use the quotient topology to describe the pushout in topological spaces.

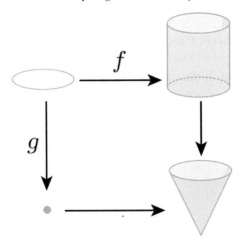

Figure 1.39 The cone can be formed by collapsing one end of a cylinder to a point.

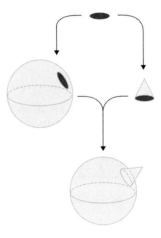

Figure 1.40 Gluing along a common subspace.

Example 1.7.15.   The pushout of $f: A \to B$ and $g: A \to C$ is the quotient of the disjoint union $B \amalg C$ given by identifying the points $f(a)$ and $g(a)$ for each $a \in A$. For example, if f and g are injective, we look at the partition of $B \amalg C$ given by the points in $B \setminus f(A)$, the points in $C \setminus g(A)$, and all subsets of the form $\{f(a), g(a)\}$ for $a \in A$.

As this last example suggests, the gluing in CW complexes can also be described in terms of pushouts (Figure 1.40). Specifically, the constructions $D^n \amalg_f X_i$ arising in the description of CW complexes (in Definition 1.5.7) are precisely pushouts.

Example 1.7.16.

1. Let B and C be the subspaces of $\mathbb{R}^3$ defined as

$$B = \{(x, y, z) \mid x^2 + y^2 + z^2 = 1, z \geq 0\}$$

and

$$C = \{(x, y, z) \mid x^2 + y^2 + z^2 = 1, z \leq 0\},$$

and let A be the circle

$$\{(x, y, 0) \mid x^2 + y^2 = 1\}.$$

Take $f\colon A \to B$ and $g\colon A \to C$ to be the evident inclusions. Then the pushout is precisely the unit sphere

$$S^2 = \{(x, y, z) \mid x^2 + y^2 + z^2 = 1\}.$$

2. More generally, we have the following pushout diagram

$$
\begin{array}{ccc}
S^{n-1} & \longrightarrow & D^n \\
\downarrow & & \downarrow \\
D^n & \longrightarrow & S^n
\end{array}
$$

3. We can do the same kind of construction with solid disks and hemispheres; see Figure 1.41.

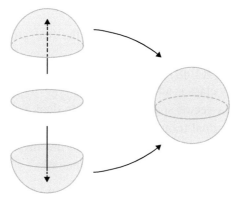

Figure 1.41 The solid sphere can be represented as the pushout of two hemispheres along a shared bounding disk.

1.7.1 Functors

For our purposes, perhaps the most important definition from category theory is the notion of a function between categories, called a *functor*. The topological invariants we study will all be functors from geometric categories to algebraic ones, for example, the function that assigns the set of path components to a topological space X.

Definition 1.7.17.   Let C and D be categories. A *functor* $F: C \to D$ is specified by

1. a function

$$F: \mathrm{ob}(C) \to \mathrm{ob}(D),$$

2. for all $x, y \in \mathrm{ob}(C)$ a function

$$F: \mathrm{Hom}_C(x, y) \to \mathrm{Hom}_D(Fx, Fy)$$

such that $F(\mathrm{id}_x) = \mathrm{id}_{Fx}$ (the maps preserve the identity) and $Fg \circ Ff = F(g \circ f)$ (the maps are compatible with the composition).

We can reinterpret and strengthen Lemma 1.3.26 in this language.

Lemma 1.7.18.   *The assignment of path components is a functor from the category* Top *to the category* Set.

Functorial constructions are ubiquitous in mathematics.

1. The functor Grp $\to$ Set that forgets the group structure is an example of a *forgetful functor*.
2. The functor Set $\to$ Grp that takes a set to the free group on generators the elements of the set is a functor.
3. The functor Top $\to$ Ho(Top) that takes each space to itself and each continuous map to its homotopy class is a functor.
4. The assignment of a vector space to its double dual and each linear transformation to its double dual transformation is a functor from Vect to itself. (The assignment of a vector space to its dual reverses the direction of the arrows, and specifies what is known as a *contravariant* functor.)

In the language of this section, we can now describe algebraic topology as the study of functors from Top to an algebraic category (e.g., Grp or Vect). For example, let Top$_*$ be the category of *based spaces*, i.e., the objects are pairs (X, x) of a

topological space and a "basepoint" $x \in X$ and a morphism $(X, x) \rightarrow (Y, y)$ is a continuous map $f: X \rightarrow Y$ such that $f(x) = y$. Then Lemma 1.6.29 can be interpreted
and strengthened as the following assertion.

Lemma 1.7.19. *For $n > 0$, the construction $\pi_n(X, x)$ specifies a functor from* Top$_*$
to Grp.

All of the invariants we study will be functorial, and in fact we will see that the
functoriality of our invariants is one of the essential facts that ensures their good
properties in algorithmic contexts.

Remark 1.7.20. Correspondingly, one might hope to cast a certain amount of
molecular biology as the study of suitable functors from genotype to phenotype.
Here the initial problem of setting up categories of genotype and phenotype, where
for instance morphisms might represent mutation and certain physical changes, is
of basic interest.

The final notion we need from category theory is the idea of a *natural
transformation*; this is a map between functors.

Definition 1.7.21. Let F and G be functors from $\mathcal{C}$ to $\mathcal{D}$. A natural transformation
$\tau: F \rightarrow G$ is specified by:

1. a map $\tau_x: F(x) \rightarrow G(x)$ for every object $x \in \mathrm{ob}(\mathcal{C})$, and
2. commuting squares

$$
\begin{array}{ccc}
F(x) & \longrightarrow & F(y) \\
\downarrow {\scriptstyle \tau_x} & & \downarrow {\scriptstyle \tau_y} \\
G(x) & \longrightarrow & G(y)
\end{array}
$$

for every morphism $x \rightarrow y$ in $\mathrm{Hom}_{\mathcal{C}}(x, y)$.

Example 1.7.22.

1. The most important example for us comes in the context of functors $\mathbb{N} \rightarrow \mathcal{C}$, for a
   category $\mathcal{C}$. A functor $F: \mathbb{N} \rightarrow \mathcal{C}$ is specified by a sequence

$$
F(0) \rightarrow F(1) \rightarrow F(2) \rightarrow \dots,
$$

and so a natural transformation $\tau\colon F \to G$ is determined by the commuting diagrams

$$
\begin{array}{ccccccc}
F(0) & \longrightarrow & F(1) & \longrightarrow & F(2) & \longrightarrow & \cdots \\
\big\downarrow{\scriptstyle \tau_0} & & \big\downarrow{\scriptstyle \tau_1} & & \big\downarrow{\scriptstyle \tau_2} & & \\
G(0) & \longrightarrow & G(1) & \longrightarrow & G(2) & \longrightarrow & \cdots
\end{array}
$$

2. For any category $\mathcal{C}$ and object $x \in \mathrm{ob}(\mathcal{C})$, there is a functor

$$
\hom(x, -)\colon \mathcal{C} \to \mathrm{Set}
$$

that takes an object y to the set of maps $\mathrm{Hom}(x, y)$ and a map $f\colon y_1 \to y_2$ to the map $\mathrm{Hom}(x, y_1) \to \mathrm{Hom}(x, y_2)$ induced by composition with f. Now, for any pair of functors $\hom(x_1, -)$ and $\hom(x_2, -)$, any map $x_2 \to x_1$ induces a natural transformation $\hom(x_1, -) \to \hom(x_2, -)$. (This is a version of the Yoneda lemma.)

1.8 Simplicial Complexes

Our most basic model of a geometric object is a topological space, which we introduced in Section 1.3. Topological spaces are too general to be feasible for algorithmic purposes, however. In Section 1.5, we introduced CW complexes, which are a more restrictive notion of a topological space; this data is a recipe for building a space from spheres and disks. Although CW complexes are an incredibly useful notion in modern algebraic topology, they are still not concise enough for algorithmic purposes. The issue is that describing the data of an attaching map $f\colon S_n \to X_n$ in general requires an infinite amount of information. That is, despite the fact that there are a limited number of building blocks, the instructions about how to glue them together are not simple enough.

We now describe an older model of topological spaces, the category of simplicial complexes, that is entirely discrete: here a space will be specified by gluing simple pieces together in a very small number of ways. As long as we are willing to work up to homotopy equivalence or weak homotopy equivalence, it will turn out that this is a general model of topological spaces. Our treatment follows the fantastic introduction given in [368].

Simplicial complexes are generalizations of graphs. And in this guise, there are many examples of simplicial complexes that are studied by systems biologists. For example, any of the networks that are described as graphs (e.g., protein interaction networks, regulatory networks, ecological interaction networks) are simplicial complexes. Thus, in a precise sense the theory we are developing here is a way to talk about higher dimensional networks.

Suppose that we are given points $\{x_0, \ldots, x_k\}$ in $\mathbb{R}^n$. We will assume that these points satisfy the condition that the set of vectors in $\mathbb{R}^n$ represented by the differences

$$\{x_1 - x_0, x_2 - x_0, \ldots, x_k - x_0\}$$

are linearly independent. For example, a set $\{x_0, x_1, x_2\}$ will satisfy this condition if the points do not all lie on the same line.

Definition 1.8.1. The *k-simplex* spanned by the points $\{x_0, \ldots, x_k\}$ is the set of all points

$$z = \sum_{i=0}^{k} a_i x_i, \qquad \sum_{i=0}^{k} a_i = 1.$$

For a given z, we refer to a_i as the *ith barycentric coordinate*.

Example 1.8.2.

1. A 0-simplex is a point.
2. A 1-simplex is a line segment (with endpoints the points x_0 and x_1).
3. A 2-simplex is a triangle with vertices the points $\{x_0, x_1, x_2\}$.

(See Figure 1.42 for examples of geometric simplices.)

The simplices are the basic building blocks for a simplicial complex; roughly speaking, a simplicial complex is a collection of simplices glued along their edges (or "edges" of their edges).

Definition 1.8.3. The *interior* of a simplex S spanned by the points $\{x_0, \ldots, x_k\}$, denoted int(S), is the subset of points where $a_i > 0$ for all the barycentric coordinates a_i. The *boundary* bd(S) is defined to be $S \setminus$ int(S). (See Figure 1.43.)

It is straightforward to check that for any *n*-simplex S, there are homeomorphisms

$$\mathrm{bd}(S) \cong S^{n-1} \qquad \text{and} \qquad S \cong D^{n+1}.$$

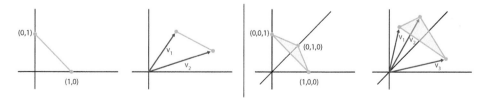

Figure 1.42 Geometric simplices specified by a set of vectors (including 0). On the left, the simplices are determined by the standard axial unit vectors; on the right, they are specified by the indicated vectors.

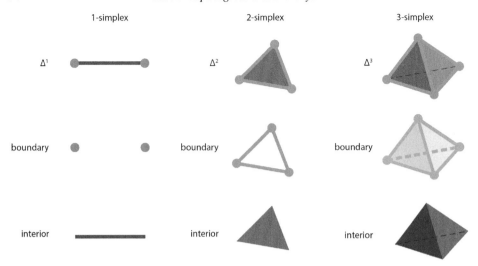

Figure 1.43 The boundary of a standard simplex is a combinatorial sphere; the interior is an open disk.

Therefore, there is a close analogy between gluing together simplices and building CW complexes. The advantage of working with simplices rather than CW complexes is that the boundaries of a simplex decompose into unions of simplices; we will be able to use a very restricted universe of attaching maps.

Definition 1.8.4.   For a simplex S spanned by the points $P = \{x_0, \dots, x_k\}$, a *face* of S refers to any simplex spanned by a subset of P.

Example 1.8.5.

1. There are no non-empty faces of a 0-simplex.
2. The non-empty faces of a 1-simplex determined by the points x_0 and x_1 are the two 0-simplices spanned by $\{x_0\}$ and $\{x_1\}$ respectively.
3. The non-empty faces of a 2-simplex determined by the points $\{x_0, x_1, x_2\}$ are the edges of the triangle and the vertices, the three 1-simplices determined by $\{x_0, x_1\}$, $\{x_1, x_2\}$, and $\{x_2, x_0\}$ and the three 0-simplices $\{x_0\}$, $\{x_1\}$, and $\{x_2\}$.

The following lemma is the key observation that allows us to glue together simplices in a simple way (Figure 1.44).

Lemma 1.8.6.   *Let S be a simplex. The union of all of the faces of S is* $\mathrm{bd}(S)$.

We now define the notion of a simplicial complex.

Definition 1.8.7.   A *simplicial complex* X in $\mathbb{R}^n$ is a set of simplices in $\mathbb{R}^n$ such that:

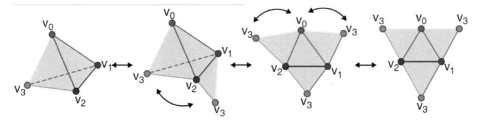

Figure 1.44 The boundary of the standard 3-simplex is a hollow pyramid; unfolding it makes clear how the 2-simplices that form the faces are glued along edges and vertices.

1. every face of a simplex in X is also a simplex in X, and
2. the intersection of two simplices in X is a face of each of them.

The zero simplices of a simplicial complex are referred to as the *vertices*. More generally, the collection of simplices of dimension at most k is referred to as the k-skeleton of the simplicial complex; we will denote the k-skeleton by X_k. For simplicity, we will restrict attention to simplicial complexes with finitely many simplices, referred to as *finite simplicial complexes*.

Definition 1.8.8. The *geometric realization* $|X|$ of a finite simplicial complex X is the topological space given by the union of simplices, given the subspace topology. (Here we regard the union as a subspace of $\mathbb{R}^n$.)

The geometric realization of a simplicial complex can be given the structure of a CW complex, where the cells correspond to the simplices and the attaching maps are determined by the faces.

Example 1.8.9. A circle can be given the structure of a simplicial complex (up to homeomorphism) in $\mathbb{R}^2$ where the 0-simplices are the points $(0,0)$, $(1,0)$, and $(1,1)$ and the 1-simplices are the line segments specified by the equations

$$x + 0y = 1, \quad 0x + y = 0, \quad \text{and } x + y = 1,$$

where $x, y \in [0,1]$. (In fact, as explained in Example 1.8.21, we can analogously model the circle with n 0-simplices and n 1-simplices connecting them for any n. See also Figure 1.45)

Remark 1.8.10. As with infinite CW complexes (recall Remark 1.5.8), we can make sense of the geometric realization of an infinite simplicial complex,

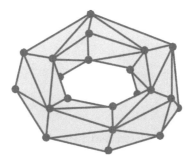

Figure 1.45 Up to homeomorphism, a torus can be triangulated as a simplicial complex.

but describing the topology is somewhat more complicated. However, all of the examples we consider in this book will be finite.

In a precise sense, a simplicial complex can be thought of as a higher dimensional generalization of a graph.

Example 1.8.11. A simplicial complex that has only 0-simplices and 1-simplices represents a graph embedded in Euclidean space, where the 0-simplices are the vertices and the 1-simplices are the edges.

We can assemble simplicial complexes into a category; for this purpose, we need an analogue of a continuous map.

Definition 1.8.12. Let X and Y be simplicial complexes. A *simplicial map* $f\colon X \to Y$ is specified by a map $X_0 \to Y_0$ such that whenever

$$\{z_0, \dots, z_k\} \subset X_0$$

span a simplex of X,

$$\{f(z_0), f(z_1), \dots, f(z_k)\}$$

span a simplex of Y.

Therefore, we can form a category with objects the simplicial complexes and morphisms the simplicial maps. It is useful to characterize the isomorphisms in this category.

Definition 1.8.13. Let X and Y be simplicial complexes. An *isomorphism of simplicial complexes* is a simplicial map $f\colon X \to Y$ that is a bijection on 0-simplices and such that for any $k > 1$, a collection of vertices $\{x_1, \ldots, x_k\}$ specifies a simplex of X if and only if $\{f(x_1), \ldots, f(x_k)\}$ is a simplex of Y.

Moreover, a simplicial map can be extended to a continuous map $f\colon |X| \to |Y|$ by linear interpolation:

$$f\left(\sum_{i=0}^{n} a_i x_i\right) = \sum_{i=0}^{n} a_i f(x_i).$$

Put another way, geometric realization is a functor.

Lemma 1.8.14. *Geometric realization specifies a functor from the category of simplicial complexes and simplicial maps to the category of topological spaces and continuous maps.*

One inconvenience with working with simplicial complexes as specified in Definition 1.8.7 is the dependence on a choice of embedding in some ambient Euclidean space $\mathbb{R}^n$. For example, ensuring that simplices intersect properly can require solving equations. Fortunately, it turns out that the data of a simplicial complex can be abstracted even further; all that is really important is the data of how many simplices there are and which faces they are glued along.

Definition 1.8.15. An *abstract simplicial complex* is a set X of finite non-empty sets such that if A is an element of X then so is every non-empty subset of A.

1. Each element of X represents a simplex; we refer to elements of X as (abstract) simplices.
2. The dimension of an abstract simplex A is $|A| - 1$, where here $|-|$ denotes the number of elements of a set.
3. Any non-empty subset of a simplex A is a face of A.
4. The vertices of X are the one-point sets in X. (Notice that any simplex of X is a union of vertices.)
5. More generally, we will denote the subset of X consisting of sets of cardinality $\leq k + 1$ as X_k, the k-skeleton.

We have a natural generalization of Definition 1.8.12 to the setting of abstract simplicial complexes.

Definition 1.8.16. A *map of abstract simplicial complexes* $f\colon X \to Y$ is specified by a map of sets $f\colon X_0 \to Y_0$ with the property that for any element $\{x_0, \ldots, x_k\}$ in X, $\{f(x_0), \ldots, f(x_k)\}$ is an element of Y.

Therefore, we have a category with objects the abstract simplicial complexes and morphisms the simplicial maps.

Definition 1.8.17. Let X and Y be abstract simplicial complexes. A simplicial map $f\colon X \to Y$ is an isomorphism if f is a bijection on 0-simplices and $\{x_0, \ldots, x_k\}$ is an element of X if and only if $\{f(x_0), \ldots, f(x_k)\}$ is an element of Y.

We now explain the relationship between abstract simplicial complexes and the simplicial complexes of Definition 1.8.7, which to be clear we will refer to as *geometric* simplicial complexes.

Lemma 1.8.18. *Let X be a geometric simplicial complex spanned by the points $x_0, \ldots, x_k \subseteq \mathbb{R}^n$. Then there is an associated abstract simplicial complex specified by the collection of subsets of the vertices of X which span a simplex in X.*

Two geometric simplicial complexes are isomorphic if and only if their associated abstract simplicial complexes are isomorphic. Moreover, every abstract simplicial complex can be uniquely associated to a geometric simplicial complex.

Theorem 1.8.19. *For every abstract simplicial complex S, there exists a geometric simplicial complex $\tilde{S}$ such that S is associated to $\tilde{S}$.*

The preceding theorem allows us to define the geometric realization of an abstract simplicial complex in terms of the geometric realization of the associated geometric simplicial complex. Once again, geometric realization is a functor.

Lemma 1.8.20. *The geometric realization of the associated simplicial complex specifies a functor $| - |$ from the category of abstract simplicial complexes and simplicial maps to the category of topological spaces and continuous maps.*

Example 1.8.21.

1. The abstract simplicial complex

$$\{\{v_0\}, \{v_1\}, \{v_2\}, \{v_0, v_1\}, \{v_1, v_2\}, \{v_2, v_0\}, \{v_0, v_1, v_2\}\}$$

   describes the 2-simplex and its faces; the geometric realization has the homotopy type of a disk in $\mathbb{R}^2$.
2. Removing the interior from the previous example, the abstract simplicial complex

$$\{\{v_0\}, \{v_1\}, \{v_2\}, \{v_0, v_1\}, \{v_1, v_2\}, \{v_2, v_0\}\}$$

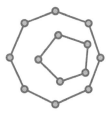

Figure 1.46 Two different models of the simplicial circle.

describes the boundary of the 2-simplex; the geometric realization has the homotopy type of a circle in $\mathbb{R}^2$. (In fact, Example 1.8.9 is homeomorphic to the geometric realization of this complex.)

3. More generally, we can make an abstract simplicial complex which models the circle using n vertices

$$\{v_0, v_1, \ldots, v_{n-1}\}$$

and n 1-simplices

$$\{\{v_0, v_1\}, \{v_1, v_2\}, \ldots, \{v_{n-2}, v_{n-1}\}, \{v_{n-1}, v_0\}\}.$$

(See Figure 1.46 for examples of this.)

4. The previous examples are all of two kinds; we can form the standard simplex Δ^n by taking a single n-simplex $[v_0, \ldots, v_n]$ and all of its subsets. The boundary $\partial\Delta^n$ is given by removing the n-simplex from the complex Δ^n.

5. Although computationally tractable, simplicial complexes describing even relatively simple surfaces can be large; see Figure 1.47 for a representation of a complex modeling a torus.

A next question one might wonder about is whether every topological space is homeomorphic or at least homotopy equivalent to a simplicial complex. In the case of homeomorphism, this kind of question turns out to be very difficult to answer. But for homotopy equivalence, there is a simple and satisfying criterion.

Proposition 1.8.22. *Let X be an abstract simplicial complex. The geometric realization $|X|$ is a CW complex with an n-cell for each n-simplex of X.*

Proposition 1.8.23. *Let X be a CW complex. Then X is homotopy equivalent to the geometric realization of a simplicial complex K. Moreover, if X is a finite CW complex, then K can be taken to be a finite simplicial complex.*

Thus a topological space is homotopy equivalent to the geometric realization of a simplicial complex if and only if it is homotopy equivalent to a CW complex.

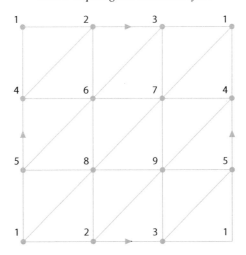

Figure 1.47 This diagram represents the vertices, 1-simplices, and 2-simplices of an abstract simplicial complex with realization homeomorphic to the torus. (Note that we identify the top edge with the bottom edge and the left edge with the right edge.)

But what about the morphisms? That is, can every continuous map $|X| \to |Y|$ be described as the geometric realization of a simplicial map? To be precise, we might ask the following question.

Question 1.8.24. Let X and Y be abstract simplicial complexes. Is every continuous map $|X| \to |Y|$ homotopic to the geometric realization of a simplicial map $X \to Y$?

As the question is posed, the answer is no.

Example 1.8.25. Let S_1 be the minimal abstract simplicial complex that models the circle; S_1 has vertices x_0, x_1, and x_2 and 1-simplices $\{x_0, x_1\}$, $\{x_1, x_2\}$, and $\{x_2, x_0\}$. If we consider simplicial maps from $S_1 \to S_1$, it is clear that there is no way to model the continuous maps $S^1 \to S^1$ given by $t \mapsto e^{kt(2\pi i)}$ for $k > 1$. That is, we cannot represent homotopy classes of maps that wrap the circle around itself more than once.

However, this deficiency can be repaired. The counterexample in Example 1.8.25 works because the "feature scale" of the domain is not fine enough. We can improve the situation using the notion of *subdivision*. In this case, if we use a model of the circle with n vertices and $n - 1$ 1-simplices, as n increases we can represent maps which wrap around the circle more and more. More generally, we can subdivide any simplicial complex by dividing the simplices into unions of smaller simplices.

The resulting complex has geometric realization homeomorphic to the original one. Since we do not need these results we do not discuss them further here, but in fact there is a fundamental result (the simplicial subdivision theorem) that guarantees that any homotopy class of maps $|X| \to |Y|$ can be represented by a simplicial map from some subdivision of X to Y.

We now turn to the discussion of algebraic invariants of topological spaces that can be computed in terms of combinatorial operations on simplicial complexes. The oldest and simplest example of such an invariant is the Euler characteristic.

1.9 The Euler Characteristic

A basic and classical combinatorial invariant associated to a CW complex or an abstract simplicial complex is the Euler characteristic.

Definition 1.9.1. Let X be a finite CW complex, with cells of dimension at most n. The *Euler characteristic* of X is defined to be the alternating sum

$$\chi(X) = \sum_{i=0}^{n} (-1)^i k_i,$$

where k_i denotes the number of i-cells.

Equivalently, we can define the Euler characteristic of a finite simplicial complex directly.

Definition 1.9.2. Let X be a finite simplicial complex, with simplices of dimension at most n. The Euler characteristic of X is defined to be the alternating sum

$$\chi(X) = \sum_{i=0}^{n} (-1)^i k_i$$

where k_i denotes the number of i-simplices (see Figure 1.48).

It is straightforward to verify that these two notions are consistent under geometric realization.

The Euler characteristic is a very appealing invariant insofar as it does not depend on any information about the way in which cells or simplices are glued together, just their counts. As a consequence, it is very easy to compute. However, it is not completely clear from the definition what sorts of equivalences the Euler characteristic is preserved by. It is easy to see that χ is an isomorphism invariant for simplicial complexes.

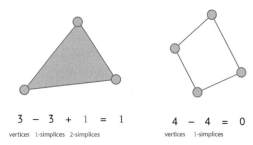

$$3 \; - \; 3 \; + \; 1 \; = \; 1 \qquad\qquad 4 \; - \; 4 \; = \; 0$$

vertices 1-simplices 2-simplices   vertices   1-simplices

Figure 1.48 The Euler characteristic of a finite simplicial complex is computed as the alternating sum of the counts of simplices.

Lemma 1.9.3. *Let $f \colon X \to Y$ be an isomorphism of simplicial complexes. Then $\chi(X) = \chi(Y)$.*

But this is not tremendously useful; as we have seen in Example 1.8.21, there are many non-isomorphic models for the circle S^1. We would like there to be a well-defined Euler characteristic for "the circle" that does not depend on the simplicial model. Direct computation is encouraging, however – all the models of the circle have n vertices and n 1-simplices, and therefore have Euler characteristic 0. It turns out that $\chi(X)$ is a homotopy invariant for CW complexes.

Another concern about the Euler characteristic is that it does not reflect simplicial maps. The issue is simply that numbers are not rich enough to support functoriality. A central motivation for constructing invariants of topological spaces that land in algebraic categories (e.g., groups or vector spaces) is to provide enough structure for them to be functors.

1.10 Simplicial Homology

In this section, we finally develop the central invariant that we will use in topological data analysis, the homology groups. The homology groups will be a collection of functors indexed on the natural numbers

$$H_n \colon \mathrm{Simp} \to \mathrm{Vect}_{\mathbb{F}}, \quad n \geq 0.$$

Let X be an abstract simplicial complex. Roughly speaking, the homology groups of X are going to encode information about the way in which the simplices in successive dimensions are glued together. For the definition, we will need to pick an *orientation* for the simplices – in the case of a 1-simplex, this amounts to picking a direction for the line segment connecting the two vertices.

Let X be an abstract simplicial complex and σ a simplex. We will pick an ordering for the set of vertices in σ. Consider the case of a 2-simplex $[v_0, v_1, v_2]$.

Then there are six possible orderings: (v_0, v_1, v_2), (v_0, v_2, v_1), (v_1, v_0, v_2), (v_1, v_2, v_0), (v_2, v_0, v_1), and (v_2, v_1, v_0). However, we want to regard the possible choices of orientation for this 2-simplex as twofold, either clockwise or counterclockwise. We can express this by identifying orderings that are given by "rotations" of the vertices.

Definition 1.10.1. An *orientation* of the vertices of a simplex σ is an equivalence class of orderings of the vertices under the equivalence relation that two orderings are the same if they differ by an even permutation. (Recall that an even permutation is one that can be written as the composite of an even number of transpositions.)

Each k-simplex can be given one of two possible orientations for $k > 0$; there is only a single orientation for a vertex. We now assume that we have chosen orientations for the k-simplices of X; this can be done arbitrarily. We let $[v_0, \ldots, v_k]$ denote the oriented simplex specified by the vertices $\{v_0, \ldots, v_k\}$, where the orientation is specified by the ordering of the vertices.

1.10.1 Chains and Boundaries

We now explain the building blocks for the homology groups, the chain groups and the boundary homomorphism. These provide algebraic encodings of the combinatorial information of a simplicial complex. We start with the case of coefficients in a field $\mathbb{F}$, as this is most relevant for topological data analysis.

Definition 1.10.2. The *k-chains* $C_k(X; \mathbb{F})$ is the vector space with basis the set of oriented k-simplices. That is, elements of $C_k(X; \mathbb{F})$ are linear combinations of generators $\{g_\sigma\}$, where σ varies over the oriented k-simplices of X.

Example 1.10.3. Consider the abstract simplicial complex

$$X = \{[v_0], [v_1], [v_2], [v_0, v_1], [v_1, v_2]\}.$$

1. The space of 0-chains $C_0(X; \mathbb{F})$ for X is a vector space which is isomorphic to $\mathbb{F} \oplus \mathbb{F} \oplus \mathbb{F}$. We think of $C_0(X; \mathbb{F})$ as having elements of the form

$$a_0 v_0 + a_1 v_1 + a_2 v_2, \quad a_0, a_1, a_2 \in \mathbb{F},$$

   where generators correspond to the vertices v_0, v_1, and v_2 respectively.
2. The space of 1-chains $C_1(X; \mathbb{F})$ for X is a vector space which is isomorphic to $\mathbb{F} \oplus \mathbb{F}$. We think of $C_1(X; \mathbb{F})$ as having elements of the form

$$a_0 g_{01} + a_1 g_{12}, \quad a_0, a_1 \in \mathbb{F},$$

   where g_{01} and g_{12} are generators corresponding to the two 1-simplices of X.

3. The space of 2-chains $C_2(X; \mathbb{F})$ (and all higher chain groups) is the trivial vector space $\{0\}$ since there are no k-simplices for $k > 1$.

We now define a linear transformation $\partial_k \colon C_k(X; \mathbb{F}) \to C_{k-1}(X; \mathbb{F})$, the *boundary map*. As we will see, this is an algebraic way to encode the boundary of a simplex.

Definition 1.10.4. The linear transformation

$$\partial_k \colon C_k(X; \mathbb{F}) \to C_{k-1}(X; \mathbb{F})$$

is specified on the generators as

$$\partial_n([v_0, \ldots, v_k]) \mapsto \sum_{i=0}^{k} (-1)^i [v_0, \ldots, \hat{v}_i, \ldots, v_k]$$

where the $\hat{v}_i$ notation means we delete that vertex. The homomorphism is then specified by extending linearly to all of $C_k(X; \mathbb{F})$. (The orientation of the image is determined by the ordering of the vertices.)

Notice that this expression has a clear geometric interpretation: the boundary map applied to a simplex is precisely the alternating sum over the faces that make up the boundary of the simplex. (See Figure 1.49.)

Example 1.10.5.

1. The boundary of the 1-simplex $[v_0, v_1]$ is $v_1 - v_0$.
2. The boundary of the 2-simplex $[v_0, v_1, v_2]$ is $[v_1, v_2] - [v_0, v_2] + [v_0, v_1]$.

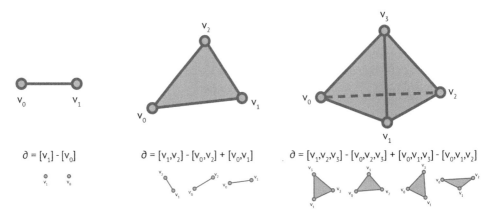

Figure 1.49 The boundary map applied to a simplex is the alternating sum of the simplices along the boundary.

The boundary map has the special property that applying it twice is 0; "the boundary of a boundary is 0."

Lemma 1.10.6. *The composite $\partial_k \circ \partial_{k+1} = 0$.*

Checking this is an easy algebraic argument; the alternating signs result in cancellation.

Example 1.10.7. We compute $\partial_1 \circ \partial_2$ applied to the 2-simplex $[v_0, v_1, v_2]$. As in Example 1.10.5 above,

$$\partial_2([v_0, v_1, v_2]) = [v_1, v_2] - [v_0, v_2] + [v_0, v_1],$$

and applying ∂_1 we obtain

$$\begin{aligned} \partial_1 \partial_2([v_0, v_1, v_2]) &= \partial_1([v_1, v_2]) - \partial_1([v_0, v_2]) + \partial_1([v_0, v_1]) \\ &= (v_2 - v_1) - (v_2 - v_0) + (v_1 - v_0) \\ &= v_2 - v_1 - v_2 + v_0 + v_1 - v_0 \\ &= 0. \end{aligned}$$

As an immediate corollary, we have the following.

Corollary 1.10.8. *For any simplicial complex X and natural number k,*

$$\mathrm{im}(\partial_{k+1}) \subseteq \ker(\partial_k).$$

1.10.2 Homology Groups

We now define the homology groups associated to the simplicial complex; the kth homology group H_k measures the failure of the inclusion of $\mathrm{im}(\partial_{k+1})$ in $\ker(\partial_k)$ to be an isomorphism. The idea of the homology groups is to take the subgroup of $C_k(X)$ of *cycles*, i.e., $\ker(\partial_k)$, and impose the equivalence relation that two chains c_1 and c_2 are *homologous* if their difference $c_1 - c_2$ is a *boundary*, i.e., if $c_1 - c_2$ is an element of $\mathrm{im}(\partial_{k+1})$.

Definition 1.10.9. The kth *homology group* with $\mathbb{F}$-coefficients $H_k(X; \mathbb{F})$ is defined to be the quotient group $\ker(\partial_k)/\mathrm{im}(\partial_{k+1})$. (In fact, this quotient group inherits the structure of a vector space.)

The zeroth homology group has a very natural interpretation.

Theorem 1.10.10. *Let X be an abstract simplicial complex. The homology group $H_0(X; \mathbb{F})$ is a vector space on generators in bijection with the path components of X.*

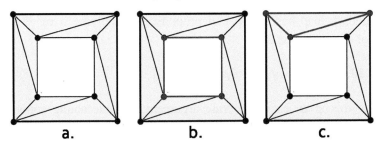

a.   b.   c.

Figure 1.50 (a) gives a simplicial complex for an annulus. The blue paths in pictures (b) and (c) are examples of cycles in the complex. The cycle in picture (b) is not the boundary of any collection of simplices in the complex; it represents a non-zero class in the first homology group. In contrast, (c) is the boundary of a simplex and therefore is 0 in the homology group.

As we make precise below in Theorem 1.10.29, the first homology group is closely related to the fundamental group and hence to loops in X (see Figure 1.50).

Crudely, we can think of homology groups as the set of cycles in $C_k(X; \mathbb{F})$ that *are not* the boundaries of elements of $C_{k+1}(X; \mathbb{F})$. Roughly speaking, the fact that an element γ in $C_k(X; \mathbb{F})$ is a cycle means that it encloses a k-dimensional region, and the fact that γ is not a boundary means that the interior of the region is not part of the space X.

More precisely, consider the simplicial complex $\partial \Delta_k$, consisting of the boundary of the standard k-simplex. There is a cycle consisting of the alternating sum of the $(k-1)$-simplices; this is the boundary of the (missing) k-simplex. But this cycle cannot be a boundary, since there are no k-simplices. Thus, it specifies a class in the homology group H_{k-1}; this class detects the "hole." But if we fill the hole in, we get the standard simplex Δ_k, and now this cycle is clearly in the image of δ_k, and so vanishes in homology. More generally, given a simplicial complex X that contains $\partial \Delta_k$ but not the k-simplex, there will be a homology class representing that hole. Of course, this analysis does not directly apply to "larger" holes (with boundaries that are the union of many $(k-1)$-simplices), but a similar analysis does apply.
    Summarizing:

1. H_0 a measure of path components of X,
2. H_1 is a measure of the one dimensional "holes" in X, and
3. more generally, H_k is a measure of k-dimensional geometric features of X, specifically, a count of the number of k-dimensional "holes" in X.

One of the advantages of simplicial homology is that it is easily computable given the data of an abstract simplicial complex. We illustrate this with some examples below.

Example 1.10.11.

1. Let S be the abstract simplicial complex $\{[v_0], [v_1], [v_0, v_1]\}$; this represents the interval. Then

$$\begin{cases} C_0(S;\mathbb{F}) \cong \mathbb{F} \oplus \mathbb{F}, \\ C_1(S;\mathbb{F}) \cong \mathbb{F}, \\ C_i(S;\mathbb{F}) = 0, \quad i > 1. \end{cases}$$

The boundary map $\partial_1 : C_1(S;\mathbb{F}) \to C_0(S;\mathbb{F})$ is specified by

$$1 \in \mathbb{F} \mapsto (1, -1) \in \mathbb{F} \oplus \mathbb{F}.$$

Then $H_0(S) = \mathbb{F}$, since $\ker(\partial_0)$ is all of $C_0(S;\mathbb{F})$ and the image of ∂_1 is $\mathbb{F}$. $H_1(S;\mathbb{F}) = 0$, as the kernel of ∂_1 is 0. And all $H_i(S;\mathbb{F}) = 0$ for $i > 1$.

Interpreting geometrically, this answer tells us that S represents a topological space that has one path component and no holes.

2. Let S be the abstract simplicial complex $\partial\Delta^2$, with vertices

$$\{[v_0], [v_1], [v_2]\}$$

and 1-simplices

$$\{[v_0, v_1], [v_1, v_2], [v_2, v_0]\}.$$

This complex is a model for the circle. Then

$$\begin{cases} C_0(S;\mathbb{F}) \cong \mathbb{F} \oplus \mathbb{F} \oplus \mathbb{F}, \\ C_1(S;\mathbb{F}) \cong \mathbb{F} \oplus \mathbb{F} \oplus \mathbb{F}, \\ C_2(S;\mathbb{F}) = 0, \quad i > 1. \end{cases}$$

Since $\partial_1([v_0, v_1]) = v_1 - v_0$, $\partial_1([v_1, v_2]) = v_2 - v_1$, and $\partial_1([v_2, v_0]) = v_0 - v_2$, it is straightforward to check that $\ker(\partial_1)$ is $\mathbb{F}$ with generator $[v_0, v_1] + [v_1, v_2] - [v_2, v_0]$. Therefore, $H_1(S) \cong \mathbb{F}$ and a similar argument shows that $H_0(S) \cong \mathbb{F}$. Specifically, $\ker(\partial_0)$ must be all of $C_0(S) \cong \mathbb{F} \oplus \mathbb{F} \oplus \mathbb{F}$. The computations of the image of ∂_1 above imply that in the quotient by $\text{im}(\partial_1)$, we have that $v_0 = v_1$ since $v_0 + \partial_1([v_0, v_1]) = v_1$. Similarly, $v_1 = v_2$. Therefore, the quotient must be $\mathbb{F}$, generated by the coincident coset of v_0, v_1, and v_2. Interpreting geometrically, this example tells us that S represents a topological space that has one path component and one one-dimensional hole (see Figure 1.51).

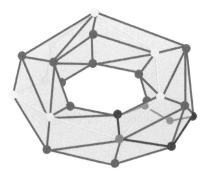

Figure 1.51 The red and yellow paths indicate representative generators for H_1 of the torus, which is $\mathbb{F} \oplus \mathbb{F}$.

3. More generally, for the simplicial complex Δ^{n+1} modeling S^n (i.e., the boundary of the standard $(n + 1)$-simplex), we compute the answer

$$
\begin{cases}
H_0(S^n; \mathbb{F}) \cong \mathbb{F}, \\
H_n(S^n; \mathbb{F}) \cong \mathbb{F}, \\
H_k(S^n; \mathbb{F}) = 0, \quad k \neq 0, n.
\end{cases}
$$

This computation makes precise the sense in which we can think of the nth homology group as capturing information about n-dimensional holes.

Of particular relevance for topological data analysis is the fact that simplicial homology is algorithmically tractable; ∂_k can be expressed as a matrix where each column specifies the image in $C_{k-1}(S; \mathbb{F})$ of a generator of $C_k(S; \mathbb{F})$. We can then compute the image and kernel using linear algebra manipulations. Specifically, using Gaussian elimination we put ∂_k and ∂_{k+1} into Smith normal form; the rank of the homology group can then be computed in terms of the ranks of ∂_k and ∂_{k+1}.

Theorem 1.10.12. *Given a simplicial complex, there exists an algorithm to compute $H_k(-; \mathbb{F})$ whose running time is polynomial (cubic) in the total number of $(k + 1)$-simplices, k-simplices, and $(k - 1)$-simplices.*

1.10.3 Homology of Chain Complexes

The impressionistic description of the homology groups as computing information about k-dimensional holes strongly suggests that the groups H_k are homotopy invariants. To provide context for stating this kind of invariance result, it is useful to describe the homology groups as functors. As we have emphasized, much of the power of the invariants of algebraic topology comes because they are functorial. We can check directly from the definition that homology is in fact a functor

$$H_n \colon \text{Simp} \to \text{Vect}_\mathbb{F}.$$

Theorem 1.10.13. *Let X and Y be abstract simplicial complexes and let $f \colon X \to Y$ be a simplicial map. Then for each $k \geq 0$ there is an induced group homomorphism*

$$f_* \colon H_k(X; \mathbb{F}) \to H_k(Y; \mathbb{F}).$$

To explain this result, we provide an algebraic category to abstract the construction underlying homology. To this end, we now define the category $\text{Ch}(\text{Vect}_\mathbb{F})$ of chain complexes of $\mathbb{F}$-vector spaces.

Definition 1.10.14. A *chain complex of vector spaces $A_\bullet$* is a collection of vector spaces $\{A_n\}$, for $n \in \mathbb{Z}$, and linear transformations

$$\partial_n \colon A_n \to A_{n-1}$$

such that $\partial_{n-1} \circ \partial_n = 0$. More succinctly, a chain complex is a functor $\mathbb{Z}^{\text{op}} \to \text{Vect}_\mathbb{F}$ satisfying the condition above on the successive composites of maps.

Having specified the objects of $\text{Ch}(\text{Vect}_\mathbb{F})$, we now need to explain the morphisms.

Definition 1.10.15. A *map of chain complexes $f \colon A_\bullet \to B_\bullet$* is a collection of linear transformations $f_n \colon A_n \to B_n$ for each $n \in \mathbb{Z}$ such that $f_{n-1} \circ \partial_n^A = \partial_n^B \circ f_n$, i.e., such that the diagrams

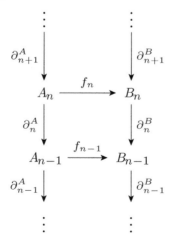

commute.

There is a natural functor $\text{Vect}_{\mathbb{F}} \to \text{Ch}(\text{Vect}_{\mathbb{F}})$ that takes a vector space V to the chain complex

$$A_\bullet = \ldots \to 0 \to 0 \to V \to 0 \to 0 \to \ldots$$

where $A_0 = V$ and $A_i = 0$ for $i \neq 0$.

As we have seen, the category of topological spaces has several useful notions of equivalence: homeomorphisms, which are categorical isomorphisms, as well as homotopy equivalences and weak equivalences. In contrast, the algebraic category $\text{Vect}_{\mathbb{F}}$ does not have a good analogue of the notion of homotopy equivalence. One of the advantages of $\text{Ch}(\text{Vect}_{\mathbb{F}})$ is precisely that is an algebraic category that enlarges $\text{Vect}_{\mathbb{F}}$ enough to have a notion of homotopy equivalence, which is called *quasi-isomorphism*.

To explain a quasi-isomorphism of chain complexes, we need to observe that the definition of homology makes sense for arbitrary chain complexes. Notice that since by definition $\partial_n \circ \partial_{n+1} = 0$, we have the evident inclusion of groups

$$\text{im}(\partial_{n+1}) \subseteq \text{ker}(\partial_n).$$

We have the following general analogue of Definition 1.10.9.

Definition 1.10.16. For a chain complex $A_\bullet$, the *n*th *homology group* H_n is defined as the quotient

$$H_n(A_\bullet) = \text{ker}(\partial_n)/\text{im}(\partial_{n+1}).$$

The construction of homology is functorial.

Lemma 1.10.17. *A map $f \colon A_\bullet \to B_\bullet$ of chain complexes induces a linear transformation of vector spaces $H_n(A_\bullet) \to H_n(B_\bullet)$. Moreover, H_n specifies a functor from the category of chain complexes to the category of vector spaces.*

We think of the homology groups of a chain complex as akin to the homotopy groups of a space, and this leads to the following definition.

Definition 1.10.18. A map $f \colon A_\bullet \to B_\bullet$ of chain complexes is a quasi-isomorphism when each induced map $H_n(A_\bullet) \to H_n(B_\bullet)$ is an isomorphism.

Of course, if each map f_n is an isomorphism, then f is a quasi-isomorphism. But there are many examples of quasi-isomorphisms that are not isomorphisms.

Example 1.10.19. Consider the chain complex where $C_3 = \mathbb{F}$, $C_2 = \mathbb{F}$, all other $C_i = 0$, and $\partial_3 = \text{id}$. Then the homology is zero for all n; this chain complex is quasi-isomorphic to the zero complex (i.e., the complex where $C_i = 0$ for all i).

For our purposes, the most interesting examples of chain complexes come from the construction of the simplicial chains. This assignment is functorial.

Lemma 1.10.20. *For a simplicial complex X, the chains $C_\bullet(X; \mathbb{F})$ form a chain complex of vector spaces. A simplicial map of simplicial complexes $f: X \to Y$ induces a chain map $C_\bullet(X; \mathbb{F}) \to C_\bullet(Y; \mathbb{F})$. That is, passage to simplicial chains induces a functor*

$$C_\bullet(-): \mathrm{Simp} \to \mathrm{Ch}(\mathrm{Vect}_{\mathbb{F}}).$$

We can immediately deduce the functoriality of homology from this construction. An isomorphism of simplicial complexes clearly induces a quasi-isomorphism of chains; in fact, so does a homeomorphism of the associated topological spaces. But the power of homology arises because it is in fact a homotopy invariant. To explain this, we need to consider the question of when two maps $f, g: A_\bullet \to B_\bullet$ induce the same map on homology.

Definition 1.10.21. We say that two maps of chain complexes $f, g: A_\bullet \to B_\bullet$ are *chain homotopic* if there exist maps $h: A_n \to B_{n+1}$ such that $f_n - g_n = \partial_{n+1} \circ h_n - h_{n-1} \circ \partial_n$.

The definition of chain homotopy is a precise analogue of the notion of homotopy of maps of spaces.

Theorem 1.10.22. *If $f, g: X \to Y$ are simplicial maps of abstract simplicial complexes such that $|f|, |g|: |X| \to |Y|$ are homotopic, then the induced maps $f, g: C_\bullet(X; \mathbb{F}) \to C_\bullet(Y; \mathbb{F})$ are chain homotopic.*

In fact, Definition 1.10.21 can be derived by considering the chain complexes $C_\bullet(X \times [0, 1]; \mathbb{F})$ and $C_\bullet(Y; \mathbb{F})$ and the conditions imposed by the existence of a homotopy $h: X \times I \to Y$.

For our purposes, the most important fact about chain homotopic maps is the following result:

Proposition 1.10.23. *If two maps $f, g: A_\bullet \to B_\bullet$ of chain complexes are chain homotopic, then they induce the same map on homology.*

The point is simply that

$$\partial_n \circ (f_n - g_n) = \partial_n \circ \partial_{n+1} \circ h_n - \partial_n \circ h_{n-1} \circ \partial_n = \partial_n \circ h_{n-1} \circ \partial_n,$$

i.e., the difference between f_n and g_n is a boundary.

Corollary 1.10.24. *If $f: X \to Y$ is a simplicial map of abstract simplicial complexes such that $|f|: |X| \to |Y|$ is a homotopy equivalence, then f induces an isomorphism on homology.*

   Put another way, we really have a functor

$$H_n: \text{Ho}(\text{Simp}) \to \text{Vect}_{\mathbb{F}},$$

where we define Ho(Simp) to be the category with objects abstract simplicial complexes and morphisms from X to Y specified by the homotopy classes of maps $|X| \to |Y|$.

Remark 1.10.25. Typically, this fact is proved using a related homology theory called singular homology, which coincides with simplicial homology but is (by definition) independent of the simplicial structure. In addition, as we mentioned in Remark 1.10.31, one can also define homology directly for CW complexes. In light of this menagerie of definitions, a basic consistency question arises: given an abstract simplicial complex X, do all the possible ways of defining its homology agree? Direct comparisons are possible, but it turns out that the collection of homology functors $H_n: \text{Top} \to \text{Vect}_{\mathbb{F}}$ can be axiomatically characterized in terms of a very simple set of axioms, the Eilenberg-Steenrod axioms. Roughly speaking, these axioms describe families of functors that have prescribed behavior on the spheres S^n and satisfy certain gluing relationships; the proof that this suffices to characterize the theories amounts to induction over a CW structure.

1.10.4 Simplicial Homology with Coefficients in an Abelian Group

In fact, simplicial homology can take values in the category of abelian groups instead of vector spaces. We consider the case of $\mathbb{Z}$ for clarity. Definitions 1.10.14 and 1.10.15 generalize immediately to the category Ch(Ab) of chain complexes of abelian groups.

Definition 1.10.26. A *chain complex of abelian groups* $A_\bullet$ is a collection of abelian groups $\{A_n\}$, for $n \in \mathbb{Z}$, and homomorphisms

$$\partial_n: A_n \to A_{n-1}$$

such that $\partial_{n-1} \circ \partial_n = 0$. More succinctly, a chain complex is a functor $\mathbb{Z} \to \text{Ab}$ satisfying the condition above on the successive composites of maps. The morphisms are the maps of chain complexes, i.e., the collections of homomorphisms $f_n: A_n \to B_n$ such that $f_{n-1} \circ \partial_n^A = \partial_n^B \circ f_n$.

We can build the simplicial chains by working with the free abelian group generated by the simplices. Specifically, we have the following definition.

Definition 1.10.27. The group of *m*-chains $C_m(X;\mathbb{Z})$ is the free abelian group with basis elements in bijection with the oriented *m*-simplices. That is, elements of $C_m(X;\mathbb{Z})$ are linear combinations (with coefficients in $\mathbb{Z}$) of generators $\{g_\sigma\}$, where σ varies over the oriented *m*-simplices of *X*.

Lemma 1.10.28. *A map of simplicial complexes $S \to S'$ determines a chain map $C_\bullet(S;\mathbb{Z}) \to C_\bullet(S';\mathbb{Z})$. That is, passage to simplicial chains induces a functor*

$$C_\bullet(-;\mathbb{Z}): \mathrm{Simp} \to \mathrm{Ch}(\mathrm{Ab}).$$

Applying homology, we get a composite functor

$$H_n(-;\mathbb{Z}): \mathrm{Simp} \to \mathrm{Ch}(\mathrm{Ab}) \to \mathrm{Ab}.$$

We refer to this as homology with coefficients in the group $\mathbb{Z}$. As in Proposition 1.10.23, homotopy classes of maps induce the same map on homology and quasi-isomorphisms of chain complexes induce isomorphisms.

In this context, the first homology group can be described in terms of something we have already seen.

Theorem 1.10.29. *Let X be an abstract simplicial complex that is connected. The homology group $H_1(X;\mathbb{Z})$ is the abelianization of the fundamental group $\pi_1(X, x)$, where here the abelianization of a group is the quotient by the subgroup generated by terms of the form $xyx^{-1}y^{-1}$.*

The advantage of working with $\mathbb{Z}$ coefficients is that the homology captures more information about the space *X*. More generally, it is possible to consider homology with coefficients in any ring *R*; the situation for $\mathbb{Z}$ is a special case. However, when working with topological data analysis, only the cases of homology with field coefficients tend to be used. We explain the reason for this in Section 2.3. Just as for the case of field coefficients, there is an efficient algorithm for computing homology with coefficients in $\mathbb{Z}$.

Theorem 1.10.30. *Given a simplicial complex, there exists an algorithm to compute $H_k(-;\mathbb{Z})$ whose running time is polynomial (cubic) in the total number of $(k + 1)$-simplices, k-simplices, and $(k - 1)$-simplices.*

Remark 1.10.31. Because of our focus on algorithmic methods, we have discussed simplicial homology in this section. As we mentioned in Remark 1.10.25, there are a number of other candidate constructions of the homology of a space; for example, one can define homology using calculus (in terms of differential forms) or infinite-dimensional functions (singular homology). Most notably, in the spirit of our discussion, it is possible to give a definition of homology that works directly from the CW complex structure on a space; one begins with a chain complex defined where $C_k(X)$ is the free abelian group on the k-cells. However, the boundary map is considerably more complicated in this case. Nonetheless, this approach has been the basis for computational work in discrete Morse theory and computational cubical homology (e.g., see [228, 280]).

1.11 Manifolds

The definition of a topological space is very general; an arbitrary topological space can be extremely complicated and have a very non-geometric flavor. For example, the Cantor set, constructed by removing the middle third from the interval $[0, 1]$, then the middle third from each of the resulting intervals, and so on (i.e., the subset of $[0, 1]$ consisting of elements whose ternary expansion does not contain 1) is an exotic topological space. When we work up to weak homotopy equivalence, we can restrict attention to simplicial complexes, which are a much nicer collection of spaces. Nonetheless, simplicial complexes still admit a very wide collection of examples with complicated local geometry.

However, in many applications (e.g., computer vision, medical imaging, physics), particularly nice examples of topological spaces tend to arise; these are spaces which admit Euclidean coordinates, at least locally, and permit the definition of a precise generalization of classical calculus. Such a topological space is called a *manifold*. A wonderful introduction to smooth manifolds is given by Milnor's classic book [355]; for more on Riemannian manifolds see [96].

In order to define a manifold, we need to explain what we mean by coordinates.

Definition 1.11.1. Let X be a topological space. Given an open set $U \subseteq X$, we say that a *chart* is a homeomorphism $\theta \colon U \to V$, where V is an open subset of $\mathbb{R}^n$. The inverse θ^{-1} equips U with a coordinate system. (See Figure 1.52 for examples of charts.)

An *atlas* for X is a collection of charts such that the $\{U_i\}$ cover X. The composites

$$\theta_\alpha \theta_\beta^{-1} \colon \theta_\beta(U_\alpha \cap U_\beta) \to \theta_\alpha(U_\alpha \cap U_\beta)$$

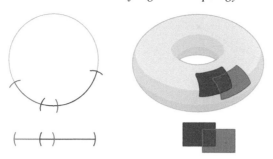

Figure 1.52 Left: Two overlapping charts on a circle. Right: Two overlapping charts on a torus. Each chart gives a little coordinate system, and transition functions connect these coordinates on the overlaps.

are referred to as transition functions. These explain how coordinates change as we move between different charts.

Definition 1.11.2. An n-dimensional *topological manifold X* is a second-countable, Hausdorff topological space equipped with an atlas where the charts are all subsets of $\mathbb{R}^n$.

(Here recall that second-countable means that the topological space has a countable base and Hausdorff means that any pair of points can be separated by enclosing open sets.)

It is often the case that examples have additional smoothness which permits the use of the methods of calculus. Since the transition functions involve maps from subsets of Euclidean space to itself, we can ask about their continuity and derivatives using the standard techniques of multivariable calculus.

Definition 1.11.3. An n-dimensional *smooth manifold* is a topological manifold where the transition functions are continuous and infinitely differentiable.

Many of the most familiar examples of topological spaces are manifolds.

Example 1.11.4.

1. Any Euclidean space $\mathbb{R}^n$ is a manifold, covered by a single chart.
2. The space $S^1 = \{(x, y) \subset \mathbb{R}^2 \mid x^2 + y^2 = 1\}$ is a manifold, covered by two charts, one covering points with $y > \frac{1}{2} - \epsilon$ and one covering points with $y < \frac{1}{2} + \epsilon$. (Here we can choose any $\epsilon > 0$.)
3. More generally, any sphere $S^n = \{(x_1, x_2, \ldots, x_{n+1}) \in \mathbb{R}^{n+1} \mid \sum_i x_i^2 = 1\}$ is a manifold covered by two charts.
4. The torus is a manifold; charts can be provided by considering a covering of the torus by little overlapping squares, for instance.

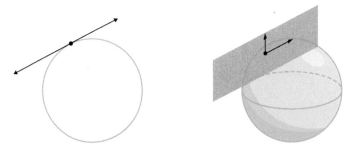

Figure 1.53 The tangent space at a point is all the directions in which a derivative of a curve could point; equivalently, it is the plane perpendicular to the normal or "outward" pointing direction.

Calculus on manifolds is expressed in terms of the notion of tangent spaces. At each point x of a manifold M, the *tangent space* $T_x M$ is simply a vector space in which the tangent vectors (i.e., derivatives) to curves through that point can lie. The derivative of a function $f: M \to \mathbb{R}$ at a point $x \in M$ is a vector which lies in $T_x M$.

Example 1.11.5.

1. The tangent space $T_x \mathbb{R}^n$ of Euclidean space $\mathbb{R}^n$ at any point $x \in \mathbb{R}^n$ is isomorphic to $\mathbb{R}^n$.
2. The tangent space $T_x S^1$ at a point $x \in S^1$ is isomorphic to $\mathbb{R}^1$; the tangent space can be viewed as the tangent line to the circle.
3. The tangent space to $T_x S^n$ to a sphere at a point $x \in S^n$ is a plane $\mathbb{R}^n$.

(See Figure 1.53 for a representation of the tangent space of spheres.)

For particularly nice manifolds (including the examples we have discussed above), the tangent spaces $T_x M$ admit inner products that vary smoothly as we move around on M. Recall that an inner product (sometimes referred to as a dot product) is a pairing of the following form.

Definition 1.11.6. An *inner product* on a vector space V over the field $\mathbb{R}$ is a function

$$\langle -, - \rangle : V \times V \to \mathbb{R}$$

such that

1. $\langle x, x \rangle \geq 0$,
2. $\langle x, y \rangle = \langle y, x \rangle$, and
3. $\langle x + y, z \rangle = \langle x, z \rangle + \langle y, z \rangle$.

The significance of an inner product is that it allows us to define the length of a vector as the *norm* $\|x\| = \sqrt{\langle x, x \rangle}$ and the (cosine of the) angle between two vectors as being proportional to their inner product. That is, manifolds with inner products on the tangent spaces admit nice notions of area and angles; such manifolds are referred to as *Riemannian manifolds*. Riemannian manifolds have a number of rich geometric properties.

1. A path-connected Riemannian manifold has a metric; a path γ in M has a length computed by integrating the norms of the tangent vectors along γ. The distance between two points p and q is computed by taking the infimum (recall Definition 1.2.17) of the lengths of all paths joining them.
2. A Riemannian manifold has a notion of area or volume of regions on the manifold, referred to as the *volume form*, coming from the determinant in the tangent spaces.
3. A Riemannian manifold M has a notion of *curvature*, which can be described in terms of the divergence of paths following the tangent vectors at a point. For example, the standard sphere has curvature 1, Euclidean space has curvature 0, and hyperbolic space has curvature -1. (Here recall that hyperbolic space is a description of the geometry that arises when Euclid's parallel postulate is modified to allow infinitely many distinct parallel lines between two points.) We will say more about this below in Section 4.7.3.

Such manifolds allow a theory of integration and sampling, and although one does not expect data to lie on such manifolds, these provide a vital source of intuition and theoretical backing for the behavior of topological data analysis algorithms; such examples play an important motivating role, as we will see in Chapters 2 and 3.

Despite their rigidity, there are an enormous number of possible manifold topologies as the dimension increases; easy estimates show the number of homeomorphism classes of manifolds grows faster than exponentially as the dimension increases [533]. We can classify manifolds in low dimensions, however.

Example 1.11.7.

1. In dimension 0, the only manifolds are disjoint unions of points.
2. In dimension 1, the manifolds are homeomorphic to disjoint unions of circles and copies of $\mathbb{R}$. For example, a compact manifold with a single path component must be a circle.
3. In dimension 2, the classification of surfaces is an early and important theorem in topology; compact manifolds can be completely described as either a sphere or a manifold classified by a pair of natural numbers, describing how the manifold is made by gluing two kinds of basic pieces (toruses and projective planes) together. (For a nice treatment, see [369, §12].)

Another important class of examples comes from matrix groups, which are examples of Lie groups.

Example 1.11.8.   Roughly speaking, a Lie group is a group which is also a topological space that is a manifold, so that the group operations are continuous. For example, the circle S^1 can be given the structure of a Lie group where the group operation is specified by adding angles. As another important example, the set $GL_n(\mathbb{R})$ of invertible matrices can be given the structure of a manifold; such manifold symmetry groups are ubiquitous in physical applications.

On the one hand, manifolds provide geometric intuition for many methods in computational topology, and provide a large and familiar class of topological spaces. On the other hand, in contrast to physics, in applications to biology and genomics we do not usually expect the metric spaces we encounter to come from Riemannian manifold structures. In many cases, we do not even expect them to come from continuous topological spaces, in the sense that for many biologically relevant metrics, there is a minimum bound such that any distance is larger than this bound – for example, the Hamming distance between strings has this property.

One potential compromise between manifolds and arbitrary topological spaces comes from the theory of *stratified spaces*. Although a precise definition is more technical than we require, roughly speaking a stratified space is a topological space that is the union of manifolds (of possibly different dimensions) that fit together nicely. (See [534] for a wide-ranging treatment.)

Example 1.11.9.

1. Any graph embedded in Euclidean space is a stratified space comprising zero dimensional manifolds (points) and one dimensional manifolds (open intervals). Notably, trees are stratified spaces.
2. The disjoint union of manifolds $\bigsqcup_i M_i$ is a stratified space.

1.12 Morse Functions and Reeb Spaces

A natural question to ask about a manifold is whether we can endow it with a CW structure which reflects the geometric structure of the manifold. A classical answer to this question is provided by *Morse theory*. Morse theory starts by considering a manifold M along with a "height function"

$$h: M \to \mathbb{R}.$$

Example 1.12.1.   Consider the standard sphere

$$S^2 = \{(x, y, z) \in \mathbb{R}^3 \mid x^2 + y^2 + z^2 = 1\}.$$

We think of this as sitting on the tangent plane $z = -1$, and we can define the height at a point (x, y, z) as simply $z + 1$. (Of course, there are many other reasonable choices of height functions.)

Given a height function, the approach of Morse theory is to study the information about M encoded in the inverse images $f^{-1}(k)$ as k varies; specifically, we consider the inverse images for $k \in \mathbb{R}$, or more generally in the inverse images $f^{-1}(I)$, where $I \subseteq \mathbb{R}$ is an open interval (a, b). The places where the inverse images change in interesting ways turn out to be precisely the critical points of the function h. That is, the idea of Morse theory is that a space can be characterized by the critical points of suitable continuous functions from $M \to \mathbb{R}$.

Example 1.12.2.   A standard example to consider is the torus "stood on its end," where the bottom has height 0. As a varies, the inverse images $h^{-1}([0, a])$ start as a disk, then become a cylinder, then the torus with a disk cut out, and then finally become the entire torus. From the perspective of homotopy theory, the process described is precisely cell attachment in a CW structure! Attaching occurs as h passes through a critical point. (See Figure 1.54.)

We do not need the full generality of Morse theory to explain the techniques of topological data analysis, so we do not give precise statements of the main theorems; for a beautiful treatment, see [354]. However, constructions inspired by this approach, the *Reeb graph* and *Reeb space*, have turned out to be incredibly useful in topological data analysis and computational geometry. We now give a brief overview of these constructions; see [459] for a more in-depth exposition.

Suppose that we are given a topological space X (e.g., a CW complex or the geometric realization of a finite simplicial complex) along with a continuous map $h: X \to \mathbb{R}^n$.

Definition 1.12.3.   We define an equivalence relation on X by stipulating that $p \simeq q$ if $p, q \in f^{-1}(k)$ for some k and moreover that p and q are in the same path component. The *Reeb space* of X is the quotient of X by this equivalence relation.

height (h)

Figure 1.54 As the height increases, the inverse image includes more and more of the torus.

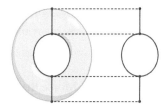

Figure 1.55 The Reeb graph of a torus.

When $n = 1$, Definition 1.12.3 yields a graph, referred to as the Reeb graph. The vertices of the Reeb graph correspond to components of the level sets, with edges connecting components that merge as k varies. See Figure 1.55 for the Reeb graph of the torus; notice the similarities to the Morse theory description above.

It is sometimes helpful in theoretical work to have a more general version of the Reeb space, referred to as the *categorical Reeb space*.

Definition 1.12.4.   Given a topological space X equipped with a continuous func-tion $h \colon X \to \mathbb{R}^n$, we specify the functor $R_{X,f}$ from the category of open sets of $\mathbb{R}^n$ with morphisms inclusions $U_1 \to U_2$ to the category of spaces as follows.

Let $R_{X,f}(I)$ be the space $f^{-1}(I)$, and let the induced map $R_{X,f}(I) \to R_{X,f}(J)$ be the evident inclusion $f^{-1}(I) \to f^{-1}(J)$.

Under good conditions, when $n = 1$ the Reeb graph of Definition 1.12.3 can be recovered from the categorical Reeb space of Definition 1.12.4 by applying π_0 to pass to components [459].

1.13 Summary

- Metric spaces, topological spaces, groups, and vector spaces are sets endowed with additional mathematical structure. These structures are the central objects upon which topological data analysis is built.
- A topological space $(X, \mathcal{U})$ is a set X endowed with a topology $\mathcal{U}$. We may describe the similarities of $(X, \mathcal{U})$ to other topological spaces by considering homeomorphisms and homotopy equivalences.
- We may construct topological spaces by gluing together simpler spaces such as cells (n-disks D^n) along their boundaries. Spaces produced in this way are called CW complexes. We may also create new topological spaces by considering the product of two or more smaller spaces (such as the torus in Example 1.5.11) or by collapsing subspaces of larger spaces, called a quotient (such as the cone in Figure 1.29).

- The fundamental group $\pi_1(X, x)$ of a topological space X is the set of homotopy classes of loops in X based at a fixed point $x \in X$. We may generalize the idea of the fundamental group to higher dimensions with the nth homotopy group $\pi_n(X, x)$ (see Definition 1.4.10). As the name suggests, $\pi_n(X, x)$ is a mathematical group under composition of rescaled maps (see Theorem 1.6.27). The homotopy groups of a space capture information about the space encoded in maps out of test spaces, namely spheres.

- Category theory provides a means of formalizing the notion of moving between different mathematical worlds. For example, we may use the language of category theory to restate the previous bulletpoint: $\pi_n(X, x)$ specifies a functor between the categories of based topological spaces and algebraic groups.

- Simplicial complexes provide a discrete, combinatorial framework for studying topological spaces. Many topological spaces arise as the geometric realizations of finite simplicial complexes.

- The combinatorial nature of simplicial complexes allows us to develop the idea of simplicial homology. For each $k \geq 0$, we may consider the group $C_k(X)$ of linear combinations of oriented k-simplices. Cycles in $C_k(X)$ may or may not form the boundaries of elements in $C_{k+1}(X)$. The kth homology group $H_k(X; \mathbb{Z})$ measures the size of the difference between $C_k(X)$ and the set of boundaries of elements in $C_{k+1}(X)$. That is, the kth homology group encodes information about the k-dimensional holes in X. The homology groups can be computed efficiently using linear algebra.

- Manifolds are topological spaces that are especially nice in that they admit Euclidean coordinates locally. Riemannian manifolds provide geometric structure: a metric, volume, and curvature.

- Given a function $f : X \to \mathbb{R}$, the Reeb space encodes information about topological changes in the level sets defined via inverse images.

1.14 Suggestions for Further Reading

The material we have covered in this section is standard, and in each of the previous sections, we have made suggestions about accessible treatments for readers who want more detail. For a reader who wants a geodesic path to the necessary background for topological data analysis, there are two sources to focus on: the first part of Munkres' book on simplicial homology [368], and Riehl's introductory textbook on category theory [428]. These are mostly self-contained (e.g., Munkres has a concise but detailed treatment of the required abstract algebra) and provide very lucid explanations.

2

Topological Data Analysis

> I prefer to express myself metaphorically. Let me stress: metaphorically, not symbolically. A symbol contains within itself a definite meaning, certain intellectual formula, while metaphor is an image. An image possessing the same distinguishing features as the world it represents.
>
> *Andrei Tarkovsky*

A central dogma of topological data analysis is that data sets have shape and that describing this shape can help explain the process generating the data. As we have outlined in the preceding chapter, from this perspective clustering techniques extract "zero dimensional" information about connected components of the data set. One of the central goals of topological data analysis is to use the methods of algebraic topology to extract higher dimensional information about the shape of the data set. For example, if we suppose that the data is sampled from a manifold, a candidate goal might be to recover the homology of that manifold. More realistically, we might simply wish to recover qualitative descriptors of the data set that are robust to perturbation and capture higher dimensional information, without necessarily postulating that there is such a clean underlying geometric description. That is, we would like to set up a pipeline

$$\{\text{data}\} \rightarrow \left\{ \begin{array}{c} \text{simplicial} \\ \text{complexes} \end{array} \right\} \rightarrow \left\{ \begin{array}{c} algebraic \\ \text{invariants} \end{array} \right\}.$$

To apply algebraic topology to discrete data, two major issues need to be tackled. First, we need a way to transform a discrete set of points into a richer topological space in order to have interesting topological invariants to compute. Second, the *feature scale* of the data must be accounted for; namely, we need to determine the relationship between the size of meaningful geometric features of the data and the distances between the sampled points. This second question is particularly interesting, since a priori the feature scale is often unknown. In this chapter, we explain approaches to these problems, with a primary focus on *persistent homology* and related constructions. The basic idea is to collect information for all feature scales at once. Persistent homology originated in the work of Frosini [187], and was independently rediscovered by Robins [433] and Edelsbrunner, Letscher, and Zomorodian [154]; in Section 2.11 at the end of the chapter we provide more comprehensive references for the interested reader.

2.1 Simplicial Complexes Associated to Data

A basic and widely applicable model for the kind of data that arises in practice is a *finite metric space*; this is simply a metric space (X, ∂_X) with finitely many points. A natural geometric example of a finite metric space is a collection of points $\{x_0, x_1, \ldots, x_k\} \subset \mathbb{R}^n$ equipped with the induced Euclidean metric $\partial_{\mathbb{R}^n}$. A natural biological example of a finite metric space is a collection of gene expression vectors in $\mathbb{R}^{20000}$, with the distance between v_1 and v_2 computed by the Pearson correlation (recall Example 1.2.6).

Recall from Example 1.3.7 that any metric space (X, ∂_X) has a natural topology where the basic open sets are the balls $B_\epsilon(x) = \{z \in X \mid \partial_X(z, x) < \epsilon\}$ for all $\epsilon > 0$. As a consequence, a first thought might be to simply regard a finite metric space (X, ∂_X) as a topological space directly. Unfortunately, such a space is not very interesting – the topology is trivial, in the sense that it is discrete.

- Every point is both open and closed.
- There are no continuous maps $\gamma\colon [0, 1] \to X$ other than the constant maps. (See Figure 2.1.)
- All homological invariants except π_0 and H_0 (which just count the number of points in X) are trivial.

In order to leverage the tools of algebraic topology to study finite metric spaces, we need a different idea for assigning a topological space to (X, ∂_X). To figure out what to do, it is useful to think about the toy model in which the sampled data X was generated by drawing from some probability distribution on a nice geometric

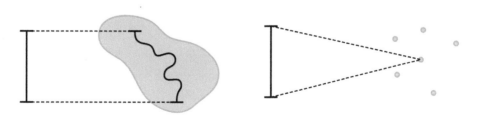

Figure 2.1 The only continuous maps from $[0, 1]$ to a discrete topological space are constant at a point.

object embedded in $\mathbb{R}^n$ (e.g., a compact smooth manifold). In this case, it is clear that we need to somehow "fill in the gaps" between the samples. If we have a rough sense of the average distance between points that are supposed to be connected, there is an evident construction: just take the union of balls around the points.

Definition 2.1.1 (Union of balls).   Let $X \subset \mathbb{R}^n$ be a finite subspace and fix $\epsilon \geq 0$. The *union of balls* is the union

$$\bigcup_{x \in X} B_\epsilon(x) \subset \mathbb{R}^n.$$

However, from a practical perspective, the union of balls is not ideal; it is not evidently algorithmically tractable, and it requires that (X, ∂_X) arise as a subspace of $\mathbb{R}^n$. To fix the first problem, we would like to produce an abstract simplicial complex that encodes the information of the union of balls. We can adapt this construction to the discrete setting by regarding the ϵ-balls around a finite set X as a cover. That is, the idea is to associate a k-simplex to a set of k points whose ϵ-neighborhoods intersect.

Definition 2.1.2 (Čech complex).   Let $X \subset \mathbb{R}^n$ be a finite subspace and fix $\epsilon > 0$. The *Čech complex* $C_\epsilon(X, \partial_X)$ is the abstract simplicial complex with

1. vertices the points of X, and
2. a k-simplex $[v_0, v_1, \dots, v_k]$ when a set of points $\{v_0, v_1, \dots, v_k\} \subset X$ satisfies

$$\bigcap_i B_\epsilon(v_i) \neq \emptyset.$$

In fact, the Čech complex (Figure 2.2) is a special case of a standard construction from algebraic topology that associates a simplicial complex to a *cover* of a space. Recall from Definition 1.3.15 that an open cover $\{U_i\}$ of a space X is a collection of open sets such that $\cup_i U_i = X$. Given a cover $\{U_i\}$ of X, we define the *nerve* of the cover as follows.

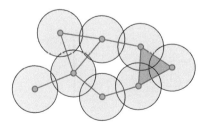

Figure 2.2   The Čech complex is a combinatorial approximation to the union of balls.

Definition 2.1.3. The *nerve* $N(\{U_i\})$ of a cover $\{U_i\}$ of X is the simplicial complex with

1. vertices corresponding to the sets $\{U_i\}$, and
2. a k-simplex $[j_0, j_1, \ldots, j_k]$ when the intersection

$$U_{j_0} \cap U_{j_1} \cap U_{j_2} \cap \ldots \cap U_{j_k} \neq \emptyset.$$

The interest of this construction is the following classical result about the relationship of the geometric realization (recall Definition 1.8.8 and Lemma 1.8.20) of this nerve to X; see e.g., [307, §15.4.3] for further discussion and a proof.

Theorem 2.1.4. *Let X be a topological space. Let $\{U_i\}$ be an open cover of X such that all non-empty finite intersections*

$$U_{j_1} \cap U_{j_2} \cap \ldots \cap U_{j_k}$$

are contractible (homotopy equivalent to a point). Then the geometric realization $|N(\{U_i\})|$ is homotopy equivalent to X.

As a corollary, we obtain the following result comparing the geometric realization of the Čech complex to the geometric Čech nerve.

Proposition 2.1.5. *Let $X \subset \mathbb{R}^n$ be a finite subspace and fix $\epsilon > 0$. There exists a homeomorphism*

$$\bigcup_{x \in X} B_\epsilon(x) \cong |C_\epsilon(X, \partial_X)|$$

between the union of balls and the geometric realization of the Čech complex.

The Čech complex provides a procedure for assigning a simplicial complex to a finite metric space embedded in $\mathbb{R}^n$. However, in order to construct the Čech complex we need to be able to decide whether the intersection of ϵ-balls is nonempty. This is a non-trivial enterprise in high dimensions. Moreover, we do not wish to assume that the data points are embedded in Euclidean space at all!

To see how to proceed, it is helpful to recall our discussion of path components and single-linkage clustering for a metric space from Section 1.3. Here, for a finite metric space (X, ∂_X) and fixed $\epsilon > 0$, we defined a graph $G = (V, E)$ with

1. vertices the points of X, and
2. edges (x_i, x_j) for each pair of points x_i and x_j such that $\partial_X(x_i, x_j) \leq \epsilon$.

Recalling that a graph is a one dimensional simplicial complex, we use a mild elaboration of this construction to define a simplicial complex associated to an arbitrary finite metric space (X, ∂_X). The Vietoris-Rips complex is the maximal simplicial complex determined by the vertices and 1-simplices specified by the graph G.

Definition 2.1.6 (Vietoris-Rips complex).   Let (X, ∂_X) be a finite metric space and fix $\epsilon > 0$. The *Vietoris-Rips complex* $\mathrm{VR}_\epsilon(X, \partial_X)$ is the abstract simplicial complex with

1. vertices the points of X, and
2. a k-simplex $[v_0, v_1, \dots, v_k]$ when

$$\partial_X(v_i, v_j) \leq 2\epsilon \qquad \text{for all} \qquad 0 \leq i, j \leq k.$$

For a point cloud in $\mathbb{R}^n$, the Vietoris-Rips complex and the Čech complex can be different; for instance, notice that there is a difference between the Čech complex in Figure 2.2 and the Vietoris-Rips complex in Figure 2.3, which are generated by the same underlying metric space. The next example highlights the kind of phenomenon that leads to such differences.

Example 2.1.7.   Consider the finite metric space $X = \{(0, 0), (1, 0), (\frac{1}{2}, \frac{\sqrt{3}}{2})\} \subset \mathbb{R}^2$. These points are the vertices of an equilateral triangle with side length 1. Choose an ϵ in the open interval $(\frac{1}{2}, \frac{\sqrt{3}}{3})$, i.e., $\frac{1}{2} < \epsilon < \frac{\sqrt{3}}{3}$. (For concreteness, $\frac{\sqrt{3}}{3} \approx 0.577$.)

1. The Vietoris-Rips complex $\mathrm{VR}_\epsilon(X, \partial_X)$ has three vertices (one for each point of X), three 1-simplices (connecting the points), and therefore has a single 2-simplex filling in the triangle.
2. In contrast, the Čech complex $C_\epsilon(X, \partial_X)$ has three vertices (one for each point of X) and three 1-simplices (connecting the points), but does not have the 2-simplex spanned by all the points since there is no point in the intersection of the balls of radius ϵ.

(See Figure 2.4 for a corresponding picture.)

The use of the Čech complex is justified by the Nerve Lemma (Theorem 2.1.4); there is no analogous result for the Vietoris-Rips complex. However, despite the

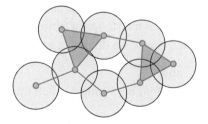

Figure 2.3 The Vietoris-Rips complex is completely determined by its 1-skeleton.

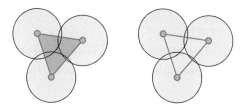

Figure 2.4 The Vietoris-Rips complex (on the left) is completely determined by its 1-skeleton, whereas the Čech complex (on the right) can potentially omit higher simplices.

fact that they are sometimes different, there is a close relationship between the Vietoris-Rips and Čech complexes.

Lemma 2.1.8. *Let $X \subset \mathbb{R}^n$ be a finite subspace and fix $\epsilon > 0$. There are natural simplicial inclusions*

$$C_\epsilon(X, \partial_X) \subseteq \mathrm{VR}_\epsilon(X, \partial_X) \subseteq C_{2\epsilon}(X, \partial_X).$$

An essential property of the constructions of the Čech complex and the Vietoris-Rips complex is that they are functorial. To be precise, these constructions are functorial in both X and ϵ. (In the following discussion, we focus on the Vietoris-Rips complex; the properties of the Čech complex are analogous.) For $\epsilon < \epsilon'$ and any metric space (X, ∂_X), there is an induced simplicial map

$$\mathrm{VR}_\epsilon(X, \partial_X) \to \mathrm{VR}_{\epsilon'}(X, \partial_X),$$

since increasing the scale parameter adds more simplices.

Next, recall that a map $f\colon X \to Y$ between metric spaces (X, ∂_X) and (Y, ∂_Y) is Lipschitz continuous with constant k if $\partial_Y(f(x_1), f(x_2)) \leq k\partial_X(x_1, x_2)$. Given a Lipschitz map $f\colon X \to Y$ with Lipschitz constant k, there is an induced simplicial map

$$f\colon \mathrm{VR}_\epsilon(X, \partial_X) \to \mathrm{VR}_{k\epsilon}(Y, \partial_Y)$$

for any ϵ. Summarizing, we have the following theorem.

Theorem 2.1.9. *The construction $\mathrm{VR}_\epsilon(-)$ specifies a functor from the category of finite metric spaces and Lipschitz maps with constant 1 to Simp. The construction $\mathrm{VR}_{(-)}(X, \partial_X)$ specifies a functor from $\mathbb{R}$ to Simp.*

This means that when we vary the scale ϵ, there is a map between the associated complexes for a given data set (X, ∂_X). And if we change a data set (X, ∂_X) to produce a new data set (Y, ∂_Y) related via a Lipschitz map, there is a map connecting the associated complexes. For example, if we add some data points, so that

$Y = X \cup A$ and the metric on Y restricts to ∂_X on $X \subset Y$, then there is a map $\mathrm{VR}_\epsilon(X, \partial_X) \to \mathrm{VR}_\epsilon(Y, \partial_Y)$.

We now turn to the question of when these constructions can recover information about the underlying geometric structure of the process that generated the data.

Question 2.1.10.   Let (X, ∂_X) be a finite metric space consisting of samples from a topological space A. When is $|\mathrm{VR}_\epsilon(X, \partial_X)|$ or $|C_\epsilon(X, \partial_X)|$ homotopy equivalent to A?

2.2 The Niyogi-Smale-Weinberger Theorem

In order to make sense of this question, we need to develop a precise model for sampling from a topological space A. We will introduce a definition of *geometric sampling* and study Question 2.1.10 in Chapter 3. However, to illustrate some of the geometric principles that motivate TDA, in this section we will explain an answer to the question in a very restricted context. Specifically, we describe a minimal sanity check: we explain the Niyogi-Smale-Weinberger result that given a finite metric space (X, ∂_X) consisting of sufficiently many points sampled "uniformly" from a compact Riemannian manifold $M \subset \mathbb{R}^n$, with high probability there is an isomorphism

$$H_* \left(\bigcup_{x \in M} B_\epsilon(x) \right) \cong H_*(M)$$

for some suitable choice of ϵ.

Going forward, we assume that we are given a compact manifold $M \subset \mathbb{R}^n$ that has a Riemannian structure. Recall from Section 1.11 that roughly speaking, this means that at each point of the manifold we can equip the tangent space with an inner product, and these inner products vary smoothly as we move on the manifold. As a consequence, M has a metric and there is a natural notion of volume of subspaces of M. In particular, there is a natural notion of what it means to sample from such a manifold, as the manifold is equipped with a probability measure called the *volume measure*.

We want to estimate how many sampled points are necessary to estimate the homology with high probability. When sampling from the volume measure on a Riemannian manifold, it is straightforward to figure out how many points to sample so that with probability $> \kappa$ (for any fixed κ) we get an ϵ-net. Therefore, we can reduce the problem to trying to understand when a finite ϵ-net $X \subset M$ has the property that for some ϵ',

$$H_*(|C_{\epsilon'}(X, \partial_X)|) \cong H_*(M).$$

When such an isomorphism occurs depends on the size of the smallest geometric features of the manifold. That is, we need to figure out how close together points need to be in order for little balls around them to capture the structure of the manifold. For a manifold M embedded in $\mathbb{R}^n$, there are two distinct but interacting factors that control how small ϵ has to be in order for the geometric nerve to have the correct topology. We need to worry about the intrinsic curvature of the manifold, and how "twisted" the embedding into $\mathbb{R}^n$ is. See Figure 2.5 for some examples of possible embeddings of familiar geometric objects into Euclidean space.

Consider the case of S^1 embedded in $\mathbb{R}^2$. In order for the Čech nerve of an ϵ-net to have the right homotopy type, we must be able to choose an ϵ' such that

1. ϵ' is large enough to cause points of the net around the circle to be connected by 1-simplices, but
2. ϵ' is small enough so that points across the circle are not connected by "cross-cutting" 1-simplices.

The relationship between the scale ϵ and ranges of suitable values for ϵ' is controlled in part by the underlying topology of the circle – sufficiently large values for ϵ' will always result in 1-simplices that connect points across the circle. On the other hand, for very twisty embeddings, we will need to choose an ϵ' that is smaller than the size of the twists.

We think of these considerations as packaged into a quantity we refer to as the *feature scale* of the manifold. A very nice way to encode the feature scale of the manifold is to use an invariant called the *condition number*. (This is sometimes also referred to as the *reach* or *feature size*.) Any manifold embedded in $\mathbb{R}^n$ can

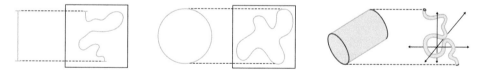

Figure 2.5 The difficulty in reconstructing a geometric object can come from both the intrinsic curvature and the twistiness of the embedding in $\mathbb{R}^n$.

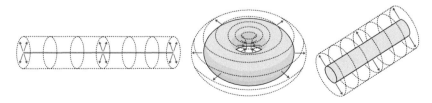

Figure 2.6 A tubular neighborhood is formed by expanding a manifold along the normal directions (perpendicular to its surface).

Figure 2.7 As the tubular neighborhood of a curve expands, eventually it self-intersects at the narrowest "pinch."

be thickened out to a *tubular neighborhood* of radius r; this is what one gets by extending out along the normal at any point. (See Figure 2.6 for some examples.)

The condition number is the minimum radius at which a tubular neighborhood of a manifold self-intersects; clearly, this can happen either because the manifold itself has small features (e.g., small holes) or because the embedding twists the manifold around on itself. (See Figure 2.7 for an example.)

The following theorem, due to Niyogi, Smale, and Weinberger [384], now provides a concrete result guaranteeing correct estimation of the homology.

Theorem 2.2.1. *Let M be a compact submanifold of $\mathbb{R}^n$ with condition number τ and let $\{x_1, \ldots, x_k\}$ be a set of points drawn from M according to the volume measure. Fix $0 < \epsilon < \frac{\tau}{2}$. Then if*

$$k > \beta_1 \left(\log(\beta_2) + \log\left(\frac{1}{\delta}\right) \right),$$

there is a homotopy equivalence

$$\bigcup_{z \in \{x_1, \ldots, x_k\}} B_\epsilon(z) \simeq M$$

between the union of balls and M (and in particular the homology groups coincide) with probability $> 1 - \delta$.

Here

$$\beta_1 = \frac{\mathrm{vol}(M)}{\cos^n(\theta_1)\mathrm{vol}(B^n_{\frac{\epsilon}{4}})}$$

and

$$\beta_2 = \frac{\mathrm{vol}(M)}{\cos^n(\theta_2)\mathrm{vol}(B^n_{\frac{\epsilon}{8}})},$$

where $\theta_1 = \arcsin\left(\frac{\epsilon}{8\tau}\right)$, $\theta_2 = \arcsin\left(\frac{\epsilon}{16\tau}\right)$, and $\mathrm{vol}(B^n_r)$ denotes the volume of the n-dimensional ball of radius r.

Remark 2.2.2. Using different techniques, one can also prove an analogous result directly for the Vietoris-Rips complex [3, 315].

To get a sense for what this means, it is helpful to do an explicit example.

Example 2.2.3. The condition number of a sphere is simply its radius. So for example, for the unit circle $S^1 \subset \mathbb{R}^2$, the condition number τ is 1. Choosing $\delta = 0.01$ and $\epsilon = \frac{1}{4}$, we compute that

$$\cos^2\left(\arcsin\left(\frac{1}{32}\right)\right) \approx 1 \quad \text{and} \quad \cos^2\left(\arcsin\left(\frac{1}{64}\right)\right) \approx 1$$

and so

$$\beta_1 = \frac{2\pi}{\pi(\frac{1}{16})^2} = 512$$

and

$$\beta_2 = \frac{2\pi}{\pi(\frac{1}{32})^2} = 2048,$$

which means that we need at least

$$512(7.6 + 4.6) \approx 6260$$

samples.

Example 2.2.4. The condition number of a torus is the minimum of r_1 and $\frac{r_2-r_1}{2}$, where r_1 and r_2 are the radii of the inner and outer bounding circles. We can repeat a similar computation as above, using the fact that the volume (surface area) of the torus is $(r_2^2-r_1^2)\pi^2$; once again, we end up with a number of points in the thousands for reasonable values of δ and ϵ.

These examples frame the application of Theorem 2.2.1 in high relief. On the one hand, this result is of critical theoretical importance, and it provides a vital consistency check for combinatorial approaches to estimating the homology of manifolds from finite data. On the other hand, the explicit bounds are useless – in practice it is difficult or impossible to estimate the condition number (although see [1]) and moreover a result of 3000 points to estimate the homology of a standard circle in $\mathbb{R}^2$ is clearly much too large. (To be sure, a direct argument can be used to obtain a much tighter bound.) In applications, we will be much more concerned about the stability of the result in the face of sampling variation and noise.

Remark 2.2.5. Theorem 2.2.1 is a statement about approximating the homotopy type of a manifold via finite sampling. One might wonder how many samples are required to estimate the homeomorphism type of M. Unfortunately, even in this very restricted setting, the problem turns out to be hopeless. Assume that M is embedded in $\mathbb{R}^n$ and the condition number is a fixed constant. Then when the dimension of M is larger than 2, the number of samples required to identify the homeomorphism type is exponential in $\text{diam}(M)^n$; see [533, §2.2] for a discussion. These concerns are relevant when studying single cell data; see Chapter 7.

2.3 Persistent Homology

The Niyogi-Smale-Weinberger theorem (Theorem 2.2.1) shows that in principle it is possible to accurately recover topological invariants of geometric objects from discrete samples. We interpret the theorem to suggest that it is reasonable to hope that in very general settings, when the distance between the samples is smaller than some *feature scale*, we can recover topological invariants of the underlying geometric object.

However, there is a key problem: the feature scale of the underlying object is usually unknowable a priori. That is, given (X, ∂_X) from M, how can we choose ϵ so that the topological invariants of $|VR_\epsilon(X, \partial_X)|$ recover information about the topological invariants of M? Moreover, choosing a single ϵ is problematic – for one thing, there might be distinct feature scales at which we can recover meaningful information, for instance if the data has regions of varying size. Another issue is that the topological invariants of $|VR_\epsilon(X, \partial_X)|$ are very unstable; small amounts of noise or sampling variation can cause large changes in the Vietoris-Rips complex and its homology. That is, at any given scale some features might not be stable with respect to noise or change of scale.

The guiding viewpoint that underlies topological data analysis is that we should simultaneously look at multiple feature scales; stable homological features that exist for a range of values of ϵ are likely to reflect the underlying signal, and this approach allows us to capture multiscale information. A naive approach to implementing this idea would simply be to vary ϵ and compute a collection of associated invariants.

1. Choose a topological invariant, e.g., the homology group $H_2(-; \mathbb{F}_p)$.
2. Select a range $[\epsilon_{\min}, \epsilon_{\max}]$, $\epsilon_{\min} < \epsilon_{\max}$. This interval reflects the smallest and largest feature scales that we will consider; a maximal choice would be to set $\epsilon_{\min} = 0$ and $\epsilon_{\max} = \text{diam}(X)$.
3. Choose values $\{\epsilon_1, \epsilon_2, \ldots, \epsilon_m\} \subset [\epsilon_{\min}, \epsilon_{\max}]$. An easy way to do this is simply to consider the equally spaced values

$$\epsilon_i = \epsilon_{\min} + i\left(\frac{\epsilon_{\max} - \epsilon_{\min}}{m}\right),$$

but it might make sense to bunch the values around regions of interest, if we have domain knowledge about interesting feature scales.

4. Compute the collection of vector spaces

$$\{H_2(|\mathrm{VR}_{\epsilon_1}(X,\partial_X)|), H_2(|\mathrm{VR}_{\epsilon_2}(X,\partial_X)|), \ldots, H_2(|\mathrm{VR}_{\epsilon_m}(X,\partial_X)|)\}.$$

5. Compare these abelian groups; for example, make a graph of the ranks of the free parts. If these are all non-zero and all the same, it suggests that there are stable topological features of M at the feature scales in the interval $[\epsilon_{\min}, \epsilon_{\max}]$. If there is a subinterval $[a, b] \subset [\epsilon_{\min}, \epsilon_{\max}]$ on which the ranks are the same, we might conclude that there are stable topological features at those ranges of scales. (Of course, there is no guarantee that we are not seeing different features at the different scales; this procedure does not really help us match topological features across scales.)

For an example of how this might work, consider the situation depicted in Figure 2.8. When ϵ is smaller than the distance between points, the Vietoris-Rips

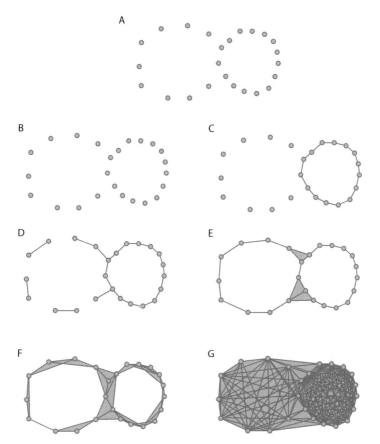

Figure 2.8 As ϵ increases, more and more simplices appear in the Vietoris-Rips complex.

complex only has 0-simplices. As ϵ increases, we first see 1-simplices appear that eventually connect the right hand circle. Then the left hand circle also appears. Finally, the circles are "filled in" by simplices crossing the circles when ϵ is large enough.

A first question is how to systematically choose the values $\{\epsilon_i\}$. Ideally, we will track places where $\text{VR}_\epsilon(X, \partial_X)$ changes. Since X is finite, there are only finitely many values $\{\epsilon_i\}$ at which the simplicial complex $\text{VR}_\epsilon(X, \partial_X)$ changes. We can see this because for $\epsilon > \text{diam}(X)$ the Vietoris-Rips complex has all $2^{|X|}$ simplices and as ϵ increases simplices are added but never removed.

Lemma 2.3.1. *Let (X, ∂_X) be a finite metric space. Then there exist at most finitely many values $\{\epsilon_i\}$ where $\text{VR}_{\epsilon_i}(X, \partial_X)$ changes, i.e., such that for all sufficiently small δ,*

$$\begin{cases} \text{VR}_\epsilon(X, \partial_X) = Z & \epsilon \in [\epsilon_i - \delta, \epsilon_i) \\ \text{VR}_\epsilon(X, \partial_X) = Z' & \epsilon \in [\epsilon_i, \epsilon_i + \delta] \end{cases}$$

and $Z \neq Z'$.

Therefore, we should choose $\{\epsilon_i\}$ to lie at these "inflection points" (and there is an upper bound on how many values we need to consider).

However, the most critical step is the last one; we need to find a systematic way to compare the various $\{H_2(\text{VR}_{\epsilon_i}(X, \partial_X))\}$. The key insight of *persistence* is that since $\text{VR}_{(-)}(X, \partial_X)$ is functorial in ϵ, for $\epsilon < \epsilon'$ we have a map of simplicial complexes

$$\text{VR}_\epsilon(X, \partial_X) \to \text{VR}_{\epsilon'}(X, \partial_X),$$

and for a collection $\epsilon_1 < \epsilon_2 < \ldots < \epsilon_m$ we obtain a sequence of simplicial maps

$$\text{VR}_{\epsilon_1}(X, \partial_X) \to \text{VR}_{\epsilon_2}(X, \partial_X) \to \ldots \to \text{VR}_{\epsilon_m}(X, \partial_X).$$

Since H_k is also a functor, applying H_k we obtain induced maps of abelian groups or vector spaces

$$H_k(\text{VR}_\epsilon(X, \partial_X)) \to H_k(\text{VR}_{\epsilon'}(X, \partial_X))$$

and

$$H_k(\text{VR}_{\epsilon_1}(X, \partial_X)) \to H_k(\text{VR}_{\epsilon_2}(X, \partial_X)) \to \ldots \to H_k(\text{VR}_{\epsilon_m}(X, \partial_X)).$$

More concisely, we can package this data as follows.

Definition 2.3.2. Given a fixed finite metric space (X, ∂_X), the Vietoris-Rips complex induces a functor

$$\text{VR}_{(-)}(X, \partial_X) \colon \mathbb{R} \to \text{Simp}$$

from $\mathbb{R}$ (regarded as the category associated to a partially ordered set) to the category of simplicial complexes. Composition with the kth homology group functor gives rise to a functor

$$H_k(\mathrm{VR}_{(-)}(X, \partial_X)) \colon \mathbb{R} \to \mathrm{Ab}.$$

It is useful to organize the resulting functors themselves into categories.

Definition 2.3.3. Let $\mathcal{C}$ be a category. The category of *filtered systems of* $\mathcal{C}$ is the category of functors $F \colon \mathbb{R} \to \mathcal{C}$ with morphisms given by natural transformations.

Clearly, any filtered system of simplicial complexes produces a filtered system of abelian groups or vector spaces. There are a variety of sources of filtered complexes that are relevant in topological data analysis, but for expositional clarity, we will focus on the Vietoris-Rips complex for the remainder of this discussion.

Example 2.3.4.

1. The Vietoris-Rips complex and Čech complex produce natural examples of filtered systems of simplicial complexes from the data of a finite metric space where we allow the scale ϵ to vary.
2. Motivated by the perspective of Morse theory, we assume the underlying data is a simplicial complex X along with a function $h \colon X \to \mathbb{R}$. There is now an induced filtered system of simplicial complexes induced by the inverse images $\{h^{-1}((-\infty, -])\}$. That is, for $b > a$, it is clear that $h^{-1}((-\infty, a])$ is a subcomplex of $h^{-1}((-\infty, b])$.

Remark 2.3.5. Note that the "Morse theoretic" perspective can be regarded as a generalization of the finite metric space approach, as follows. Given a compact subset $K \subseteq \mathbb{R}^n$, define the distance function

$$\partial_K(z) = \inf_{k \in K} \partial_{\mathbb{R}^n}(k, z).$$

Then for a finite set of points $X = \{x_1, \ldots, x_n\} \subseteq \mathbb{R}^n$, the filtered system of Čech complexes associated to the level sets of ∂_X is isomorphic to the filtered simplicial complex $\{C_*(X, \partial_X)\}$.

The functor $H_k(\mathrm{VR}_{(-)}(X, \partial_X))$ provides a means of addressing our problem about comparisons between the homology of the complexes as ϵ varies:

1. an element $\gamma \in H_k(\mathrm{VR}_{\epsilon_i}(X, \partial_X))$ is a k-dimensional feature at scale ϵ_i, and
2. we can determine the significance and stability of γ by finding the maximum $j > i$ such that the image of γ under the group homomorphism

$$\theta_{ij} \colon H_k(\mathrm{VR}_{\epsilon_i}(X, \partial_X)) \to H_k(\mathrm{VR}_{\epsilon_j}(X, \partial_X))$$

is non-zero.

Roughly speaking, an element $\gamma \in H_k(\mathrm{VR}_{\epsilon_i}(X, \partial_X))$ represents a k-dimensional hole in the geometric realization of the Vietoris-Rips complex at ϵ_i. If γ does not exist for $\epsilon' < \epsilon_i$, we think of this feature as being "born" at ϵ_i. When $\theta_{ij}(\gamma) = 0$, it means that the hole has been filled in by a collection of simplices with boundary γ. This suggests that it makes sense to try to figure out the "lifespan" of a particular element in homology, i.e., when it first appears and when it vanishes. More precisely, for a filtered simplicial complex $X_\bullet$, an element $\gamma \in H_k(X_i; \mathbb{F})$ is

1. *born* at i if it is not in the image of $H_k(X_{i-q}; \mathbb{F}) \to H_k(X_i; \mathbb{F})$ for any $q > 0$, and
2. *dies* at $\ell > i$ if it becomes zero in $H_k(X_\ell; \mathbb{F})$ or its image in $H_k(X_\ell; \mathbb{F})$ coincides with the image of another class that was born earlier.

Thus, we can think of the information contained in the filtered system of vector spaces as a series of elements with intervals representing their lifetime. Precisely, the persistent homology of a finite metric space can be described via a "barcode," a collection of intervals. Each interval represents the lifespan of a homological feature. (See Figure 2.9 for a simple representative example.)

Definition 2.3.6. A barcode is a multiset of non-empty intervals of the form either $[x, y) \subset \mathbb{R}$ or $[x, \infty)$. (A multiset is a generalization of a set where repeated elements are allowed, e.g., $\{1,1,2\}$.)

To be precise about the connection between persistent homology and barcodes, we require some finiteness hypotheses that always hold in practice, since we only have finitely many data points. We fix a field $\mathbb{F}$ for the remainder of this section.

Definition 2.3.7. A filtered simplicial complex is *tame* if the homology groups $H_i(-; \mathbb{F})$ are always of finite rank and change at only a finite number of indices.

By Lemma 2.3.1, the filtered complexes produced by applying the Vietoris-Rips complex construction to a finite metric space are always tame.

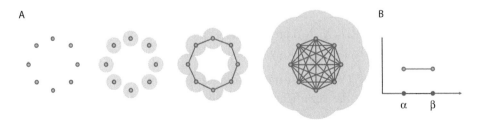

Figure 2.9 In (A), we have an idealized Vietoris-Rips filtration: when $\epsilon = \alpha$, the circle appears, and when $\epsilon = \beta$, the circle is filled in. In (B), the barcode has a single bar (representing a $\mathbb{Z}$ in homology) that appears at α and vanishes at β; this is the homology of the circle, for as long as it lasts.

Lemma 2.3.8. *Let $X\colon \mathbb{R} \to$ Simp be a tame filtered simplicial complex. The filtered vector space produced as $H_i(X(-); \mathbb{F})$ has the property that*

1. *each vector space $H_i(X(\epsilon); \mathbb{F})$ is of finite rank and*
2. *there exists N such that $H_i(X(\epsilon_1); \mathbb{F}) \to H_i(X(\epsilon_2); \mathbb{F})$ is an isomorphism for $\epsilon_2 > \epsilon_1 > N$.*

We say such a filtered vector space is of finite type.

Remark 2.3.9. A filtered vector space of finite type can be regarded as indexed on $\mathbb{Z}$, where the integral indices correspond to values in $\mathbb{R}$ where the homology changes.

The key classification result of Zomorodian and Carlsson [551] is then the following.

Theorem 2.3.10. *Let $\mathbb{F}$ be a field. There is a bijection between the set of finite barcodes and the set of isomorphism classes of filtered $\mathbb{F}$-vector spaces of finite type.*

The basic idea of this classification is quite simple; we define *interval modules*, which are filtered systems I_{ab} of $\mathbb{F}$-vector spaces $\{V_i\}$ where for $i \in [a,b]$, $V_i = \mathbb{F}$, and all the maps $\mathbb{F} \to \mathbb{F}$ are the identity (and the others are necessarily zero). Then any filtered system of $\mathbb{F}$-vector spaces is a direct sum of interval modules; the interval modules correspond to the bars in the barcode representing the lifetime of particular elements in homology.

Theorem 2.3.10 tells us that all of the information in the filtered system of vector spaces can be encoded as barcodes. It is often useful to think of a barcode as a collection of points in $\mathbb{R}^2$, specified by the endpoints of the intervals. Such a set is referred to as a *persistence diagram*, and often it is regarded as containing the entire diagonal (consisting of size zero bars).

In conclusion, we have the "persistent homology pipeline"

$$\begin{Bmatrix} \text{finite} \\ \text{metric} \\ \text{spaces} \end{Bmatrix} \to \begin{Bmatrix} \text{filtered} \\ \text{simplicial} \\ \text{complexes} \end{Bmatrix} \to \begin{Bmatrix} \text{barcodes / persistence} \\ \text{diagrams} \end{Bmatrix}.$$

We now turn to some examples of the use of barcodes to describe shape. When $k = 0$, the persistent homology is describing a standard hierarchical clustering construction.

Example 2.3.11. Recall from Theorem 1.10.10 that for a simplicial complex X, $H_0(X)$ is computing the free abelian group on the components. In the case of $VR_\epsilon(X, \partial_X)$ for a finite metric space (X, ∂_X), $H_0(\mathrm{VR}_\epsilon(X, \partial_X))$ computes the single-linkage clustering at scale ϵ of (X, ∂_X).

When considering the persistent homology, observe that each cluster at time $p+i$ can be thought of as resulting from the merger of clusters at i. This is clearly closely related to the information encoded in the hierarchical clustering dendrogram associated to single-linkage clustering. (See Figure 2.10 for comparison of the barcode and dendrogram for a synthetic data set.)

In Figure 2.11, we see an idealized situation involving sampling from an object in $\mathbb{R}^2$. In practice, however, the barcodes are often not so easy to interpret. Even for geometrically simple situations, complications can arise. In Figure 2.12, we illustrate how the barcode can change due to perturbation of the data by considering a sequence of nested circles.

In Figure 2.13, persistent homology of genomic sequence data generated by coalescent simulation is shown. As explained in Section 5.7, this is a way of modeling evolutionary phenomena. Typically, one fits phylogenetic trees to the finite metric space of sequences; here, we compute the persistent homology instead. Computing the first persistent homology group detects when "non-tree-like" events are occurring, i.e., when there is genetic recombination. Another example of this kind of application of persistent homology in studying recombination rates in the evolution of bacteria is discussed in Section 5.6.3; see Figure 2.14. In both of these applications, increased recombination can be detected by a large number of bars in the PH_1 barcode.

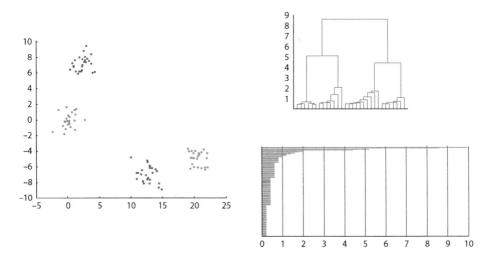

Figure 2.10 For the data set on the left, both the dendrogram and the zeroth persistent homology barcode capture how clusters merge as ϵ increases.

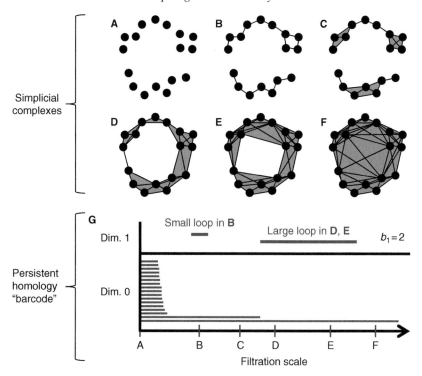

Figure 2.11 The points in panel A form a circle, with a horizontal gap separating upper and lower points. Panels A-F show the Vietoris-Rips filtration on these points as ϵ increases. Panel G shows the barcode. PH_0 (dimension 0) shows clustering of the data at different scales; each horizontal bar in the barcode is a cluster. In panel A (filtration scale 0), no points are connected; each is its own cluster (represented as 17 horizontal bars). As the scale increases, points in the simplicial complex connect, represented in the barcode as termination of a bar. There are two distinct clusters through panels B and C and one cluster in panels D, E, and F. PH_1 (dimension 1) shows loops in the data at different scales. Each bar in this part of the barcode identifies a different loop. There are two loops in this data: a short-lived loop in the top-right of the simplicial complex at scale B, and a long-lived loop appearing in panel D and persisting through panel E – this loop is represented as the long bar in the dimension 1 barcode. Robust features of the data set are captured in the barcode: the data clusters into two groups (two dimension 0 bars through scale C), and forms a loop (one long dimension 1 bar). The persistent first Betti number (b_1) is the total number of dimension 1 bars; here it is equal to 2.

In Section 8.3, we discuss an application of persistent homology to study the physical structure of DNA. Modeling DNA as a sequence of repeated units that have prescribed interaction points, persistent homology can be used to extract information about loops in the strands from a similarity matrix encoding the contact of sites with other sites. (See Figure 2.15.)

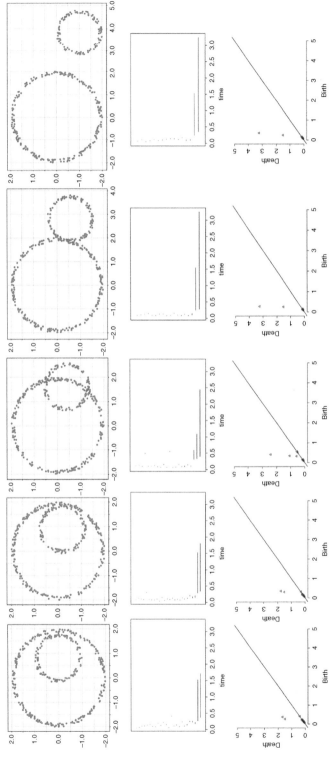

Figure 2.12 With two disjoint circles, we expect to see a barcode with two long bars, one significantly longer than the other to reflect the difference in radius. But when the circles are nested, the bars are nearly the same length as the inner circle interferes with the outer circle. Moreover, little loops connecting the two circles generate a lot of short bars. When the circles intersect non-trivially, we see an extra bar representing the loop formed by the intersection. And finally when the circles are disjoint and separated, we see the expected two bars, one longer and one shorter, corresponding to each circle.

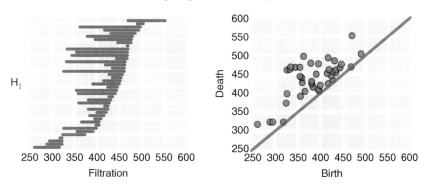

Figure 2.13 Two representations of the persistent homology of data from an evolutionary simulation; see Section 5.7 for discussion. On the left, a barcode diagram. On the right, a persistence diagram. Rather than identifying specific bars with geometric features, in this case the count of the bars conveys important information about the underlying process.

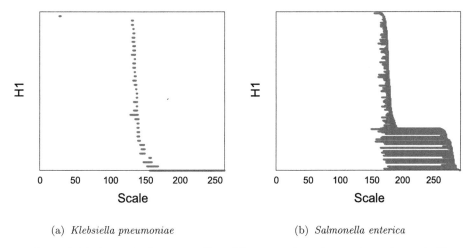

(a) *Klebsiella pneumoniae*   (b) *Salmonella enterica*

Figure 2.14 Barcode diagrams reflect different scales of genomic exchange in *K. pneumoniae* and *S. enterica*. Source: [161].

There are algorithms to compute the barcodes with running time cubic in the number of simplices. See Section 2.7 and Appendix A for discussion of the computational aspects of computing persistent homology.

2.4 Stability of Persistent Homology under Perturbation

In order to use topological invariants to describe data, it is essential that small perturbations of the data give rise to small changes in the resulting invariants.

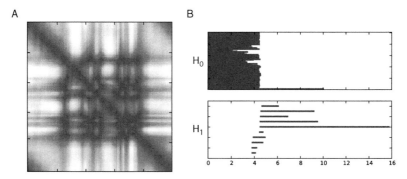

Figure 2.15 DNA can be simulated as a long polymer consisting of a large number of monomeric units interacting at specific places. Here, we show the data of a 50 Mb polymer with 10 fixed loops at random positions in the genome consisting of 1000 monomeric units. (A) The average of 5000 simulations allows us to construct a contact map. (B) Using persistent homology in a similarity matrix derived from the contact map one can clearly identify the ten loops as ten long bars in dimension one persistent classes. Source: [163].

One of the very useful aspects of persistent homology is that the set of barcodes forms a metric space; the distance between barcodes allows us to be precise about measuring changes in the output of topological data analysis. For the input, it turns out to be very useful to adopt a metric on the space of finite metric spaces, the Gromov-Hausdorff distance. These metric space structures make it possible to prove *stability theorems* that relate perturbation of the input data in the Gromov-Hausdorff metric to perturbation of the output barcodes in the barcode metric [105, 117].

These stability results are the most important theorems in the subject. In order to understand what they really say, we need to explain

1. what it means for two finite metric spaces to be close in the Gromov-Hausdorff metric, and
2. what it means for two barcodes to be close in the barcode metric.

Definition 2.4.1. Let A and B be non-empty subsets of a metric space (X, ∂_X). Then we define the *Hausdorff distance* between A and B to be

$$d_H(A, B) = \max \left(\sup_{a \in A} \inf_{b \in B} \partial_X(a, b), \sup_{b \in B} \inf_{a \in A} \partial_X(a, b) \right).$$

It is sometimes convenient to consider the equivalent formulation of the Hausdorff distance as

Figure 2.16 The Hausdorff distance is determined by the point in *A* with the largest distance to the closest point in *B* (and vice versa).

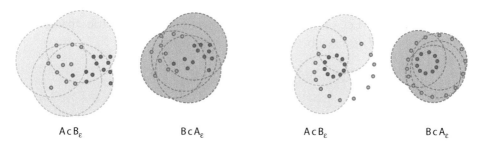

$A \subset B_\varepsilon$ $B \subset A_\varepsilon$ $A \subset B_\varepsilon$ $B \subset A_\varepsilon$

Figure 2.17 The Hausdorff distance can be computed by considering the smallest ϵ fattening of each set that contains the other.

$$d_H(A, B) = \inf_{\epsilon>0}\{B \subseteq A_\epsilon, A \subseteq B_\epsilon\},$$

where A_ϵ and B_ϵ denotes the sets of all points within distance ϵ of *A* and *B*, respectively (see Figures 2.16, 2.17).

Example 2.4.2.

1. Let $A \subset X$ and suppose that *B* is generated from *A* by perturbing each point $a \in A$ by at most ϵ; i.e., the points of *B* are in bijection with those of *A* and (denoting the bijection by θ) we have $\partial_X(a, \theta(a)) \leq \epsilon$. For instance, consider $A = \{[0,0,0],[1,2,3],[-1,0,5]\} \subset \mathbb{R}^3$ and $B = \{[\epsilon,0,0],[1,2+\epsilon,3],[-1,0,5-\epsilon]\}$. Then $d_H(A, B) \leq \epsilon$.
2. The Hausdorff distance is heavily influenced by the single most extreme point; given $A \subset X$, let $A' = A \cup \{x\}$. Then $d_H(A, A') = \min_{a \in A} \partial_X(x, a)$.

Lemma 2.4.3. *The Hausdorff distance imposes a metric on the set of non-empty subsets of a metric space* (X, ∂_X).

However, we cannot in general assume that the metric spaces we consider are given as subsets of a common ambient metric space. A key insight of Gromov is to circumvent this issue by considering the infimum of the Hausdorff distance over all isometric embeddings of the two metric spaces into a larger ambient metric space. Here an isometric embedding

$$\phi \colon (X, \partial_X) \to (Y, \partial_Y)$$

is an injective map $X \to Y$ such that

$$\partial_X(x_1, x_2) = \partial_Y(\phi(x_1), \phi(x_2)).$$

That is, an isometric embedding identifies X with a submetric space of Y.

Definition 2.4.4.   Let (X_1, ∂_{X_1}) and (X_2, ∂_{X_2}) be compact metric spaces. The *Gromov-Hausdorff distance* between X_1 and X_2 is defined to be

$$d_{GH}((X_1, \partial_{X_1}), (X_2, \partial_{X_2})) = \inf_{\substack{\theta_1 \colon X_1 \to Z \\ \theta_2 \colon X_2 \to Z}} d_H(X_1, X_2).$$

Here θ_1 and θ_2 are isometric embeddings of (X_1, ∂_{X_1}) and (X_2, ∂_{X_2}) in (Z, ∂_Z) respectively (see Figure 2.18 for an example); the infimum is taken over all such (Z, ∂_Z) and embeddings θ_1 and θ_2.

We will say that two metric spaces are *isometric* if there exists an isomorphism $f \colon X \to Y$ that preserves all distances. This clearly defines an equivalence relation on the set of metric spaces.

Theorem 2.4.5.   *The Gromov-Hausdorff distance is a metric on the set of isometry classes of compact metric spaces.*

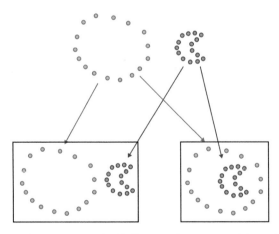

Figure 2.18 The Gromov-Hausdorff distance is computed by minimizing over all embeddings; here, the embedding on the right has a much smaller Hausdorff distance between the two image sets than the embedding on the left.

As defined above, it is hard to see how one might ever compute the Gromov-Hausdorff distance in practice. For this purpose, an alternative formulation is useful; it is also conceptually helpful in understanding what d_{GH} is measuring. Let $\mathcal{R}$ be a correspondence between X_1 and X_2, i.e., a subset of $X_1 \times X_2$ such that there exists a tuple with first coordinate x for each $x \in X_1$ and a tuple with second coordinate y for each $y \in X_2$.

The Gromov-Hausdorff distance can now be described by the formula

$$d_{GH}((X_1, \partial_{X_1}), (X_2, \partial_{X_2})) = \inf_{\mathcal{R} \subseteq X_1 \times X_2} \frac{1}{2} \left(\sup_{(x,x') \in \mathcal{R}, (y,y') \in \mathcal{R}} |\partial_{X_1}(x, y) - \partial_{X_2}(x', y')| \right).$$

Roughly speaking, the Gromov-Hausdorff distance measures the maximum distortion in the best matching between the two metric spaces.

Example 2.4.6.

1. Suppose that X' is an ϵ-net in X (recall that this means that for each $x \in X$, there exists a point $x' \in X'$ such that $\partial_X(x, x') < \epsilon$). Then $d_{GH}((X', \partial_X), (X, \partial_X)) < \epsilon$.
2. Let (X, ∂_X) be a metric space and suppose that $(X', \partial_{X'})$ is formed by adding a single point $\{z\}$ to X such that $\partial_{X'}(z, x) = \kappa > \operatorname{diam}(X)$ for any $x \in X$. (That is, we are adding a single point to X which is "far away" from the rest of the points.) Then $d_{GH}((X, \partial_X), (X', \partial_{X'})) > \frac{\kappa}{2}$.
3. Suppose that (X, ∂_X) and (Y, ∂_Y) are isometric metric spaces. Then $d_{GH}((X, \partial_X), (Y, \partial_Y)) = 0$.

There is an interesting body of work on the topology induced on the set of isometry classes of compact metric spaces by d_{GH}. For our purposes, one thing to observe is that any compact metric space can be approximated as the Gromov-Hausdorff limit of finite metric spaces. (See Figure 2.19 for an example of this kind of convergence.)

Lemma 2.4.7. *Given a compact metric space (X, ∂_X), let $\{X_n\}$ denote a sequence of finite $\frac{1}{n}$-nets in X. Then*

$$\lim_{n \to \infty} d_{GH}((X, \partial_X), (X_n, \partial_X)) = 0.$$

The Gromov-Hausdorff distance is a suitable means for capturing perturbations of data sets that involve bounded changes in each point, and therefore for measuring the impact of certain kinds of noise. On the other hand, Example 2.4.6 makes it clear that arbitrary changes in a constant number of points can cause arbitrary changes in the Gromov-Hausdorff distance. We will return to a discussion of this phenomenon in Chapter 3; see Section 3.4 in particular.

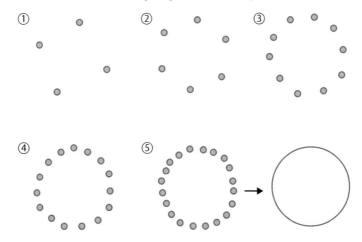

Figure 2.19 Samples of points that lie on a circle converge to the circle in the Gromov-Hausdorff distance as the sampling density increases.

We now turn to the description of various metrics on the set of barcodes. We begin with the *bottleneck distance*. Given two intervals $[a_1, b_1)$ and $[a_2, b_2)$, define

$$d_\infty([a_1, b_1), [a_2, b_2)) = \max(|a_1 - a_2|, |b_1 - b_2|).$$

We extend d_∞ to include $\emptyset$ by defining

$$d_\infty([a, b), \emptyset) = \frac{|b - a|}{2}.$$

Now given two barcodes B_1 and B_2, we define a matching between B_1 and B_2 as follows. Without loss of generality, assume that $|B_1| < |B_2|$. Then a matching is specified by a bijection $\phi: A_1 \to A_2$, where A_1 is a multi-subset of B_1 and A_2 is a multi-subset of B_2. We formally add $\emptyset$ to B_1 and B_2, and we regard the elements of $B_1 \setminus A_1$ and $B_2 \setminus A_2$ as matched with $\emptyset$.

Definition 2.4.8. Let B_1 and B_2 be barcodes. The *bottleneck distance* is defined to be

$$d_B(B_1, B_2) = \inf_\phi \sup_{Z \in B_1} d_\infty(Z, \phi(Z)),$$

where ϕ varies over all matchings between B_1 and B_2 and the supremum is taken over bars in B_1.

Roughly speaking, the bottleneck distance measures the worst discrepancy in the best matching between the two barcodes. Note that two barcodes which are a distance ϵ apart in the bottleneck distance could differ in an essentially arbitrary number of short bars of length less than $\frac{\epsilon}{2}$. Put another way, two barcodes are close

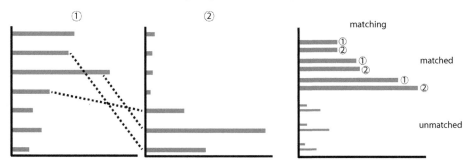

Figure 2.20 The bottleneck distance on barcodes is computed by matching long bars. Figure from experiment performed by Elena Kandror, Abbas Rizvi, and Tom Maniatis at Columbia University, with permission.

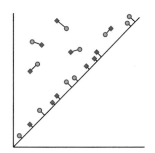

Figure 2.21 The bottleneck distance when expressed in terms of persistence diagrams is computed by matching nearby points and assigning points close to the diagonal to the nearest diagonal point.

in the bottleneck distance if after ignoring "short" bars, the endpoints of matching "long" bars are close (see Figures 2.20 and 2.21 for examples.)

There are other sensible metrics on the space of barcodes, most notably including mass transportation (Wasserstein) metrics. Since it will be convenient for later use, we will also introduce the *Wasserstein metric* here.

Definition 2.4.9. Let B_1 and B_2 be barcodes. For $p > 0$, the *p-Wasserstein distance* is defined to be

$$d_{W_p}(B_1, B_2) = \left(\inf_{\phi} \sum_{Z \in B_1} d_\infty(Z, \phi(Z))^p \right)^{\frac{1}{p}}.$$

We can now state the stability theorem for persistent homology, arguably the most important theorem in the subject [117]. (See Figure 2.22 for an illustration.)

Theorem 2.4.10. *Let (X, ∂_X) and (Y, ∂_Y) be finite metric spaces. Then for all $k \geq 0$,*

$$d_B(\text{PH}_k(\text{VR}(X, \partial_X)), \text{PH}_k(\text{VR}(Y, \partial_Y))) \leq d_{GH}((X, \partial_X), (Y, \partial_Y)).$$

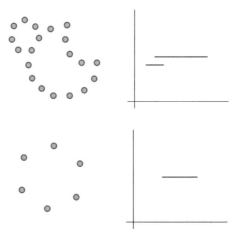

Figure 2.22 The two samples are close together in the Gromov-Hausdorff distance; although at various ϵ the homology groups are different, the barcodes are close together.

Remark 2.4.11. Analogous results hold when using the Čech complex or using the Wasserstein metric.

There are versions of the stability theorem expressed in terms of the "Morse filtration" approach to persistent homology as well. The set of functions $\{f : X \to \mathbb{R}\}$ can be endowed with a metric specified as

$$d_\infty(f, g) = \sup_{x \in X} |f(x) - g(x)|.$$

We say that a function $f : X \to \mathbb{R}$ is admissible if $H_k(f^{-1}(-\infty, t]; \mathbb{F})$ is finite rank for all $t \in \mathbb{R}$.

Theorem 2.4.12. *Let X be a topological space. Let $f, g : X \to \mathbb{R}$ be admissible functions. Then for all $k \geq 0$,*

$$d_B(\mathrm{PH}_k(X, f), \mathrm{PH}_k(X, g)) \leq d_\infty(f, g).$$

Using the observation of Remark 2.3.5 and the relationship between the Čech and Vietoris-Rips complex, we can regard Theorem 2.4.12 as a generalization of Theorem 2.4.10.

Remark 2.4.13. Theorems 2.4.10 and 2.4.12 are incarnations of an algebraic stability theorem, which says that for persistence modules that are κ-interleaved (which is a precise way of expressing the notion of being approximately isomorphic), the resulting barcodes are within κ in the bottleneck metric [42, 107]. This formulation of the stability theorem allows us to substantially weaken the hypotheses necessary to apply it and also extends its reach.

2.5 Zigzag Persistence

Persistent homology is defined in situations where we have a filtered system of complexes. As we have described above, these filtrations typically arise by varying a scale parameter of some sort. Sometimes, however, we might not expect to have a filtration but rather some kind of more general diagram. That is, a natural question that arises is whether other "filtration shapes" could be used as input. We now discuss an answer to the following specific form of this question [91].

Question 2.5.1. Does a construction like persistent homology make sense when considering "filtrations" in which not all the arrows go in the same direction?

This more general kind of diagram can easily arise in practice. For example, suppose we consider many sets of samples X_i from each fixed metric space (X, ∂_X). We then can form the sequence

$$X_1 \longrightarrow X_1 \cup X_2 \longleftarrow X_2 \longrightarrow X_2 \cup X_3 \longleftarrow X_3 \longrightarrow \ldots$$

where the maps are the obvious inclusions (Figure 2.23).

Applying the composite of $H_k(-; \mathbb{F})$ and the Vietoris-Rips complex functor (for some fixed ϵ) to this sequence yields a corresponding diagram of $\mathbb{F}$-vector spaces

$$H_k(\text{VR}_\epsilon(X_1)) \longrightarrow H_k(\text{VR}_\epsilon(X_1 \cup X_2)) \longleftarrow H_k(\text{VR}_\epsilon(X_2))$$

$$H_k(\text{VR}_\epsilon(X_2 \cup X_3)) \longleftarrow H_k(\text{VR}_\epsilon(X_3)) \longrightarrow \ldots$$

In order to study these sorts of "filtrations" more carefully, we need to develop some notation for describing the pattern of arrows. To do this, we consider *zigzag diagrams* of shape S, where S is a string on the alphabet L, R.

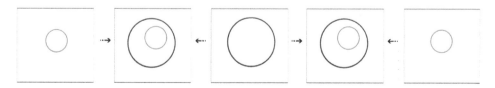

Figure 2.23 We get a natural zigzag by taking unions of samples.

Definition 2.5.2.   A *zigzag diagram* (or *zigzag module*) of shape S is defined to be a sequence of linear transformations between $\mathbb{F}$-vector spaces:

$$X_1 \xrightarrow{\ f_1\ } X_2 \xrightarrow{\ f_2\ } \dots \xrightarrow{\ f_{k-1}\ } X_k,$$

where each map f_i has its direction specified by the ith letter of the string S. (This is also referred to as a zigzag module.)

The definition of a zigzag diagram is a strict generalization of the notion of a filtration. When the shape S is $RRRRRRRRR\dots R$ or $LLLLLLL\dots L$, a zigzag diagram is simply a filtered $\mathbb{F}$-module.

Example 2.5.3.

1. Let $S = RRR$. Then a zigzag diagram of shape S is a diagram

$$M_1 \to M_2 \to M_3 \leftarrow M_4$$

   of vector spaces.
2. Let $S = RLRLRL$. Then a zigzag diagram of shape S is a diagram

$$M_1 \to M_2 \leftarrow M_3 \to M_4 \leftarrow M_5 \to M_6 \leftarrow M_7$$

   of vector spaces.

In the original setting for persistent homology, it was intuitively clear that the "lifespan" of a homological feature was an interesting topological invariant associated to a filtration. When working with zigzag diagrams, the corresponding idea is that of a homological feature that is "consistent" across the zigzag. For example, if we are considering a zigzag of shape RL,

$$M_1 \xrightarrow{\ f_1\ } M_2 \xleftarrow{\ f_2\ } M_3,$$

then a zigzag feature should represent a collection of elements $m_1 \in M_1, m_2 \in M_2, m_3 \in M_3$ consistent in the sense that $f_1(m_1) = m_2 = f_2(m_3)$. In the context of the sampling example we started with, a zigzag feature should represent some kind of geometric property that is stable across different samples.

To work with this notion, one would again hope for an analogue of Theorem 2.3.10 that allows us to characterize homological invariants of zigzag diagrams in terms of some kind of numerical invariant like barcodes. We now switch to using the zigzag module terminology.

Definition 2.5.4.   A *zigzag submodule* N of a zigzag module M of shape S is a zigzag module of shape S such that each N_i is a subspace of M_i and the maps are determined by the restrictions of the f_i.

Example 2.5.5. Let $\mathbb{F} = \mathbb{R}$, and suppose we are given the zigzag module

$$\mathbb{R} \longrightarrow \mathbb{R}^2 \longleftarrow \mathbb{R},$$

where the first map is $x \mapsto (x, 0)$ and the second map is $x \mapsto (0, x)$. Then there is a zigzag submodule

$$\mathbb{R} \longrightarrow \mathbb{R} \longleftarrow \mathbb{R},$$

where the $\mathbb{R}$ in the middle comes from the first coordinate of $\mathbb{R}^2$; the maps are now $x \mapsto x$ and $x \mapsto 0$.

We say that a zigzag submodule M is *decomposable* if it can be written as the direct sum of non-trivial submodules $\{N_j\}$ (recall Definition 1.6.40); otherwise, we say it is *indecomposable*.

Lemma 2.5.6. *Any zigzag module M of shape S can be written as a direct sum of indecomposables in a way that is unique up to permutation.*

Indecomposable zigzag modules have a very constrained form.

Definition 2.5.7. An *interval zigzag module* of shape S is a zigzag module

$$X_1 \xrightarrow{\ f_1\ } X_2 \xrightarrow{\ f_2\ } \dots \xrightarrow{\ f_{k-1}\ } X_k,$$

where for fixed $a \leq b$,

$$\begin{cases} X_i = \mathbb{F}, & 1 \leq a \leq i \leq b \leq k \\ X_i = 0, & \text{otherwise} \end{cases}$$

and the maps between the $\mathbb{F}$ are the identity map, and the zero map otherwise.

We can now state the main theorem that gives rise to zigzag barcodes.

Theorem 2.5.8. *The indecomposable zigzag modules are precisely the interval zigzag modules.*

As a consequence, we can obtain a barcode multiset which is referred to as the zigzag persistence, and tends to be represented the same way as persistence barcodes (Figure 2.24).

In Section 5.4.3, zigzag persistence is used to study HIV in tissue samples taken from central nervous system (CNS) and non-CNS regions. See Figure 2.25 for an indication of the data.

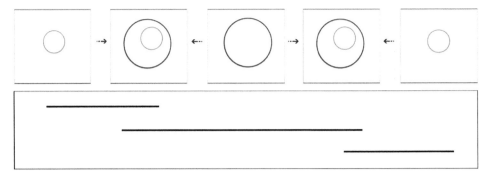

Figure 2.24 The bars represent features that persist across zigzags.

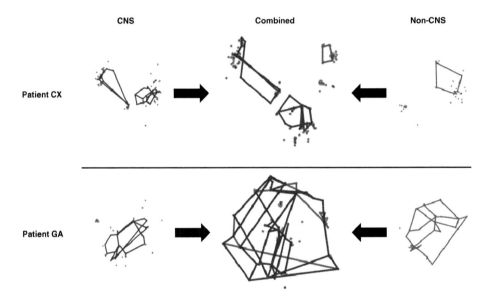

Figure 2.25 Phylogenetic networks of HIV-1 gp120 sequences obtained from Patients CX and GA. Each node represents one sequence; larger nodes show sequences that were sampled multiple times. Blue nodes were sampled from the CNS; red nodes were sampled from elsewhere in the body. The position of each node is determined by the first two principal components (computed via MDS) of genetic distance (Hamming distance). The network backbone (thin gray edges) is a minimum spanning tree, and the thick red and blue edges are generators of cycles identified by persistent homology. Red cycles denote putative recombination events that involve sequences sampled fully outside the CNS; blue cycles denote events that involve some sequences from the CNS.

We now discuss a basic zigzag that arises from metric data. In what follows, let (X, ∂_X) be a finite metric space, and choose an ordering for the points – we will denote the ordering as

$$X = \{x_1, x_2, x_3, \ldots\}.$$

Let X_k denote the subset of X consisting of the first k points in the ordering, i.e.,

$$X_1 = \{x_1\}, \quad X_2 = \{x_1, x_2\}, \quad X_3 = \{x_1, x_2, x_3\},$$

and so on. We can then define a series of distinguished scales $\epsilon_i = d_H(X_k, X)$. Notice that $\epsilon_i \geq \epsilon_{i+1}$; X_{i+1} will always be at least as close to X as X_i in the Hausdorff distance.

Definition 2.5.9. Choose real numbers $\alpha > \beta > 0$. The *Rips zigzag* consists of the zigzag module specified by the diagram of simplicial complexes

$$\cdots \longleftarrow \mathrm{VR}_{\beta\epsilon_{i-1}}(X_{i-1}) \longrightarrow \mathrm{VR}_{\alpha\epsilon_{i-1}}(X_i) \longleftarrow \mathrm{VR}_{\beta\epsilon_i}(X_i)$$

$$\downarrow$$

$$\mathrm{VR}_{\alpha\epsilon_i}(X_{i+1}) \longleftarrow \mathrm{VR}_{\beta\epsilon_{i+1}}(X_{i+1}) \longrightarrow \cdots$$

Notice that the constituent complexes in this zigzag have size controlled by the limits α and β; it was originally proposed by Morozov for the purpose of computational efficiency. Work of Oudot and Sheehy [393] provides theoretical validation for the use of this zigzag, showing that when $X \subseteq \mathbb{R}^n$ is close in Hausdorff distance to a well-behaved compact subset $Y \subseteq \mathbb{R}^n$, then there are long zigzag intervals in the Rips zigzag that permit recovery of the homology of X for suitable α and β. (As with Theorem 2.2.1, the actual numerical bounds extracted from these results are much larger than needed in practice.)

Finally, given the central importance of the stability theorem for persistent homology, one would hope for something similar in the context of zigzag persistence. In [91], stability results were proved in the context of a particular construction of zigzags from finite metric spaces, the *level set zigzag diagram*. In general, the specific form of stability results depends on the particulars of the process of constructing the zigzag. Nonetheless, theoretical considerations [67] show that essentially any reasonable procedure for producing zigzag modules will have some kind of stability theorem.

2.6 Multidimensional Persistence

The underlying idea of persistence, namely that a sensible way to cope with uncertainty about parameter settings is simply to aggregate information as the parameter changes, is a powerful and general one. But why limit ourselves to just the feature scale? There are often many parameters which we might like to apply persistence to: for example, in the motivating example for zigzag persistence, it would make sense to vary both the samples and the feature scale ϵ. And in many probabilistic settings we want to simultaneously vary a density parameter as well as ϵ. In this section we discuss two approaches to considering persistence in multiple directions. First, we explain a systematic framework for multidimensional persistence. Then we discuss a closely related idea, the persistent homology transform.

2.6.1 Multidimensional Persistence

In many situations, it is natural to consider multiple filtrations on a data set; e.g., for a finite metric space (X, ∂_X) one filtration will come from the distance scale parameter and another from an additional property of the data. A key motivating example arises when the density of the data is not uniform: it often makes sense to consider one filtration direction generated by the distance scale and another generated by density.

Provided that these filtrations interact in a natural way, we can define multidimensional persistent homology as a generalization of the definition of persistent homology given above. Specifically, we regard $\mathbb{R}^n$ as a partially ordered set and hence a category by setting $\{a_1, \ldots, a_n\} \leq \{b_1, \ldots, b_n\}$ when each $a_i \leq b_i$.

Definition 2.6.1. A *multifiltered system of simplicial complexes* is a functor from $\mathbb{R}^n$ to simplicial complexes. A *multifiltered vector space* is a functor from $\mathbb{R}^n$ to $\mathbb{F}$-vector spaces.

Explicitly, for $n = 2$, a multifiltered complex $\{X_{\alpha,\beta}\}$ is specified by a commutative diagram

$$
\begin{array}{ccc}
X_{a,b} & \longrightarrow & X_{x_1,y_1} \\
\downarrow & & \downarrow \\
X_{x_2,y_2} & \longrightarrow & X_{c,d}
\end{array}
$$

for any $x_1, x_2 \in [a, c]$ and $y_1, y_2 \in [b, d]$.

Example 2.6.2. Suppose we have a finite metric space (X, ∂_X) and a codensity function $\gamma: X \to \mathbb{R}$, where γ is small at higher density points and large at sparse points. For example, γ could be a normalized count of the distance to the kth-nearest neighbor. (Here k is a parameter that has to be chosen.) Then we define a functor

$$\mathbb{R} \times \mathbb{R} \to \mathrm{Simp}$$

via the formula

$$(\epsilon, \delta) \mapsto \mathrm{VR}_\epsilon(\gamma^{-1}(-\infty, \delta]).$$

Given any multifiltered complex, by passing to homology, we obtain a multifiltered vector space, the multidimensional persistent homology. There is again a structure theorem for multifiltered vector spaces, but in contrast to the one dimensional case, the irreducible objects are not easily described. As a consequence, there is no tractable analogue of the barcode in this context which completely describes the isomorphism type of the multifiltered vector space, and so no easy summarization of the results of computing multidimensional persistent homology.

A number of possible solutions to this problem have been proposed: even though there is no complete invariant, there are many interesting invariants which capture partial information that are relevant to data analysis.

1. Zomorodian and Carlsson proposed the rank invariant: this is the numerical invariant obtained by taking the ranks of the maps in the filtration [92].
2. Lesnick and Wright studied in detail the "fibered barcode," a version of the rank invariant (introduced under a different name in [99]), which is the collection of invariants obtained by choosing lines through the filtrations and computing the one dimensional persistence in that direction [325]. They have developed a tool, Rivet [324], that supports exploratory data analysis in this context, displaying the rank invariant as well as the bigraded Betti numbers. See Figures 2.26 and 2.27 for examples.

2.6.2 The Persistent Homology Transform

In the general spirit of persistence, one approach to choosing lines through the filtration is to consider the collection of all of them at once. We now discuss an implementation of this idea in the restricted context of surfaces embedded in Euclidean space.

Beyond difficulties in computing persistent homology, as we discuss in detail in Chapter 3, it can be difficult to interpret the results of persistent homology computations even for data embedded in comparatively low-dimensional Euclidean spaces $\mathbb{R}^n$ for $n > 3$. One approach to this issue is to restrict attention to spaces embedded in $\mathbb{R}^2$ or $\mathbb{R}^3$; such examples arise when considering surfaces, for instance. In the setting of cancer genomics, motivating examples arise from the imaging of tumors, as we discuss in a bit more detail in Section 3.8.

When working in $\mathbb{R}^2$ or $\mathbb{R}^3$, filtrations generated by a height function seem particularly useful. However, one issue with filtrations generated by height functions is that they depend on a choice of orientation – along which direction do we measure height? Just as the basic idea of persistence is to consider all scales at once, a simple approach is to consider all possible orientations at once. We now explain a direct approach to considering a kind of multidimensional persistence in this setting [510].

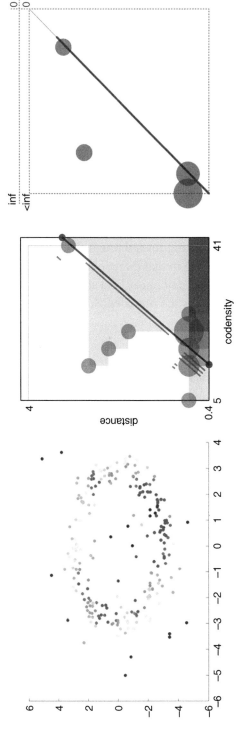

Figure 2.26 Persistence along a line that increases codensity and feature scale; in dimension zero, there is a single long bar that appears in this direction. In dimension 1, we see a single point away from the diagonal; simultaneously filtering by scale and density reduces the impact of noise. The size of the dots indicates multiplicity of the point, and the shading reflects the dimension of the vector space.

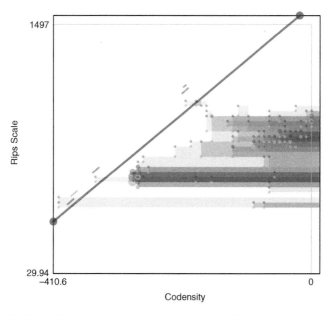

Figure 2.27 Multidimensional persistence for the HIV data set. From Monica Nicolau, Arnold J. Levine, Gunnar Carlsson, *Proceedings of the National Academy of Sciences* Apr 2011, 108 (17), 7265–7270. Reprinted with Permission from *Proceedings of the National Academy of Sciences*.

Suppose that our data is presented as a finite simplicial complex M embedded in $\mathbb{R}^d$. For each direction, represented by a point $v \in S^{d-1}$, we define a filtration of M as

$$M(v)_\epsilon = \{x \in M \mid x \cdot v \le \epsilon\}.$$

We can now consider summarizing M by considering the persistent homology of each of these filtrations in aggregate. Specifically, we have the following definition.

Definition 2.6.3. The *persistent homology transform* of $M \subseteq \mathbb{R}^d$ is the function

$$\mathrm{PHT} \colon S^{d-1} \to \mathcal{B}^d$$

specified by the assignment

$$v \mapsto [\mathrm{PH}_0(M(v)_\bullet), \mathrm{PH}_1(M(v)_\bullet), \ldots, \mathrm{PH}_d(M(v)_\bullet)].$$

The main theorem of [510] shows that in dimensions 2 and 3, we can use the collection of persistent homologies here to uniquely characterize the input object.

Theorem 2.6.4. *Let $d = 2$ or $d = 3$. Then* PHT *specifies an injective function from the set of finite simplicial complexes $M \subset \mathbb{R}^d$ to the set of functions from S^{d-1} to $\mathcal{B}^d$.*

2.7 Efficient Computation of Persistent Homology

In order for topological data analysis to be useful in practice, it must be possible to efficiently compute invariants like PH of real data sets. For example, one reason for the ubiquity of linear regression, PCA, and clustering in data analysis is the ease of computation, even for large data sets. Moreover, since many applications of TDA are in the context of exploratory data analysis, it is important that repeatedly recomputing with different parameters be feasible. In this section, we give an overview of the source of computational difficulty in applying TDA; Appendix A has a more detailed discussion of specific software packages.

As a baseline for comparison, we note the following.

1. Computing the single-linkage clustering dendrogram for a finite metric space (X, ∂_X) where $|X| = n$ can be done in time proportional to $n \log n$.
2. Similarly, Mapper (described in Section 2.8) can also be computed very efficiently.

Persistent homology is another matter. As is evident from the discussion of the computation of homology, persistent homology cannot be computed much more efficiently than matrix multiplication on matrices with dimensions given by the number of simplices – and for non-sparse matrices, practical algorithms for matrix multiplication are roughly cubic.

To compute persistent homology, we proceed as follows. Suppose that we have a filtered simplicial complex X. We choose a total ordering of the simplices of X that is compatible with the filtration on X; i.e., $\sigma < \tau$ if σ appears in a lower filtration than τ. (The order of simplices within a given filtration level is arbitrary.) Let n denote the number of simplices of X. We now form the $n \times n$ matrix D defined by the formula

$$D_{i,j} = \begin{cases} 1, & \text{if } \sigma_i \text{ is a codimension 1 face of } \sigma_j \\ 0, & \text{otherwise.} \end{cases}$$

We now define $\text{low}(j)$ to be the row number of the last 1 in column j; we set $\text{low}(j) = 0$ if column j consists only of zero entries. We will say that the matrix D is *reduced* if $\text{low}(j_1) \neq \text{low}(j_2)$ for $j_1 \neq j_2$. The following algorithm reduces the matrix D:

```
for j = 1 to n
  while there exists k < j such that low(k) = low(j) != 0:
    add column k to column j
```

The algorithm clearly terminates, since each step decreases low in a given column. We can extract the persistence diagram from the reduced form of D by observing that the pairs $(j, \text{low}(j))$ specify persistence intervals.

The serious issue that arises here is the dependence of the running time on the number of simplices. For example, for the Vietoris-Rips complex (or the Čech complex), this can be a problem when the feature scale ϵ approaches the maximum distance between any pair of points in the data set.

Lemma 2.7.1. *Let (X, ∂_X) be a finite metric space, and choose $\epsilon > \text{diam}(X)$, i.e.,*

$$\forall x, y \in X, \qquad \epsilon > \partial_X(x, y).$$

Then $\text{VR}_\epsilon(X, \partial_X)$ has $2^{|X|}$ simplices.

The inexorable conclusion of Lemma 2.7.1 is that in order to efficiently compute persistent homology, it will be necessary to control the number of simplices. One way to do this is to only work with low-dimensional homology; state of the art implementations (see Appendix A) can handle thousands of points when computation is limited to H_1. A general approach to this problem is simply to study the Vietoris-Rips complex over a range $[0, \epsilon_{\max}]$ that ensures a tractable number of simplices at $\epsilon_{\max}$. Another technique is to take many subsamples from (X, ∂_X) such that each subsample results in tractable persistent homology computations, and then combine the persistent homology of the subsamples in some way to estimate the persistent homology of X. This idea is part of the motivation for zigzag persistence, notably the Rips zigzag of Definition 2.5.9. Because zigzag persistence can be used in contexts where we control the size of the maximal complex, modern implementations can be used on data sets with thousands of points. Moreover, techniques for combining such subsamples in a systematic way along with methods for understanding error and variability in the results lead us naturally into the domain of statistical methods; we explore this in detail in the next chapter.

Another possibility is to construct a smaller complex. An early approach to this is the weak witness (or weak Delaunay) complex [458]. The idea is to choose as vertices a set of *landmarks* but use all of the data points to determine the complex.

Definition 2.7.2. Let (X, ∂_X) be a finite metric space. Consider a set of points

$$A = \{x_0, x_1, \ldots, x_k\} \subset X.$$

Then a point $w \in X$ is a *weak witness* for A if $\partial_X(w, x_i) \le \partial_X(z, x_i)$ for all i and $z \in X - A$.

Roughly speaking, the witness complex will only include simplices for which weak witnesses exist.

Definition 2.7.3. Let $L \subset X$ be a subset of the finite metric space (X, ∂_X). The *witness complex* is the simplicial complex specified by the rule that a simplex $[x_0, x_1, \ldots, x_k]$ for $x_i \in L$ is in the complex if all subsimplices admit weak witnesses.

In practice, the landmarks are often picked randomly or using an algorithm to maximize dispersion (Figure 2.28).

Although very attractive from the perspective of efficiency, the witness complex has problematic theoretical properties:

1. There do not appear to be good stability theorems for the witness complex,
2. the dependence on choice of landmarks is not well understood [105], and
3. the witness complex can fail to reconstruct the homotopy type even in simple examples [215].

In light of these issues, we believe that the only way to extract information from witness complexes is by using the statistical techniques outlined in the next chapter.

For points embedded in $\mathbb{R}^n$, other "small" complexes come from consideration of the Voronoi tesselation of $\mathbb{R}^n$. For example, the Delaunay complex is the simplicial complex obtained as the nerve of the cover of $\mathbb{R}^n$ given by the sets U_x for $x \in X$, where

$$U_x = \{z \in \mathbb{R}_n \mid \partial_X(x, z) \le \partial_X(x', z) \; \forall x' \in X\}.$$

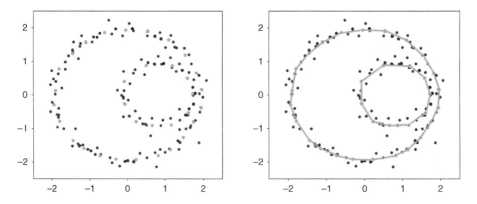

Figure 2.28 The landmark points give rise to concise simplicial circles that capture the topology of the data.

In low dimensions, the Delaunay complex can be computed very efficiently and faithfully recovers the homotopy type of X, although the dependence on the ambient dimension is exponential in general. A variant of this is called the α-complex, which again can be computed efficiently in low dimensions. (Both of these complexes can be computed for data sets consisting of thousands of points via state the art packages.)

As another example, using techniques from the theoretical computer science literature about approximation of metric spaces, the paper [455] explores how to build a hierarchical collection of approximations to suitable finite metric spaces such that for any given accuracy the computation time is linear in the number of points X. Here, suitable means that the metric space has constant doubling dimension, which is a measure of how the volume of balls changes as the radius changes. Note however that metric spaces with doubling dimension d admit low distortion embeddings into $\mathbb{R}^d$; from a practical perspective, it is not clear when these complexes are useful.

2.8 Multiscale Clustering: Mapper

For very large data sets, the techniques of topological data analysis described above can be computationally infeasible. For example, the number of simplices in the Vietoris-Rips complex can grow too rapidly for computation of higher (persistent) homology to be practical. (See Section 2.7 and Section 3.4 for various ways to address this problem.) Another issue is that the output of persistent homology can be hard to interpret for large high-dimensional data sets. An approach to answering these questions when handling very large data sets is to consider integration of ideas of persistence with clustering.

In this section, we describe a method for multiscale clustering: this is the Mapper algorithm of Singh, Mémoli, and Carlsson [462]. Roughly speaking, the idea of Mapper is to define a function on the data set, for example a measure of local density, and then perform clustering at different ranges of values of this function and keep track of how the clusters change as these ranges vary.

The basic framework assumes the data is presented as a finite metric space (X, ∂_X) and we choose

- a *filter function* $f: X \to \mathbb{R}^n$, and
- a cover $\mathcal{C} = \{U_\alpha\}$ of the range of f in $\mathbb{R}^n$; typically this cover is taken to be a collection of overlapping closed boxes. In the case of $n = 1$, a typical cover is a collection of closed intervals.

We now proceed as follows. This algorithm amounts to a discretization of the Reeb graph (see Section 1.12) at each scale.

1. Cluster each inverse image $f^{-1}(U_\alpha) \subseteq X$, regarded as a metric subspace of X, for all $U_\alpha \in C$; denote by $C_{\alpha,i}$ the ith cluster. (Any clustering algorithm can be used that takes as input only the interpoint distances and does not require specification of the number of clusters; single-linkage clustering is a standard choice.)

2. Form a graph where the vertices are given by the clusters $C_{\alpha,i}$ as α and i vary and there is an edge between $C_{\alpha,i}$ and $C_{\alpha',j}$ when

$$C_{\alpha,i} \cap C_{\alpha',j} \neq \emptyset \qquad \text{(clusters overlap)}.$$

3. Finally, we assign a color to each vertex in the graph corresponding to a particular cluster $C_{\alpha,i}$ according to the average value of f on $x \in C_{\alpha,i}$.

The results are of course dependent on the choice of filter function and the cover; this algorithm is well adapted to the methodology of *exploratory data analysis*, where we are trying to understand the data without an explicit hypothesized model to describe it. For the cover, it is standard to try successive refinements of the range of f, sometimes equally spaced, but often with increased resolution in areas where we expect more interesting behavior to occur. Standard filter functions include density measures and eccentricity measures; these depend on the data, and we will see in the examples and applications many different useful choices of filter function.

Example 2.8.1.

1. Let (X, ∂_X) be any finite metric space, $f: X \to \mathbb{R}$ an arbitrary function, and $C = \{(-\infty, \infty)\}$. Then the output of Mapper is simply the graph consisting of a point for each cluster of (X, ∂_X), no edges, and the clusters colored with the average value of f on the cluster. (See Figure 2.29 for an example.)

2. Let (X, ∂_X) be any finite metric space, $f: X \to \mathbb{R}$ an arbitrary function, and $C = \{[0, 1], [2, 3]\}$. Writing

$$X_{[0,1]} = f^{-1}([0, 1]) \qquad \text{and} \qquad X_{[2,3]} = f^{-1}([2, 3])$$

the output of Mapper is the union of a collection of vertices for the clusters of $X_{[0,1]}$ and a collection of vertices for the clusters of $X_{[2,3]}$. Again, there are no edges, since the cover does not overlap, and the colors represent the average values of f on the cluster corresponding to the vertex.

3. Now consider the previous example, but modify the cover to be $C = \{[0, 1], [0.5, 3]\}$. In this case, there are potentially edges between the vertices for clusters that overlap.

4. Consider points sampled densely from a unit circle in $\mathbb{R}^2$, let $f: X \to \mathbb{R}$ be the function $(x, y) \mapsto x$ that takes a point to its x-coordinate, and take C to be a series of overlapping subsets of $[-1, 1]$. (Specifically, we take ten intervals which overlap by 25% on each side.) Then the Mapper graph recovers the circle; see Figure 2.29.

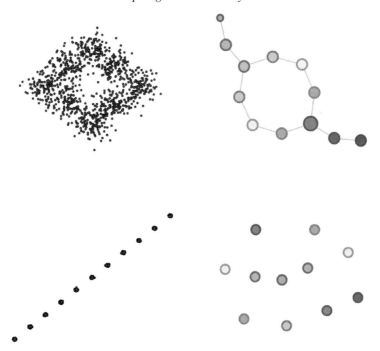

Figure 2.29 Top: The filter function is a projection onto the *x*-axis and there are 10 overlapping charts; the Mapper graph recovers the topology of the circle. Bottom: When there is a single chart that covers the domain, Mapper just returns the results of clustering, colored by the filter function (in this case, distance from the mean of the data). From Abbas H. Rizvi et al., *Nature Biotechnology* 35, 551–560 (270). © 2017 Nature. Reprinted with Permission from Springer Nature.

In practice, Mapper has turned out to be very useful for identifying clinically significant subsets of the data that are hard to find with traditional clustering methods. It has also been an effective way to represent the structure of the data set across feature scales. To give a sense of what this means, we illustrate with some examples of the use of Mapper on real data.

Example 2.8.2.   An early and celebrated example of the application of Mapper was work on a breast cancer data set, by Nicolau, Levine, and Carlsson [383]. The data here is presented as a finite metric space comprising vectors of expression data in $\mathbb{R}^n$. Expression-based classification of tumors is a well-studied problem and has been the subject of a vast number of papers (e.g., see [236, 512]); clustering is the standard technique here. However, there is reason to worry about the efficacy of basic clustering techniques: for example, different tumors activate or suppress pathways with varying strengths, and there is widely variable infiltration of healthy cells into tumor samples. As a consequence, one expects clinically significant features to appear at varying scales.

The analysis used samples from 295 breast cancers as well as additional samples from normal breast tissue; see Figure 2.30. The original expression vectors were in $\mathbb{R}^{24479}$, but a preprocessing step projected them into $\mathbb{R}^{262}$.

- The distance metric was given by the correlation between (projected) expression vectors.
- The filter function used was a measure taking values in $\mathbb{R}$ of the deviation of the expression of the tumor samples relative to normal controls.
- The cover was overlapping intervals in $\mathbb{R}$.

In the Mapper graph, the samples divide into two branches. The lower right branch itself has a subbranch (referred to as c-MYB+ tumors), which are some of the most distinct from normal and are characterized by high expression of genes including c-MYB, ER, DNALI1 and C9ORF116. Interestingly, all patients with c-MYB+ tumors had very good survival and no metastasis. These tumors do not correspond to any previously known breast cancer subtype; the grouping seems to be invisible to classical clustering methods – for example, hierarchical clustering fails to identify this particular subset of tumors (see bottom left of Figure 2.30). We will study this example in detail in Section 6.7.

Example 2.8.3.   Another interesting application of Mapper is to the study of the differentiation process from murine embryonic stem cells to motor neurons. The process is demonstrated in Figure 2.31; over time, undifferentiated embryonic cells become differentiated motor neurons when retinoic acid and sonic hedgehog (a differentiation-promoting protein) are applied.

The data generated corresponds to RNA expression profiles from roughly 2000 single cells.

- The distance metric was provided by correlation between expression vectors.
- The filter function used was multidimensional scaling (MDS) projection into $\mathbb{R}^2$; as we review in Section 4.2, this is a procedure for embedding an arbitrary metric space in a lower dimensional Euclidean space.
- The cover was overlapping rectangles in $\mathbb{R}^2$.

As can be seen in Figure 2.32, the Mapper diagram neatly identifies various regions characterized by their state in the differentiation process; in contrast, conventional clustering directly applied to the raw metric data does not produce clusters that encode information about the progress of differentiation. We will study this example in Section 7.3.

One potential concern for applications is the fact that the Mapper algorithm is not stable in the sense that we have described for persistent homology. For one thing, choice of parameters for the clustering algorithm can lead to unstable results; for example, when hierarchical clustering is used, the results are very sensitive to the choice of cutoff parameter. Worse, it is possible to construct examples of metric spaces (X, ∂_X) and a cover $\mathcal{C}$ such that two very similar filter functions give rise to very different results.

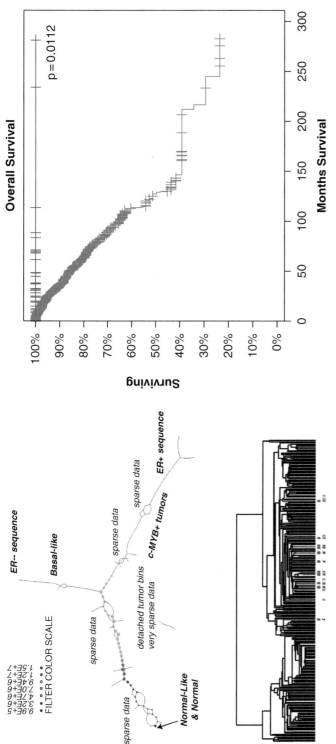

Figure 2.30 **Mapper applied to breast cancer expression data**. Top left: Mapper representation of the gene expression data from 295 breast tumors. Blue color indicates samples similar to normal tissue. Tumors with expression profiles that deviate significantly from those of normal tissue appear in the two arms on the right side. The upper arm is characterized by low expression of estrogen receptor (ER−). The lower right branch contains samples with high expression of c-MYB+ and cannot be identified using standard clustering techniques, as indicated on the lower left. Independent validation using 960 breast invasive carcinomas from "The Cancer Genome Atlas," of two of the highest expressed genes in c-MYB+ tumors, DNALI1 and C9ORF116, show very good prognosis for these tumors. Source: [383].

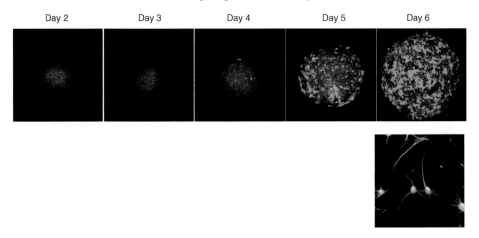

Figure 2.31 Over time, embryonic stem cells differentiate into distinct cell types. These pictures capture the in vitro differentiation of mouse embryonic stem cells into motor neurons over the course of a week. Embryonic stem cells are marked in red, and fully differentiated neurons in green. Figure from experiment performed by Elena Kandror, Abbas Rizvi and Tom Maniatis at Columbia University.

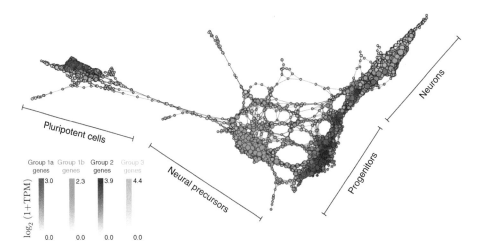

Figure 2.32 The different regions in the Mapper graph nicely line up with different points along the differentiation timeline. Source: [431].

Effectively, the issue is that a mismatch between the scale of change in the data and the width of the overlap of inverse images can give rise to dramatic changes in the Mapper graph in response to small shifts in filter function or cover. (See Figure 2.33 for a representative example of this phenomenon.)

There are various different approaches to handling this instability in practice.

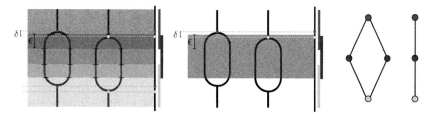

Figure 2.33 Small perturbation of the data relative to the cover can lead to large changes in the Mapper graph.

1. As we discuss in Section 3.9 below, various approaches motivated by standard considerations in statistics give us tools to establish confidence in the robustness of Mapper output.
2. Another possibility is to reintroduce persistence in the cover direction: the idea is to consider a tower of successive refinements of covers. With a suitable metric on such towers of covers, one can prove a stability theorem in this context [143].

The notion of refinement of covers also gives rise to a way to make precise the connection between Mapper and the Reeb graph. Specifically, consider the sequence of covers $\mathcal{C}_\epsilon$ consisting of all intervals of size ϵ. Then as $\epsilon \to 0$, the resulting Mapper graph converges to the Reeb graph [366].

2.9 Towards Persistent Algebraic Topology

In this chapter, we have focused primarily on ways of associating homological invariants to data sets; our focus reflects the majority of existing work on topological data analysis. From a pragmatic perspective, this choice of emphasis is very natural. Homology groups are distinguished in part by being computable; as we have seen, given a topological space presented as the geometric realization of a simplicial complex, there is an efficient algorithm for computing its homology.

In contrast, computing homotopy groups is an intractable problem. Computing the homotopy groups of spheres is a basic and unsolved problem in algebraic topology. From an algorithmic standpoint, we have the following hardness results.

1. Even for a finite complex X, $\pi_1(X)$ is uncomputable in general. (This problem is equivalent to solving the "word problem" in groups, which asks for an algorithm to determine whether two expressions in a generators and relations presentation of a group are equal.)
2. For simply connected finite complexes X and fixed k, computing $\pi_k(X)$ can be done in time polynomial in the number of simplices of X [85], although the complexity is completely infeasible for realistic use.

3. If k is allowed as part of the input (i.e., not fixed at the outset), even computing the ranks of $\pi_k(X)$ is a #P-complete problem [14] (and therefore likely to be exponential, provided that current beliefs about computational complexity are true).

Notwithstanding, one can define and study persistent homotopy groups. This is an interesting endeavor for several reasons. For one thing, it is possible to use partial computations of such persistent homotopy groups to distinguish topological features of data [59]. From a theoretical perspective, consideration of persistent homotopy groups leads to efforts to understand *persistent algebraic topology*.

In classical algebraic topology, homology groups are homotopy invariants and thus capture information about the homotopy type of the space. In fact, a version of Whitehead's theorem (Theorem 1.6.31) shows that a map $f \colon X \to Y$ between simply connected CW complexes that is an isomorphism on homology groups is a homotopy equivalence. There are corresponding questions about the relationship between persistent homology and some kind of persistent homotopy equivalence.

1. What is the right notion of persistent homotopy equivalence and persistent weak equivalence? Is there an analogue of the Whitehead theorem (Theorem 1.6.31)?
2. Can we axiomatically characterize persistent homology in an analogous fashion to the way we can axiomatically characterize ordinary homology?
3. How should we think about the stability theorem (Theorem 2.4.10) in these terms?

Although it is not totally clear what candidate answers for these questions might look like, the stability theorem and the importance of the metric structure on barcodes suggests that what we are seeing is the outline of some kind of "approximate algebraic topology." See [58] for the beginnings of foundations for such a theory.

2.10 Summary

- We may assign mathematical structure to a data set by viewing the points of the set as points in a suitable metric space (X, ∂_X).
- This chapter focuses on two ways to assign a simplicial complex to a finite metric space (X, ∂_X). For a given $\varepsilon > 0$, we have the Čech complex $C_\varepsilon(X, \partial_X)$ (see Definition 2.1.2) and the Vietoris-Rips complex $VR_\varepsilon(X, \partial_X)$ (see Definition 2.1.6). These complexes are functorial in ϵ.
- Given a finite metric space (X, ∂_X) uniformly sampled from a compact Riemannian manifold M, the Niyogi-Smale-Weinberger Theorem (see Theorem 2.2.1) shows that it is possible to recover topological invariants of the underlying

geometric object M, provided the distance between sampled points is smaller than some feature scale.

- The feature scale of data is unknown a priori. The idea of persistent homology is to keep track of how homological features change as the scale parameter varies.
- To investigate persistence, we examine filtered systems of simplicial complexes (see Definition 2.3.3), which arise via the functoriality of $VR_\varepsilon(X, \partial_X)$ in ε.
- In order to use topological invariants to describe data, we must guarantee that small perturbations in the data correspond to commensurately small changes in the resulting invariants. To measure the size of these changes in the data, we use the Gromov-Hausdorff distance (see Definition 2.4.4). To measure changes in the barcodes, we use the bottleneck distance (see Definition 2.4.8). The stability theorem for persistent homology (Theorem 2.4.10) bounds the size of changes in barcodes by the size of changes in the data.
- Zigzag persistence is the study of persistent homology considering filtrations of different shapes where the arrows have different orientations. This approach may be helpful in controlling the number of simplices, allowing efficient computation of persistent homology.
- In some cases, a single data set may give rise to multiple filtrations. For example, we might filter by both scale and density. This is the focus of multidimensional persistence.
- The Mapper algorithm is a method for multiscale clustering that has been effectively applied to identify clinically significant information in data sets that traditional clustering may miss. Mapper performs clustering at different scales, keeping track of changes in the clusters as the scale varies.

2.11 Suggestions for Further Reading

Topological data analysis is a young field, and for many aspects of it the original papers remain the best reference. However, there have been a number of excellent introductory articles, ranging from brief treatments (e.g., [193, 326, 535]) to more comprehensive (and technical) overview articles [90, 103, 156, 157]. There are also now a number of good books [111, 155, 194, 392, 550], with slightly different areas of emphasis.

3

Statistics and Topological Inference

O! it is pleasant with a heart at ease,
Just after sunset, or by moonlight skies,
To make the shifting clouds be what you please ...
Samuel Coleridge

Our central goal in this book is to explain how to use topological data analysis as a tool for scientific inference in biology. In the previous chapter, we described a strategy for assigning topological invariants to experimental data presented as a finite metric space. Moreover, we have presented theoretical justification that in ideal cases these topological invariants encode information about the shape underlying the data. But when trying to understand how to extract answers to specific scientific questions from the shape of real experimental data, many methodological questions immediately arise.

1. How confident can we be that the results of TDA applied to sampled data correctly reflect something about the underlying process generating the data?
2. How stable are the results of TDA in the face of noise and differing choices of parameters?
3. What does a particular value of a topological invariant tell us about the shape of the data?

These questions are not unique to this setting, but arise pervasively in data analysis. But the last of these questions is particularly acute in the context of topological data analysis. The geometric significance of clustering is fairly clear; the data breaks up into groups which are made up of similar points. This is not to say that it is always easy to make use of clustering for inference, but we feel like we understand the information about the shape of the data that it provides. In contrast, suppose that you compute the homology of a data set at scale 0.75 and discover that H^6 has rank 15. What then?

In this chapter, we describe answers to the three questions above using statistical techniques to analyze topological invariants computed from data. It is easy to engage in self-deception with incautious use of statistical techniques. As a consequence, our focus is on trying to understand how to sensibly and reliably use these tools to analyze data.

3.1 What Can Topological Data Analysis Tell Us?

In order to understand the use of statistics in topological data analysis, it is useful to draw a contrast with the basic approaches in classical statistics. Consider the most fundamental problem.

1. We are given a finite collection of samples $\{x_1, \ldots, x_n\} \subset \mathbb{R}$ which have been drawn independently from a Gaussian with mean μ and standard deviation σ. The probability density function of this distribution is

$$p(x) = \frac{1}{\sqrt{2\pi\sigma^2}} e^{-\frac{(x-\mu)^2}{2\sigma^2}}.$$

2. We know that the data has come from some Gaussian, and we want to estimate μ and σ.

This is a *parametric* problem; we know the answer lies in a family of unknown distributions in which each member is described by a collection of numbers. To recover the distribution, we would estimate μ and σ using the *sample mean*

$$\hat{\mu} = \frac{1}{n} \sum_i x_i$$

and the *sample variance*

$$\hat{\sigma}^2 = \frac{1}{n-1} \sum_i (x_i - \hat{\mu})^2,$$

which are unbiased estimators of the mean and variance of the underlying distribution.

Deep theoretical results provide confidence in this procedure. The *law of large numbers* tells us that as n increases, the sample mean converges to μ in a suitable sense. Since $\hat{\mu}$ depends on the particular sample, it will vary, and the *central limit theorem* describes the distribution of $\hat{\mu}$; specifically, it tells us that this quantity itself has a Gaussian distribution. We can summarize the information we obtain about $\hat{\mu}$ in terms of a *confidence interval*; this is an interval $[a, b] \subset \mathbb{R}$, defined in terms of the samples, that contains the true parameter value with a specified probability. For example, the 95% confidence interval for the mean of a Gaussian is centered around $\hat{\mu}$ and has width that depends on the standard deviation σ.

In general parametric settings, we cannot always assume that the underlying distribution is Gaussian or that we know a closed form expression for the distribution of the parameter we are estimating. As a consequence, in practice we often form confidence intervals using *the bootstrap*: this procedure estimates the distribution of the parameter by repeatedly generating samples (with replacement) from the given samples and computing the test statistic from them.

Sometimes we do not want to assume that we know a parametric family of distribution that generated the samples; this is the domain of *non-parametric statistics*. Even in these cases, the law of large numbers and central limit theorem tell us a great deal about how to estimate various summary statistics of the underlying distribution. For example, the law of large numbers tells us that the *empirical distribution* on a sample $\{x_i\}$, which assigns probability to each value proportional to its frequency, converges to the underlying distribution. For more general summary statistics, the bootstrap remains a powerful way to estimate confidence intervals in this setting. Another possibility is to try to describe the distribution using *density estimation*; for example, we could solve the optimization problem of fitting the observed samples to a mixture of Gaussians and regard this result as an approximation of the underlying distribution.

In topological data analysis, we have access to many fewer tools. As we discuss below, it is very hard to algorithmically specify the underlying geometric object, even if we assume it is a manifold, except under very restrictive hypotheses. This implies it will be hard to recover it as well. Moreover, for distributions on general metric spaces, we do not necessarily expect many of the analogues of classical statistics to hold. And writing down parametric distributions is generally difficult. As a consequence, statistical inference in topological data analysis immediately focuses on estimating distributions of summary statistics, often generated by persistent homology barcodes.

In the literature on topological data analysis, there is often an implicit (and sometimes explicit) view of *topological inference* as a process in which some sort of underlying geometric "ground truth" can be recovered. In a setup where we have access to samples which we regard as coming from a probability distribution on an underlying space, there are a number of ways of formalizing what we mean.

1. A first goal might be simply to recover the persistent homology of the underlying space (or rather, the support of the sampling distribution) from computation of persistent homology of the samples, the *empirical persistent homology*.
2. A more sophisticated version of the preceding goal would be to recover information about both the persistent homology of the underlying space and the

probability measure generating the samples. For instance, a natural way to proceed is to try to recover the persistent homology of the *level set filtration*. Given a suitable probability density ρ on $A \subseteq \mathbb{R}^n$, the super level sets

$$\Gamma_\rho(z) = \{x \in A \mid \rho(x) > z\}$$

induce a filtration as z varies.

3.1.1 Persistent Homology and Sampling

We begin by considering the first question above: can we recover the persistent homology of the underlying space from the empirical persistent homology? An initial consistency check, described in Section 3.4, is that with large enough samples we can always recover the persistent homology of the support of the probability distribution from the empirical persistent homology (Figure 3.1). The basic observation is simply that with sufficiently many samples, even regions of low probability density will be well sampled.

However, in practice we will usually not know how many samples are enough; the feature scale of the underlying object is often unknown and even when we have some estimates of the scale, experimental realities may limit the number of data points available. Thus, we need to understand the behavior of the empirical persistent homology as it converges, i.e., when we do not necessarily assume the number of samples is large. The kind of situation we might worry about is represented in Figure 3.2; an anomalous sample leads to misleading results.

Thus, we need to understand sampling variability and decide how to assemble an estimate that aggregates the empirical persistent homology from different samples; for example, we might hope to build a confidence region for the population value of the parameter. Figures 3.3 and 3.4 indicate sampling variability in persistent homology at different sample sizes. We study questions of convergence properties and confidence intervals for estimates in Section 3.5. Thinking about summaries of collections of barcodes raises interesting questions about what it means to compute the "average" barcode or to think about the variance or spread of a collection of barcodes; we will discuss these issues throughout the chapter, notably in Sections 3.3 and 3.6.

Summarizing collections of empirical barcodes is an interesting endeavor from another perspective: we might regard the probability distribution generating the samples as itself worthy of investigation, and so want to have an invariant or collection of invariants for persistent homology which explicitly encodes information about the distribution. For example, Figure 3.5 illustrates that different distributions on the same underlying space can result in very different barcodes at small sample sizes.

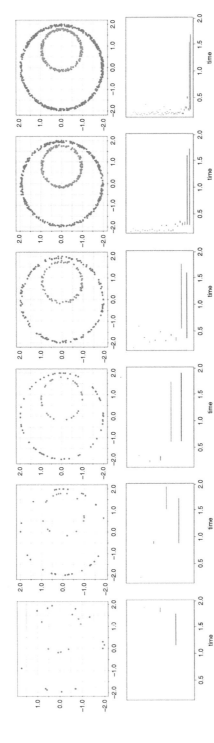

Figure 3.1 As the sample size increases, the persistent homology of the sample converges to the persistent homology of the support of the distribution, in this case the underlying space.

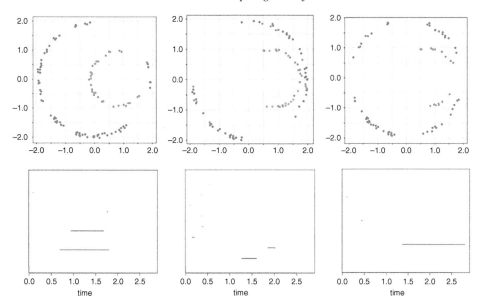

Figure 3.2 These samples were all generated from a uniform distribution on nested circles, and underneath we graph the PH$_1$ barcode. The barcode on the left is consistent with our expectations. But in the sample in the middle (which was a particularly anomalous sample among the many we generated), the two bars are very short and do not coexist, and on the right, there is only a single bar.

In the limiting cases where we have many samples, regions of low probability mass can make the same contribution to the topology as regions of high probability mass. And we might not regard this insensitivity to the density as a feature!

A closely related question is to understand the impact of noise in the data. One might expect the empirical persistent homology to behave well with respect to noise. After all, part of the original intuition behind persistent homology is to make homology computations robust to perturbation by integrating information across various feature scales; and this intuition is confirmed by Theorem 2.4.10, the stability theorem for persistent homology. And indeed, persistent homology is relatively stable in the face of noise concentrated around the real data; see Figure 3.6 for an example.

However, even in this case, the barcode has an increasing number of short "noise bars." The difficulties are exacerbated when we deal with data coming from a low-dimensional space embedded in a higher dimensional Euclidean space; then the noise is often the same dimension as the ambient space, which can lead to very complicated topological signals arising from the noise. These considerations motivate the study of the topology of "random" geometric complexes, which we review in Section 3.7.

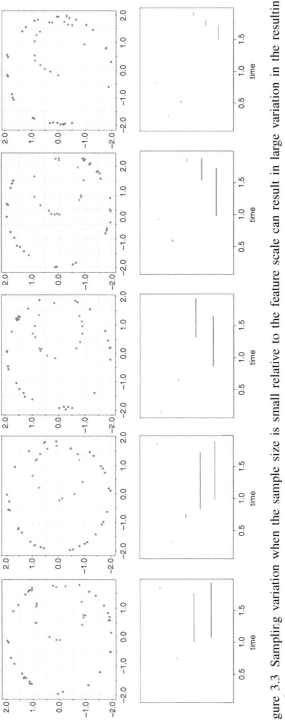

Figure 3.3 Sampling variation when the sample size is small relative to the feature scale can result in large variation in the resulting barcodes.

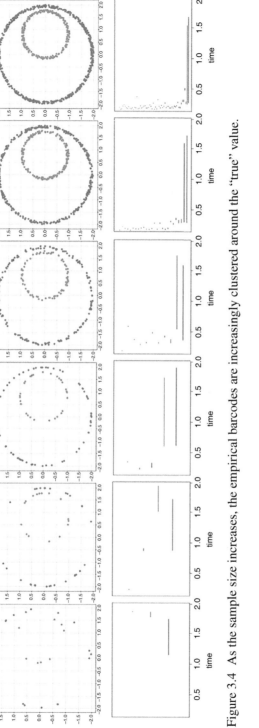

Figure 3.4 As the sample size increases, the empirical barcodes are increasingly clustered around the "true" value.

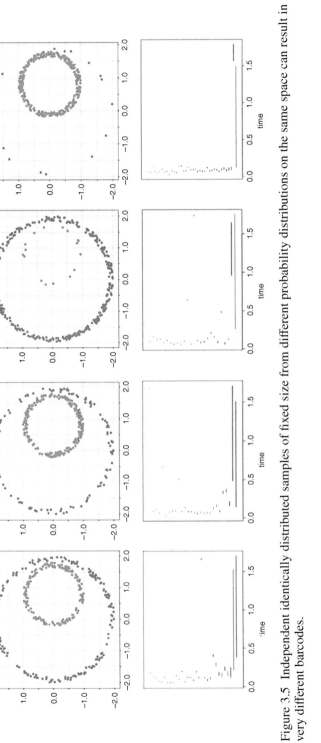

Figure 3.5 Independent identically distributed samples of fixed size from different probability distributions on the same space can result in very different barcodes.

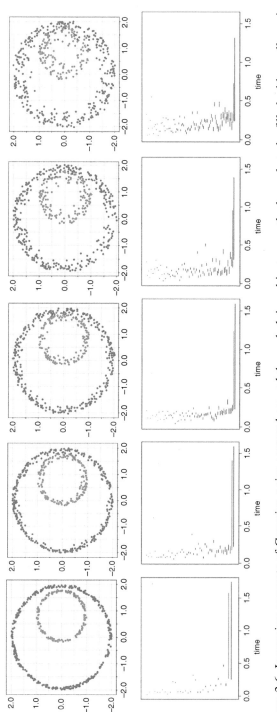

Figure 3.6 Increasing amounts of Gaussian noise centered around the underlying object cause the barcode to be filled with small spurious bars.

An even more serious problem is that not all noise is concentrated around the data. And the stability theorem has basically nothing to say about the presence of arbitrary outliers (i.e., noise points that are far from the data points). Adding a single point to a metric space (X, ∂_X) can perturb it in the Gromov-Hausdorff distance arbitrarily (recall Example 2.4.6). And we can perturb PH_i arbitrarily by adding "synthetic i-spheres" far away from the points of X. For instance, when $i = 1$, we can add 4 points at the vertices of a square with side-length k; this adds an interval $\left[\frac{k}{2}, \frac{k\sqrt{2}}{2}\right)$. Using more points, we can control the size of the interval and introduce additional intervals. (See Figure 3.7 for an example of the effect of outliers.)

This kind of instability is a well-known problem that arises even in very basic statistical inference.

Example 3.1.1.   Consider computing the mean of a set of points $\{x_1, \ldots, x_n\} \subset \mathbb{R}$. Specifically, let us take $\{1, 2, 3, 4, 5\}$; we find the mean is $\frac{1+2+3+4+5}{5} = 3$. Now change the point 5 to 10^{50}. To first approximation, the mean is now very close to 10^{49}. Put another way, given a set $\{x_1, \ldots, x_n\}$, one can make the mean any arbitrary value by suitably modifying a single data point! (See Figure 3.8 for a picture of this phenomenon.)

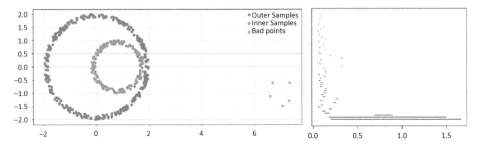

Figure 3.7 Adding a small number of points to create a circle far away from the real data can make a significant change in the barcode; a tiny number of "bad points" creates a noticeable third bar.

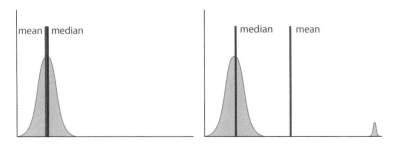

Figure 3.8 A small amount of outlying probability mass can have a large effect on the mean but cannot affect the median very much.

Traditionally, this phenomenon is the purview of *robust statistics* [251]; the mean is not robust. In contrast, the median is the classic example of a robust replacement for the mean. Changing 5 to 10^{50} does not affect the median of $\{1, 2, 3, 4, 5\}$ at all. (More generally, one needs to change at least 50% of the points in order to achieve arbitrary change in the median.) The situation with persistent homology turns out to be even worse, since whereas the mean is stable with respect to small perturbations of the distribution, barcodes of samples are not.

There are various ways to try to make persistent homology invariants more robust. One possibility is to simply preprocess the data to remove "outliers"; when there are a small number of points that are very far away from the bulk of the points, it is easy to identify them. A more principled way to do this is to consider filtering the data by density; we discuss one approach to using density estimators in Section 3.5.1, and we discuss the use of the density filtration with Mapper in Section 3.9. Another version of this strategy involves subsampling; if the number of outliers is small, most subsamples will not contain many outliers. We explain in more detail in Sections 3.4 and 3.5 how to use these ideas to bound the impact of parts of the data set with small probability mass. Finally, in Section 3.6, we discuss how to use real-valued invariants of the data and techniques from robust statistics.

3.1.2 Topological Inference

In contrast to the relative success of procedures for trying to recover information about the persistent homology of the underlying space, we cannot hope in general to identify the homotopy type of the underlying space. Any effort to identify topological spaces runs up against the fact that there is no algorithmic classification of topological spaces up to homeomorphism or homotopy type; the problem is provably uncomputable in dimensions ≥ 4, even if we restrict attention only to manifolds. (See [550, §4.1] for a nice review of these results.)

Theorem 3.1.2. *The problem of determining whether two manifolds M and N of dimension ≥ 4 presented as finite simplicial complexes are homeomorphic is undecidable.*

This result is proved by constructing a manifold whose fundamental group $\pi_1(M)$ encodes the word problem (recall Example 1.6.22 and Remark 1.6.23). Weakening homeomorphism to homotopy equivalence does not help.

Corollary 3.1.3. *The problems of determining whether two manifolds M and N of dimension ≥ 4 presented as finite simplicial complexes are homotopy equivalent or weakly homotopy equivalent are undecidable. (Even the problems of determining* *whether a given manifold is homeomorphic or homotopy equivalent to a fixed manifold Z are undecidable.)*

Worse, as the allowable diameter grows, there are exponentially many possible homeomorphism types of manifolds arising as submanifolds of Euclidean space of dimension greater than 2 [533, §1.2]. Similar bounds hold for the number of possible homotopy types. As a consequence, it is not in general plausible to parametrize hypothesis classes of spaces except when imposing strong restrictions or using coarse invariants. Moreover, the explicit sample bounds for recovery of persistent homology (from Section 2.2 above and Section 3.5 below) are exponential in the intrinsic dimension of the data.

Even if we restrict ourselves to the seemingly easy problem of distinguishing spheres of different dimensions (i.e., S^{50} versus S^{51}), basic results about *concentration of measure* in high-dimensional Euclidean spaces imply that under an oblivious sampling model (i.e., when samples are drawn independently of one another) most of the mass on a sphere S^n is concentrated around a radial region which is homeomorphic to S^{n-1}. This shows that this problem requires an exponentially large number of points [533, §1.3]. More generally, as we discuss below in Section 4.6, it is very difficult to successfully estimate the dimension of very high-dimensional manifolds.

These constraints place sharp limits on the kind of geometric inference that we can expect. We have basically three options: work with low-dimensional topological features of the data and perform exploratory data analysis, work with low-dimensional data where exact topological inference is reasonable, or treat the results of topological data analysis as signals about shape that are potentially uninterpretable except as input to statistical inference or machine learning procedures. In more detail, TDA provides the following.

1. A methodology for exploratory data analysis via description and visualization of low-dimensional shape information. Arguably the most widely used TDA technique is Mapper, and indeed the standard usage pattern for Mapper is to search for meaningful clusters in the data which can then guide further experiments. We discuss this at a high level in Section 3.9. In the second part of the book, we will explain many examples of this approach, including applications to tumor classification and cell differentiation (see Sections 6.7, 7.3, and 7.4).

2. Exact information about data that truly does lie in low-dimensional topological spaces. In these cases, topological data analysis can be interpreted to provide specific geometric information about the data and is often applied in a "hypothesis testing" framework. For instance, in dimension 1, specific hypotheses about the process generating the data are reasonable, and analogues of parametric

statistics make sense. We will discuss an example of this kind of approach in phylogenetics in Section 5.2, where persistent H_1 is used to detect divergence from the "tree hypothesis" for evolutionary data and estimate recombination rates.

3. Robust "topological signals" to use as features for classification, inference, and supervised learning algorithms. Although many topological features cannot be interpreted directly (e.g., "$H^{15}(X)$ is approximately 39"), they still convey discriminative information about the data. Ideally, this approach permits integration of information from topological data analysis with other sources of information (e.g., standard parametric statistical models). Two examples of this approach that we will discuss are surface recognition via the persistent homology transform (see Section 3.8 for a general discussion and Section 9.3 for specific applications) and the use of persistent homology information to fit parameters for population genetics models (see Section 5.7).

We now explain how to integrate topological data analysis with suitable statistical techniques in order to carry out these three kinds of analyses.

3.2 Background: Geometric Sampling and Metric Measure Spaces

At the most basic level we access geometry through a metric. Therefore, we want to work with metric spaces equipped with probability measures that are compatible with the metric. We do this using the machinery of *metric measure spaces*. This framework makes it possible to extend intuitive and familiar ideas from ordinary statistics in Euclidean space to a very broad class of geometric objects.

3.2.1 Metric Measure Spaces

To express the compatibility of metric and probability measure in a precise fashion, we work with the notions of measurable spaces and measures. A *measurable space* is a set along with a collection of subsets to which we can assign "area." A *measure* on a measurable space is a rule for assigning area, i.e., a theory of integration. We rapidly review these definitions here; we recommend [56, 57] for more in-depth treatments.

Definition 3.2.1. For a set X, a *σ-algebra* is a collection Σ of subsets of X such that

1. $\emptyset \in \Sigma$,
2. given a countable set $\{U_i\}$ such that $U_i \in \Sigma$, then the union $\bigcup_i U_i$ is also in Σ, and
3. if $U \in \Sigma$, then the complement $X \setminus U$ is in Σ.

As we noted above, these closure properties are motivated by the perspective that the elements of a σ-algebra have area; for instance, given a collection of sets that have area, we should be able to measure the area of their union. Given an arbitrary collection S of subsets of X, the σ-algebra generated by this collection is the smallest σ-algebra containing S; roughly speaking, we simply add all missing unions, intersections, and complements.

Example 3.2.2.   The most important example of a σ-algebra is the *Borel σ-algebra* associated to a topological space X; this is just the σ-algebra generated by the collection of open sets of X. (Equivalently, it is generated by the collection of closed sets of X.)

In fact, when (X, ∂_X) is separable (i.e., contains a countable dense subset; recall Definition 1.2.14), then the Borel σ-algebra is generated by the collection of open balls $\{B_\epsilon(x)\}$ as ϵ varies over $\mathbb{R}^{>0}$ and x over the points of X.

Definition 3.2.3.   A *measurable space* is a pair (X, Σ) consisting of a set X and a σ-algebra Σ.

Example 3.2.4.

1. Let X be a countable set; the power set of X forms a σ-algebra, which we refer to as the *counting σ-algebra*. This σ-algebra is generated by the points $x \in X$.
2. Euclidean space $\mathbb{R}^n$ with the σ-algebra generated by the boxes $(a_1, b_1) \times \cdots \times (a_n, b_n)$ is a measurable space.
3. More generally, any topological space is a measurable space with the Borel σ-algebra.
4. It turns out to be technically advantageous to equip Euclidean space with a more sophisticated σ-algebra, the Lebesgue σ-algebra. This is an enlargement of the Borel σ-algebra; it includes more measurable sets, in order to force every subset of a set of measure zero to be measurable. (This enlargement is referred to as the "completion" of a σ-algebra.)

Functions between measurable spaces are defined in analogy with continuous functions.

Definition 3.2.5.   Let (X, Σ) and (X', Σ') be measurable spaces. A map of sets $f \colon X \to Y$ is a *measurable function* if $f^{-1}(A) \in \Sigma$ for all $A \in \Sigma'$. A measurable function is a *measurable isomorphism* when f is an isomorphism of sets and f^{-1} is also a measurable function.

Measurable spaces support the computation of area; a *measure space* is a measurable space that has been equipped with a specific area function, a *measure*.

Definition 3.2.6. A *measure* μ on a measurable space (X, Σ) is a function

$$\mu \colon \Sigma \to \mathbb{R}^{\geq 0}$$

such that

1. $\mu(\emptyset) = 0$, and
2. for $X_i \in \Sigma$ such that $X_i \cap X_j = \emptyset$ for all i and j,

$$\mu\left(\bigcup_{i=1}^{\infty} X_i\right) = \sum_{i=1}^{\infty} \mu(X_i).$$

A basic theorem that allows us to construct measures is that for a σ-algebra generated by a collection of subsets S, it suffices to define the measure on the sets in S. This result is closely related to the construction of the Riemann integral. (See Figure 3.9 for an example of the process.)

Example 3.2.7.

1. For a finite set X with the counting σ-algebra, the *counting measure* on X assigns to each subset $A \subseteq X$ the cardinality of A, i.e.

$$\mu(A) = \#A, \quad A \subseteq X.$$

    This can be regarded as the measure determined by setting each point $x \in D$ to have measure 1.
2. For $\mathbb{R}^n$ with the box σ-algebra, the standard measure is determined by assigning to each rectangle its area, i.e.,

$$\mu([a_1, b_1] \times \ldots \times [a_n, b_n]) = \prod_{i} (b_i - a_i).$$

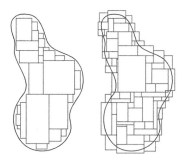

Figure 3.9 In favorable cases, the measure of an arbitrary region is bounded by the area of inner and outer covers by generating sets; the actual area is given by taking limits as the inside sum increases and the outside sum decreases. In general, the inner measure is defined in terms of the outer measure of the complement.

Given a measure μ on X, we can integrate any measurable function $f: X \to \mathbb{R}$ over a region $A \subseteq X$ as follows. Assuming temporarily that $f \geq 0$, we set

$$\int_A f d\mu = \sup_{\substack{B_1 \cup B_2 \cup ... B_\ell = A \\ B_i \cap B_j = \emptyset, i \neq j}} \sum_i \left(\inf_{x \in B_i} f(x) \right) \mu(B_i).$$

Here the sup is computed over all decompositions of A into finitely many disjoint subsets $\{B_i\}$ (in particular, ℓ will vary). If f takes both positive and negative values, we define the integral in terms of the expression above for the positive part and negative part separately and take the sum.

We are most interested in *probability measures*, for which we require that $\mu(X) = 1$. An important class of examples of probability measures are determined by *probability density functions*. Given a probability measure μ and a measurable function f, the integral $\int_X f$ is called the *expectation* of X.

Definition 3.2.8. Let μ be a measure on (X, Σ) and f be a measurable function $f: X \to \mathbb{R}$ so that $\mu(\{z \mid f(z) < 0\}) = 0$. Then there is an induced measure on X

$$\nu(A) = \int_A f d\mu, \quad A \subseteq X.$$

We say that the measure ν has *density* f with respect to μ.

Remark 3.2.9. It is standard to describe measures via probability densities when working with a basic reference measure for integration, e.g., the Lebesgue measure on $\mathbb{R}^n$ or the counting measure on a finite set. In the following discussion, we will sometimes omit specification of the measure when working with densities.

When (X, ∂_X) is a metric space, we can now use the topology induced by the metric and the Borel σ-algebra to express compatibility of metric and measure. A *Borel measure* is a measure with respect to the Borel σ-algebra.

Definition 3.2.10. A *metric measure space* with a probability measure is a metric space (X, ∂_X) that is complete and separable, equipped with a Borel probability measure μ_X. The *support* of a metric measure space is the subset supp(X) of X consisting of points x for which every neighborhood U of x satisfies $\mu_X(U) > 0$.

Remark 3.2.11. More generally, we can consider metric measure spaces where the measure is not a probability. We will not use such examples in this chapter, however.

Definition 3.2.10 provides a theoretical framework for describing data sampled from some kind of geometric object.

> Our working hypothesis throughout this chapter is that we have data presented as samples from an underlying metric measure space (X, ∂_X, μ_X).

Example 3.2.12.

1. A finite metric space (X, ∂_X) with the normalized counting measure

$$\mu(A) = \frac{\#A}{\#X}, \qquad A \subseteq X$$

   is a metric measure space.
2. For any subset $A \in \mathbb{R}$ and a measure μ (not necessarily a probability measure) such that $\mu(A) < \infty$, A becomes a metric measure space via the uniform measure

$$\mu'(S) = \frac{\mu(S)}{\mu(A)}, \qquad S \subseteq A.$$

3. More generally, the standard probability distributions on $\mathbb{R}$ and $\mathbb{R}^n$ equip them with the structure of metric measure spaces. For example, $\mathbb{R}$ with a Gaussian density gives rise to the Gaussian measure when integrated with regard to the Lebesgue measure.
4. Manifolds also provide natural geometric examples of metric measure spaces – any compact Riemannian manifold M is a metric measure space under the volume measure [144]. Samples from the volume measure on a manifold have the property that any small region has a number of points proportional to its volume; this is a version of the uniform distribution.
5. Given any metric measure space (X, ∂_X, μ_X), any measurable subset $A \subset X$ is itself a metric measure space, where

$$\mu_A(V) = \frac{\mu_X(V)}{\mu_X(A)}$$

   for $V \subset A$.

We can describe finite independent identically distributed (i.i.d.) samples as follows.

Definition 3.2.13. Let (X, ∂_X, μ_X) be a metric measure space. The *product measure* $\mu_X^{\otimes n}$ makes the metric space $(\prod_{i=1}^{n} X, \prod_{i=1}^{n} \partial_X)$ into a metric measure space, where

$$\mu_X^{\otimes n}(A_1 \times A_2 \times \ldots \times A_n) = \mu_X(A_1)\mu_X(A_2)\ldots\mu_X(A_n),$$

$$A_1 \times A_2 \times \ldots \times A_n \subseteq \prod_{i=1}^{n} X = X \times X \times \ldots \times X.$$

Thus, an i.i.d. sample of size n from (X, ∂_X, μ_X) can be described as a draw from the distribution $\mu_X^{\otimes n}$.

We will be interested in measures induced by the application of functions (e.g., persistent homology). To be precise about this, we need the notion of the *pushforward* of a measure.

Definition 3.2.14. Let $f: (X, \partial_X) \rightarrow (Y, \partial_Y)$ be a measurable function between the Borel measure spaces X and Y. Then given a probability measure μ_X, the *pushforward measure $f_*\mu_X$* on Y is specified by the formula

$$f_*\mu_X(A) = \mu_X(f^{-1}(A)),$$

for A a measurable set in Y.

Another useful way of generating new measures is by combining old ones.

Definition 3.2.15. Let μ and ν be finite Borel measures on $\mathbb{R}^n$. Then the convolution $\mu * \nu$ can be defined as

$$\mu * \nu = +_*(\mu \times \nu),$$

the pushforward of the product measure along the addition map $+: \mathbb{R}^n \times \mathbb{R}^n \rightarrow \mathbb{R}^n$.

Explicitly, the convolution is given by the formula

$$(\mu * \nu)(A) = \int_{\mathbb{R}^n} \int_{\mathbb{R}^n} \mathbf{1}_A(x + y)d\mu(x)d\nu(y),$$

where $\mathbf{1}_A$ is the indicator function for the measurable set $A \subset \mathbb{R}^n$. Convolution with a Gaussian affords a useful general technique for smoothing distributions with complicated local structure; the width of the Gaussian controls the degree of smoothing.

Finally, we note that it is frequently useful to have a notion of size for real-valued functions on a metric measure space. To this end, we quickly recall the definition of the L_p and L_∞ norms.

Definition 3.2.16. Let (X, ∂_X, μ_X) be a metric measure space and let $f: X \rightarrow \mathbb{R}$ be a measurable function such that $\int_X f^p d\mu < \infty$. Then the L_p *norm* of f for $1 \leq p < \infty$ is

$$\|f\|_p = \left(\int_X |f|^p d\mu \right)^{\frac{1}{p}}.$$

When $p = \infty$, we define

$$\|f\|_\infty = \inf\{k \in \mathbb{R} \mid \mu(\{x \in X \mid f(x) > k\}) = 0\}.$$

Remark 3.2.17. When X is a finite set and μ_X is the *counting measure* (i.e., the measure that assigns probability mass $\frac{1}{|X|}$ to each point), these norms reduce to the pth root of the sum of pth powers and the max, respectively.

Remark 3.2.18. Geometric sampling on non-Euclidean metric measure spaces can be very subtle, even when dealing with the volume measure on a compact Riemannian manifold [144]. For example, consider the problem of sampling from the surface of the sphere $S^2 \subseteq \mathbb{R}^3$. In this case, there is a natural parametrization of the points of the sphere arising from spherical coordinates (r, θ_1, θ_2). A naive approach is to use the spherical coordinates to sample: sample uniformly θ_1 and θ_2 from $[0, 2\pi]$ and $[0, \pi]$ respectively and consider the map $\sigma \colon [0, 2\pi] \times [0, \pi] \to \mathbb{R}$ specified by

$$x = \sin(\theta_2)\cos(\theta_1)$$
$$y = \sin(\theta_2)\sin(\theta_1)$$
$$z = \cos(\theta_2).$$

Denoting by U the uniform distribution on $[0, 2\pi] \times [0, \pi]$, we have the pushforward $\sigma_* U$ which is supported on $S^2 \subseteq \mathbb{R}^3$. However, $\sigma_* U$ is concentrated at the poles and is not the distribution associated to the area form. In this case, we can simply sample uniformly in a cube around the origin in $\mathbb{R}^3 \setminus \{0\}$, discard points further than 1 from the origin, and divide by the norm. More generally, one needs to use either rejection sampling or Markov chain Monte Carlo (MCMC) techniques. These methods can be applied to general manifolds, provided one has access to an explicit and computationally tractable parameterization; of course, this is often a serious problem.

3.2.2 The Fréchet Mean and Variance of a Metric Measure Space

Most practical applications of statistics involve the use of summary statistics. As such, it is natural to look for notions of mean and variance that apply in the general context of metric measure spaces. The standard approach to this problem is the theory of the Fréchet mean and variance of a probability measure μ on a metric measure space (e.g., see [487] for an introduction to this theory). Although it turns out that this theory is not particularly useful in barcode space (as we explain below), we nonetheless quickly review it here since understanding the pathological behavior of the Fréchet mean motivates the techniques used in practice. We restrict our attention to probability measures μ satisfying a finiteness condition.

Definition 3.2.19.   Let (X, ∂_X, μ_X) be a metric measure space. Then the *Fréchet variance* as a function of $z \in X$ is the integral

$$v_\mu(z) = \int_X \partial_X(z, x)^2 d\mu(x).$$

We will assume that $v_\mu < \infty$. Then the *Fréchet mean* is defined as follows.

Definition 3.2.20.   The *Fréchet mean* is the set

$$e_\mu = \operatorname{argmin}\left(\inf_z v_\mu(z)\right) \subseteq X,$$

i.e., the values $z \in X$ that achieve the infimum.

   When dealing with a finite sample $\{x_1, x_2, \ldots, x_n\}$ from (X, ∂_X, μ_X), the Fréchet mean and variance of the underlying distribution are approximated using the empirical measure which assigns probability $\frac{1}{n}$ to each point in the sample. See Figure 3.10 for a simple example.
   It is not at all clear that the Fréchet mean exists for general metric measure spaces; in practice, we rely on the following result.

Theorem 3.2.21.   *Let (X, ∂_X, μ_X) be a metric measure space. If μ_X has compact support (e.g., if X is compact), then the Fréchet mean exists.*

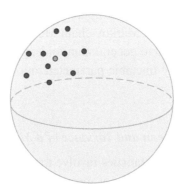

Figure 3.10 The Fréchet mean (green) of a finite sample (red) from the uniform distribution on a sphere is the point on the sphere that is the "centroid" of the sample.

More generally, the Fréchet mean can be shown to exist as long as the "tails" of μ decay sufficiently rapidly. (See [285] for a precise statement.)

The general theory of the Fréchet mean and variance provides laws of large numbers; given finite samples from μ equipped with the empirical measure, the Fréchet means of the samples converge to the Fréchet mean of μ.

Theorem 3.2.22. *Let (X, ∂_X, μ) be a metric measure space. Let $\{Z_k\}$ be a collection of i.i.d. samples $Z_k \subset X$ drawn according to μ, such that $|Z_k| \to \infty$ as $k \to \infty$. Let μ_k denote the empirical measure on Z_k. Then almost surely $e_{\mu_k} \to e_\mu$ (i.e., the probability of convergence is 1).*

The problem of understanding the convergence of derived quantities of distributions for increasing finite samples suggests that we should put a topology on the set of probability measures. We now turn to a discussion of how to construct metrics on probability distributions and on metric measure spaces.

3.2.3 Distances on Measures and Metric Measure Spaces

In order to state Theorem 2.4.10, the stability theorem for persistent homology of finite metric spaces, we used a metric on the set of isometry classes of finite metric spaces. To state the analogous stability theory describing the interaction of sampling and persistent homology, we will use a metric on the set of isomorphism classes of compact metric measure spaces. Recall that the Gromov-Hausdorff metric is defined in terms of a metric on subspaces of a fixed metric space, the Hausdorff metric. To define a metric on metric measure spaces, we will start with a metric on probability measures on a fixed metric space.

To motivate this definition, we quickly explain the notion of *weak convergence* of probability measures. For a metric space (X, ∂_X), let $\mathcal{P}(X)$ denote the set of Borel probability measures on X.

Definition 3.2.23. Let (X, ∂_X) be a metric space. A sequence $\{\mu_n\} \subset \mathcal{P}(X)$ *weakly converges* to $\mu \in \mathcal{P}(X)$ if for all bounded continuous functions $f : X \to \mathbb{R}$,

$$\int_X f d\mu_n \to \int_X f d\mu.$$

The idea of weak convergence is that a sequence of distributions converges when the average value of any function f converges; i.e., weak convergence means that the expectation of any random variable converges. This notion of convergence is of particular importance because it is the kind of convergence that arises in the central limit theorem.

Warning 3.2.24. Weak convergence is very different from requiring that the measure of each set converge!

Since convergence of sequences can be defined in terms of a metric (recall Definition 1.2.7), it is natural to look for a metric that controls weak convergence. We now introduce several such metrics that are useful in topological data analysis, starting with the *Prohorov distance*.

Definition 3.2.25. Let (X, ∂_X) be a metric space equipped with two Borel measures μ_1 and μ_2. Then we define the *Prohorov distance* between μ_1 and μ_2 to be

$$d_{Pr}(\mu_1, \mu_2) = \inf\{\epsilon > 0 \mid \mu_1(A) \leq \mu_2(B_\epsilon(A)) + \epsilon \quad \text{and} \quad \mu_2(A) \leq \mu_1(B_\epsilon(A)) + \epsilon\},$$

where A varies over all closed sets in X and

$$B_\epsilon(A) = \{z \in X \mid \exists a \in A, \partial_X(z, a) \leq \epsilon\}.$$

To understand what the Prohorov distance means, it can be convenient to use an alternative formulation. For this, we need the notion of a *coupling*, which is a probability distribution θ on $X \times X$ such that $\theta(A \times X) = \mu_1(A)$ and $\theta(X \times B) = \mu_2(B)$ for arbitrary measurable subsets $A, B \subseteq X$.

Lemma 3.2.26. *Let (X, ∂_X) be a metric space equipped with two Borel measures μ_1 and μ_2. Then we can compute the Prohorov distance as*

$$d_{Pr}(\mu_1, \mu_2) = \inf_C \inf\{\epsilon > 0 \mid C\{(x, x') \in X \times X \mid \partial(x, x') \geq \epsilon\} < \epsilon\},$$

where C varies over all couplings.

Roughly speaking, two measures are within ϵ in the Prohorov metric when there is a matching of the space with itself such that on a region of probability mass $1 - \epsilon$ matched points are within ϵ and can vary arbitrarily on the remainder.

Proposition 3.2.27. *The distance d_{Pr} is a metric on $\mathcal{P}(X, \partial_X)$, the space of probability measures on X. If X is complete and separable, then given a sequence of probability measures $\{\mu_i\}$ that converges to a measure μ in d_{Pr}, μ_i weakly converges to μ.*

A complete and separable metric space is called a *Polish space*. In general, Polish spaces are a good setting for probability theory: not only is weak convergence metrizable, but in addition certain pathologies with product measures do not arise.

Example 3.2.28.

1. Let μ_1 and μ_2 be distributions determined by δ-functions, i.e., μ_1 has mass 1 on a point x_1 and μ_2 has mass 1 on a point x_2. Then

$$d_{Pr}(\mu_1, \mu_2) = \min(\partial_X(x_1, x_2), 1).$$

2. Let (X, ∂_X, μ_X) be a metric measure space and $Y \subset X$ have measure $> 1 - \epsilon$. Then μ_X regarded as a distribution on Y has Prohorov distance $< \epsilon$ from μ_X.

In fact, there are many metrics on $\mathcal{P}(X)$ that metrize weak convergence [195]. Optimal transport theory suggests the use of the Wasserstein or "earth-mover" metric [517]. Here, the rough idea is to imagine distributions modeled by piles of dirt; the Wasserstein distance is the minimal amount of energy (dirt times distance) that must be expended to transform one distribution into another.

Definition 3.2.29. Let (X, ∂_X) be a compact metric space equipped with two Borel measures μ_1 and μ_2. For $p \geq 1$, the *p-Wasserstein distance* between μ_1 and μ_2 is

$$d_{W_p} = \left(\inf_C \int_{X \times X} \partial_X(x, y)^p dC(x, y) \right)^{\frac{1}{p}},$$

where C varies over all couplings.

Any of the Wasserstein distances metrize weak convergence of probability measures on metric spaces with bounded diameter (i.e., where the maximum distance between $x_1, x_2 \in X$ is bounded).

Lemma 3.2.30. *The distance d_{W_p} is a metric on $\mathcal{P}(X, \partial_X)$, the space of probability measures on X. Let (X, ∂_X) have bounded diameter. Then given a sequence of probability measures $\{\mu_i\}$ that converges to a measure μ in d_{W_p}, then μ_i weakly converges to μ.*

Example 3.2.31.

1. Let μ_1 and μ_2 be distributions specified by δ-functions; μ_1 has mass 1 on $x_1 \in X$ and μ_2 has mass 1 on $x_2 \in X$. Then $d_{W_p}(x_1, x_2) = \partial_X(x_1, x_2)$.
2. Let μ_1 and μ_2 be empirical distributions on finite subsets $\{x_i\} \subset X$ and $\{x_i'\} \subset X$ such that $|\{x_i\}| = |\{x_i'\}|$. Then the Wasserstein distance can be computed as

$$d_{W_p} = \min_{\theta: \{x_i\} \to \{x_i'\}} \left(\sum_i (\partial_X(x_i, \theta(x_i)))^p \right)^{\frac{1}{p}},$$

where θ varies over all bijections.

Remark 3.2.32. It is common in information theory and Bayesian statistics to measure the difference between distributions μ_1 and μ_2 in terms of the Kullback-Leibler divergence. Taking p and q to be probability mass functions on a discrete space X where $q(x) = 0 \implies p(x) = 0$, the Kullback-Leibler divergence is computed as

$$\sum_{x \in X} p(x) \log \frac{p(x)}{q(x)},$$

where we interpret the contribution of a term with $p(x) = 0$ to be 0. (An analogous definition can be given in the setting of measure spaces, but setting it up is sufficiently complicated that we do not pursue it here; see [195] for a discussion, where it is referred to as relative entropy.)

The Kullback-Leibler divergence has many interesting properties, but it is not a metric; it is neither symmetric nor satisfies the triangle inequality.

The Wasserstein distance and the Prohorov distance are related, in the sense that

$$d_P(\mu_1,\mu_2)^2 \le d_{W_1}(\mu_1,\mu_2) \le (\text{diam}(X) + 1)d_P(\mu_1,\mu_2)$$

(and $d_{W_1}(\mu_1,\mu_2) \le d_{W_p}(\mu_1,\mu_2) \le Cd_{W_1}(\mu_1,\mu_2)$ for a suitable constant C) [195]. We can convert the Prohorov and Wasserstein distances into metrics on isomorphism classes of compact metric measure spaces. The approach is to use an analogue of the technique that converts the Hausdorff distance into the Gromov-Hausdorff metric on isometry classes of compact metric spaces.

Definition 3.2.33. Let (X,∂_X,μ_X) and (Y,∂_Y,μ_Y) be compact metric measure spaces. The *Gromov-Prohorov distance* is defined as

$$d_{GPr}((X,\partial_X,\mu_X),(Y,\partial_Y,\mu_Y)) = \inf_{\phi_X,\phi_Y,Z} d_{Pr}((\phi_X)_*\mu_X,(\phi_Y)_*\mu_Y),$$

where here $\phi_X\colon X \to Z$ and $\phi_Y\colon Y \to Z$ are isometric embeddings into a metric space Z.

Definition 3.2.34. Let (X,∂_X,μ_X) and (Y,∂_Y,μ_Y) be compact metric measure spaces. The *Gromov-Wasserstein distance* is defined as

$$d_{GW_p}((X,\partial_X,\mu_X),(Y,\partial_Y,\mu_Y)) = \inf_{\phi_X,\phi_Y,Z} d_{W_p}((\phi_X)_*\mu_X,(\phi_Y)_*\mu_Y),$$

where (ϕ_X,ϕ_Y,Z) is as in the previous definition.

Lemma 3.2.35. *The Gromov-Prohorov and Gromov-Wasserstein distances are metrics on the set of isomorphism classes of compact metric measure spaces.*

Remark 3.2.36. Although we do not review this here, there is very interesting work on the details of the topology induced on the set of isomorphism classes of metric measure spaces by these metrics [207, 346, 347].

3.3 Probability Theory in Barcode Space

The foundation of any statistical approach to persistent homology is the notion of a probability distribution of barcodes. The set $\mathcal{B}$ of barcodes is a metric space under the bottleneck distance d_B (Definition 2.4.8) or the p-Wasserstein distance d_{W_p} (Definition 2.4.9). Therefore, $\mathcal{B}$ endowed with the Borel σ-algebra becomes a measurable space: we can work with the collection of Borel probability measures on $\mathcal{B}$. Proposition 3.2.27 shows that the Prohorov metric on the set of Borel probability measures metrizes weak convergence of probability measures when the underlying metric space is complete and separable. We begin this section by constructing subspaces of barcode space that are complete and separable.

3.3.1 Polish Spaces of Barcodes

A first thought is to consider the set of finite barcodes. It is easy to see that this barcode space is separable for either the bottleneck or Wasserstein distance; an arbitrary "bar" $[a, b)$, with $a, b \in \mathbb{R}$, can be approximated arbitrarily well by choosing rational approximations a' for a and b' for b. However, the set of finite barcodes is not complete.

Example 3.3.1. Consider a sequence of barcodes $\{X_i\}$ where $X_0 = \emptyset$ and X_i is obtained from X_{i-1} by adding a disjoint bar $[0, \frac{1}{n})$. That is,

$$X_i = \{[0, 1), [0, 1/2), [0, 1/3), \ldots, [0, 1/i)\}.$$

Working with the bottleneck distance, it is easy to check that $\{X_i\}$ is a Cauchy sequence (recall Definition 1.2.9),

$$d_B(X_i, X_j) \leq \frac{1}{\max(i, j)},$$

as the distance between X_i and X_j is bounded by the longest bar present in X_j and not in X_i (assuming that $j > i$). But $\{X_i\}$ does not converge to any element of $\mathcal{B}$; the sequence is clearly converging to a barcode with infinitely many bars! (See Figure 3.11 for a picture of this sequence.)

Instead, we can consider countable barcodes, although certain finiteness conditions are still required.

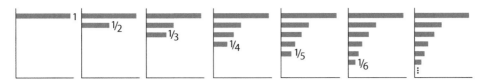

Figure 3.11 By adding shorter and shorter bars, this sequence eventually converges to a barcode with infinitely many bars!

Definition 3.3.2. Let $\overline{B}$ denote the subspace of B consisting of those barcodes such that for all $\epsilon > 0$, the number of bars of length $> \epsilon$ is finite. We regard $\overline{B}$ as a metric space with the bottleneck metric (recall Definition 2.4.8).

When working with the p-Wasserstein metric, it turns out that we need to use a slightly different finiteness condition.

Definition 3.3.3. Let B_P denote the subspace of B consisting of those barcodes B for which

$$d_{W_p}(B, \emptyset) < \infty.$$

We regard B_P as a metric space with the p-Wasserstein metric (recall Definition 2.4.9).

These finiteness conditions rule out phenomena like that exhibited in Example 3.3.1: we can now show that $\overline{B}$ and B_P are complete metric spaces [60, 352].

Theorem 3.3.4. *The metric spaces* $(\overline{B}, d_B)$ *and* (B_p, d_{W_p}) *are complete and separable.*

In order to summarize distributions in $\overline{B}$ and B_p, we need to define summary statistics. In light of the discussion in the preceding section, one might hope to use the Fréchet mean and variance. Unfortunately, the Fréchet mean of a distribution of barcodes is not that useful in practice.

1. Computing the Fréchet mean is computationally expensive. An algorithm for computing an approximation to the Fréchet mean for finite sets of barcodes equipped with the empirical measure is given in [511]; however, the algorithm involves gradient descent (and so only finds local minima of the variance expression) and the rate of convergence is not well understood.
2. The Fréchet mean of a distribution μ is not necessarily unique; barcode space is positively curved [511], which means that unique geodesics do not connect all points; see Section 4.7.3. (In fact, most pairs of points are not connected by

unique geodesics, in a precise sense.) In particular Fréchet means may not be unique.

3. The Fréchet mean is very unstable; small perturbations in the sample distribution can cause the mean to jump around. To handle both this and the preceding problem, the paper [367] proposes using a distribution-valued variant of Fréchet means. Nonetheless, computation is still basically intractable.

As a consequence, the Fréchet mean and variance of distributions on barcode space are primarily of theoretical interest; in Section 3.6 below, we discuss various practical summary statistics.

3.3.2 Sampling and Hypothesis Testing in Barcode Space

We now describe our formalization of sampling problems in persistent homology using the analysis of the barcode space above. Specifically, we work with the following assumptions.

Hypothesis 3.3.5.

1. The data consists of independent samples from a metric measure space (X, ∂_X, μ_X).
2. For any k, the function assigning the kth persistent homology barcode to a sample $\{x_1, \ldots, x_n\} \subset X$ drawn from μ_X is a measurable map. (For example, in the case of the Vietoris-Rips complex, the stability theorem for persistent homology (Theorem 2.4.10) implies that persistent homology is continuous and hence measurable.)
3. Therefore, taking the product measure $\mu_X^{\otimes n}$ on X^n and then computing persistent homology, we obtain an induced measure $\mathrm{PH}_* \mu_X^{\otimes n}$ on $\overline{\mathcal{B}}$. This distribution represents the distribution of barcodes associated to PH_k computed from samples of size n.

A standard statistical approach would now be to assume that the distribution μ on (X, ∂_X) is parametrized by values $(z_1, z_2, \ldots, z_k)$. We might then hope to compute a joint density function in terms of a likelihood function. In this way, in principle, one could use a maximum likelihood method to estimate the parameters. However, in general, these kinds of statistical procedures are not really feasible, as we explained above in Section 3.1.2. The problem is that without stringent constraints there is no reasonable way to come up with sensible "topological hypotheses," for the following basic reasons. Theorem 3.1.2 and Corollary 3.1.3 show that the problem of specifying a topological hypothesis is ill posed. Only in certain special cases (e.g., the data is known to be low dimensional or known to be contractible) is it at all reasonable to imagine producing a guess about the underlying topological type of the process generating the data or a parametric distribution for sampling from this topological space.

Even in the situation where a specific topological hypothesis is reasonable, it is often a challenging problem to provide an efficient algorithm for sampling from the null hypothesis. There are not natural parametric families of distributions for most metric spaces (X, ∂_X). Even in the case of a manifold, the most naive approach to specifying a distribution involves choosing coordinate charts and sewing together distributions on each chart – parametric inference and sampling is complicated in this setting. As an example of the difficulties, recall from the discussion in Section 3.2 above (notably Remark 3.2.18) that even correctly sampling from the volume measure on a compact Riemannian manifold defined by specific systems of equations requires some care. It is possible to compare the homology of observed data against samples generated from some standard random distribution on a compact geometric region bounding the empirical support. See Section 3.7 below for discussion of recent progress on theoretical understanding of the resulting distributions of barcodes; of course, simulation can also produce empirical estimates of these distributions. But more general topological hypotheses are out of reach except under stringent hypotheses about the dimension or complexity of the underlying space.

As a consequence, we focus on how to reliably estimate barcodes from samples and how to produce tractable features from barcodes. We can now reformulate more precise versions of the questions from the introduction to this section; we pose the problems in terms of how to estimate the persistent homology of a metric measure space (X, μ_X) from a sample $\{x_1, x_2, \ldots, x_k\}$ and use this estimate for inference. (For expositional convenience, we assume that $\mathrm{supp}(\mu_X) = X$.)

1. If k is large enough, does the sample faithfully represent the persistent homology of the underlying space X? To be precise, if we take a sequence of finite samples S_n of increasing size from a metric measure space (X, ∂_X, μ_X), does the sequence $\{PH_k(S_n)\}$ converge to $PH_k(X)$?
2. Under the conditions for which the first question has a positive answer, how fast is the rate of convergence? Can we construct confidence intervals controlling the expected error in the estimate $PH_k(S_n)$?
3. Analogously, if our points are sampled from a density ρ on $A \subseteq \mathbb{R}^n$, can we recover the persistent homology of the level set filtration associated to the super level sets

$$\Gamma_\rho(z) = \{x \in A \mid \rho(x) > z\}.$$

Can we understand the rate of convergence and construct confidence intervals?

4. Given a collection of barcodes generated by samples of size k, how do we produce summaries of these barcodes? The discussion in Section 3.3 above suggests that the Fréchet mean is not useful in practice. A related question is how to produce numerical summaries that can be used as input to standard machine learning algorithms.

5. In the presence of noise, how can we ensure reliable estimation of barcodes? The stability theorem for persistent homology (Theorem 2.4.10) implies that if the noise is concentrated in the Gromov-Hausdorff metric, we can expect good behavior. But suppose the noise consists of "outliers" that are far from the data. How can we ensure that the estimates of persistent homology are not arbitrarily disrupted?

3.4 Stability Theorems for Persistent Homology of Metric Measure Spaces

We begin with analogues of the stability theorem in the context of metric measure spaces. We describe two related approaches to such a theorem. First, we consider distributions of samples. The idea is to consider the induced distributions on barcode space associated to the empirical persistent homology of subsamples of a fixed size (Figure 3.12). For samples of size n, we define the associated *distributional persistent homology* of a metric measure space (X, ∂_X, μ_X) as follows.

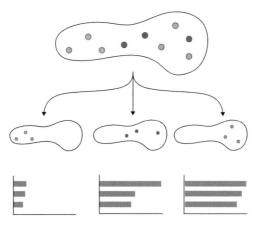

Figure 3.12 The distribution of barcodes is induced by taking many samples of a fixed size and computing their persistent homology.

Definition 3.4.1. For n and k, we define the *distributional persistent homology*

$$\Phi_k^n(X, \partial_X, \mu_X) = (\mathrm{PH}_k)_*(\mu_X^{\otimes n}),$$

the distribution on $\overline{B}$ induced by pushforward along PH_k of the product measure on the Cartesian product X^n.

In practice, we approximate Φ_k^n by sampling many blocks of size n and computing the empirical distribution; as the number of blocks approaches ∞, the law of large numbers guarantees that these approximations converge to the underlying distribution Φ_k^n. We might also subsample these blocks of size n from a larger sample from μ_X; see Figure 3.13 for an example of this.

In order for Φ_k^n to recover the persistent homology of X, the size n must be sufficiently large so that the samples can capture topological features of X; selecting n large enough requires information about the feature scale. However, even when n is too small, we can regard Φ_k^n as containing geometric information about the data, because of the following stability theorem [60].

Theorem 3.4.2. *Let (X, ∂_X, μ_X) and $(X', \partial_{X'}, \mu_{X'})$ be metric measure spaces. Fix n and k.*

$$d_{Pr}(\Phi_k^n(X, \partial_X, \mu_X), \Phi_k^n(X', \partial_{X'}, \mu_{X'})) \le n d_{GPr}((X, \partial_X, \mu_X), (X', \partial_{X'}, \mu_{X'})).$$

Interestingly, this bound is tight (and the n is unavoidable). One way of understanding the role of n is that as n increases the invariants become finer and finer and better approximate the support of the measures, which can be far apart even though the Gromov-Prohorov distance of the metric measure spaces is small. Theorem 3.4.2 implies that the distributional invariants Φ_k^n are robust invariants, in the sense that changing X on an ϵ-probability mass arbitrarily can perturb Φ_k^n by at most $n\epsilon$. One can also formulate a Gromov-Wasserstein version of this result.

However, note that there is some subtlety to the behavior of these invariants in n; having a smaller n can make the results less sensitive to outliers since fewer noise points turn up in any given sample. On the other hand, smaller n means less resolution for detecting actual topological features of the data. Compare Figures 3.14 and 3.15.

Of course, as we have discussed, working with distributions of barcodes directly is difficult, and so in practice we will rely on ways of approximating these by distributions on $\mathbb{R}$; we will describe ways to do this in Section 3.6. Before moving on, we note two pragmatic benefits to using distributional invariants: the parameter n can be chosen to accommodate the computational power available, and the computation of Φ_k^n can be evidently parallelized with linear speedup.

We now turn to another approach to a probabilistic stability theorem which is similar in spirit to Theorem 3.4.2. We suppose we are given a data set X embedded

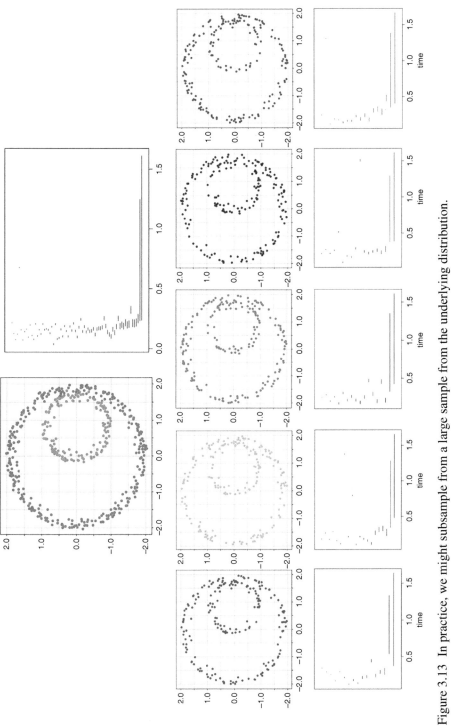

Figure 3.13 In practice, we might subsample from a large sample from the underlying distribution.

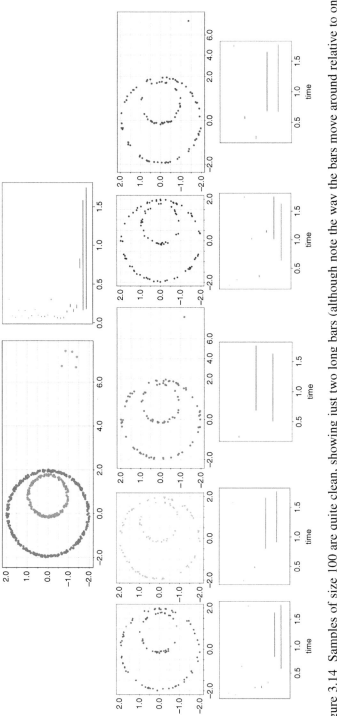

Figure 3.14 Samples of size 100 are quite clean, showing just two long bars (although note the way the bars move around relative to one another).

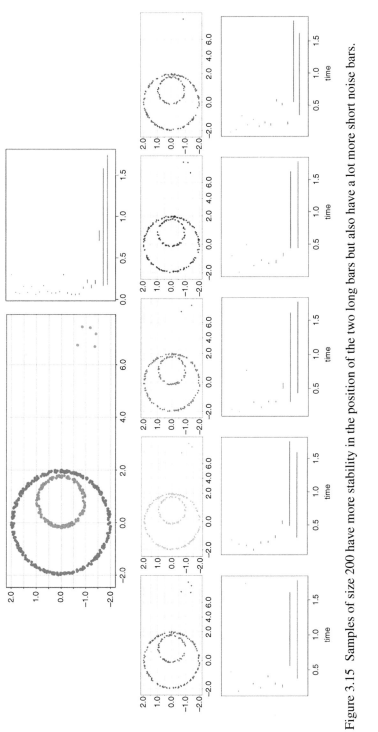

Figure 3.15 Samples of size 200 have more stability in the position of the two long bars but also have a lot more short noise bars.

in $\mathbb{R}^n$. Recall from Remark 2.3.5 that the filtered complex associated to the Čech complexes on X can be alternatively described in terms of the filtration imposed by the distance function. Specifically, let C be a compact subset of $\mathbb{R}^n$. The distance function $D\colon \mathbb{R}^n \to \mathbb{R}$ is defined as

$$D(x) = \inf_{z \in C} \partial_{\mathbb{R}^n}(x, z).$$

The sublevel sets $\{x \mid D(x) \le \epsilon\}$ as ϵ varies are precisely the filtration imposed by the geometric Čech complexes of C. (When working with a finite metric space (X, ∂_X), the inf is replaced by the minimum.)

   Estimating the persistent homology of the filtration for $X \subset \mathbb{R}^n$ via samples from some distribution on X is very sensitive to outliers. The work of [104, 108] proposes to handle this by replacing the distance function D (which captures the distance to the support of X) by a generalization that incorporates the measure on X. This generalization is referred to as the *distance to a measure*. For a continuous distribution, we have the following definition.

Definition 3.4.3.   Let (X, ∂_X, μ_X) be a compact metric measure space. Let $F_x(t) = \mu_X(\{z \mid \partial_{\mathbb{R}^n}(x, z) \le t\})$. Then for $0 < m < 1$ we define the *distance to a measure* to be

$$\delta_{\mu_X, m}(x) = \sqrt{\frac{1}{m} \int_0^m F_x^{-1}(u)^2 du},$$

where here

$$F_x^{-1}(u) = \inf_t \{t \mid F_x(t) \ge u\}.$$

   Here m is a resolution parameter that is a measure of the feature scale; choice of suitable values of m is once again an issue in practical use. The idea of the parameter m is that we are averaging density-biased approximations to the distance over a range controlled by m. Along these lines, for finite samples, the distance to a measure has a much simpler expression.

Lemma 3.4.4.   *Given a finite sample $Y = \{x_1, x_2, \ldots, x_n\} \subseteq X$, the distance to a measure function for the empirical distribution on Y for m is*

$$\delta_m(x) = \sqrt{\frac{1}{k} \sum_{z_\alpha \in N_k(x)} \partial_{\mathbb{R}^n}(z_\alpha, x)^2},$$

where here k is the smallest integer $\ge mn$ and $N_k(x)$ denotes the k nearest neighbors of x in Y.

   Notice that when m is very small, the distance to a measure function is very close to the distance function D. The advantage of the distance to a measure is that it is Wasserstein stable, in the sense that the L_∞ norm distance is bounded by the 2-Wasserstein distance. Specifically, we have the following theorem.

Theorem 3.4.5. *Suppose that μ_1 and μ_2 are two probability measures on $\mathbb{R}^n$. Then for mass parameter $0 < m < 1$, we have*

$$\|\delta_{\mu_1,m} - \delta_{\mu_2,m}\|_\infty \le \frac{1}{\sqrt{m}} d_{W_2}(\mu_1, \mu_2).$$

In turn, the bottleneck distance between the persistence diagrams associated to the distance filtrations on these two functions is bounded by the L_∞ norm.

Corollary 3.4.6. *Suppose that μ_1 and μ_2 are two probability measures on $\mathbb{R}^n$. Then for mass parameter $0 < m < 1$, we have*

$$d_B(P_{\delta_{\mu_1,m}}, P_{\delta_{\mu_2,m}}) \le \|\delta_{\mu_1,m} - \delta_{\mu_2,m}\|_\infty \le \frac{1}{\sqrt{m}} d_{W_2}(\mu_1, \mu_2).$$

As a consequence, we can conclude that the persistent homology estimate associated to the distance to a measure filtration is robust to outliers having low probability mass. (See Figure 3.16 for an example demonstrating robustness in the face of outliers.)

Furthermore, one can show [108] that the distance to a measure is statistically well behaved in the sense that a uniform law of large numbers applies to establish that it can be approximated by finite samples. Moreover, there are natural confidence intervals describing how well it is approximated by empirical estimates. Another interesting aspect of the distance to a measure is that, since for small m it approaches the ordinary distance function to X, in principle it can be used for geometric inference. On the other hand, computing the distance to a measure is difficult in practice due to problems associated to estimating level sets. See [79] for recent work that provides better algorithms and also extends the methodology to arbitrary metric spaces.

We now turn to the issue of understanding the way that the empirical persistent homology converges to the persistent homology of the underlying space.

3.5 Estimating Persistent Homology from Samples

Suppose we take a sequence of finite samples S_n of increasing size from a metric measure space (X, ∂_X, μ_X). It is straightforward to see that in fact $\{PH_k(S_n)\}$ does converge to $PH_k(\text{supp}(\mu_X))$ almost surely, provided that X is bounded: Lemma 1.2.20 shows that for any compact metric measure space (X, ∂_X), there

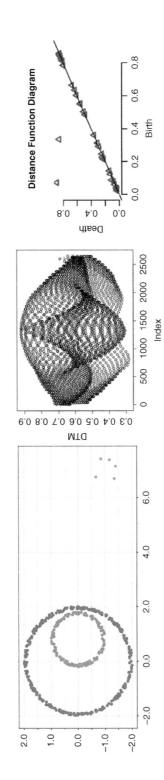

Figure 3.16 The level set persistence of the distance to a measure recovers the persistent homology of the two circles and ignores the outliers. (The middle panel plots the distance function.)

exists a finite ϵ-net X_ϵ for each $\epsilon > 0$. If we were given a sequence $\{X_n\}$ such that as $n \to \infty$, X_n is an $\frac{1}{n}$-net,

$$\{X_n\} \longrightarrow X$$

in the Gromov-Hausdorff metric and so

$$\{PH_k(X_n)\} \longrightarrow PH_k(X)$$

in the barcode metric. The point now is that for any ϵ, there exists an n sufficiently large so that any sample of size $> n$ is with high probability an $\frac{1}{n}$-net. This implies the following result.

Theorem 3.5.1. *Let (X, ∂_X, μ_X) be a metric measure space. Let $\{S_n\}$ be a sequence of finite samples drawn from μ_X such that $|S_n| \to \infty$. Then almost surely $PH_k(S_n)$ converges to $PH(\mathrm{supp}(\mu_X))$ in the barcode metric (or Wasserstein metric).*

Theorem 3.5.1 focuses attention on the rate of convergence of $\{PH_k(S_n)\}$. The key issue is to analyze the number of samples needed to obtain an ϵ-net with high probability (for some fixed ϵ). Such estimates require knowledge of the feature scale; we need to be able to compute how likely we are to sample in a ball around any given point. Estimates for compact Riemannian manifolds were given by Niyogi-Smale-Weinberger [384] (as explained in our discussion of Theorem 2.2.1), and elaborated on and extended by [170]. We describe the problem in the framework of the latter, which is more general and is expressed explicitly in terms of the language of *confidence regions*.

A confidence region is the multivariate analogue of the basic statistical notion of a confidence interval, which we now review. Returning to our example of estimating parameters of a Gaussian, we suppose that we have a sample $\{x_1, \ldots, x_n\}$ from a Gaussian distribution with mean μ and standard deviation σ. As discussed above, to estimate μ we compute the empirical mean $\hat{\mu}$ from the samples. We know that as n increases, it is very likely that $\hat{\mu}$ will be a good approximation of μ. One way to make that precise is to talk about a confidence interval.

Definition 3.5.2. A confidence interval $[a, b]$ with confidence level α for the parameter θ is specified by two random statistics a and b such that the probability that $\theta \in [a, b]$ is α.

For example, we know that $\hat{\mu}$ is distributed according to the t-distribution around μ with parameters determined by $\hat{\sigma}$, and using this fact we can derive the confidence interval for μ

$$\left[\hat{\mu} - \frac{c\hat{\sigma}}{\sqrt{n-1}}, \hat{\mu} + \frac{c\hat{\sigma}}{\sqrt{n-1}} \right],$$

where c is chosen such that the probability in the tail of the distribution larger than c has mass $\frac{1-\alpha}{2}$.

We now turn to the analogous notions for persistence diagrams. Associated to a specific c, the confidence set around a barcode B is a subset of the set of barcodes within a distance c of B. We can visualize this as the union of squares with side-length $2c$ is centered at each point of the persistence diagram. Points where the bounding box intersects the diagonal can be interpreted as noise. (Alternatively, we can put a band of width $(\sqrt{2})c$ around the diagonal.) See Figure 3.17 for an example.

To define a confidence set with probability α, we need to find c such that the true parameter is within c of the empirical barcode with probability larger than α. To formulate this, it turns out to be useful to talk about asymptotic confidence sets, defined as follows.

Definition 3.5.3. Fix a reference barcode $\mathcal{B}$ and denote by $\widehat{\mathcal{B}}_n$ the empirical barcode computed from a sample of size n. For $0 < \alpha < 1$, the *asymptotic* $1 - \alpha$ *confidence set* is the collection of regions determined by a (usually decreasing) sequence $c_n > 0$, where

$$\limsup_{n \to \infty} \Pr(d_B(\mathcal{B}, \widehat{\mathcal{B}}_n) > c_n)) < \alpha.$$

(Recall that lim sup denotes the limit of the supremums of the remaining terms in the sequence.)

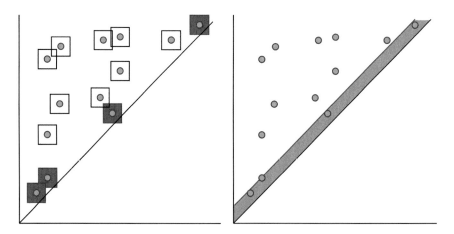

Figure 3.17 The confidence interval around the persistence diagram (in blue) is given by the boxes; a band around the diagonal contains "noise."

As one would expect, the rate of convergence (i.e., how large n has to be in order to obtain sufficiently small c_n) depends on the details of the density and the feature scale of the underlying manifold space. Going forward, we will assume that M is a compact manifold of dimension d embedded in $\mathbb{R}^k$ ($k > d$), that the condition number (recall Section 2.2) of M is positive, and that the samples are drawn from a probability density on $\mathbb{R}^k$ which is supported on M, smooth, and bounded away from 0.

Remark 3.5.4. More generally, it suffices for M to be a compact and rectifiable (piecewise smooth) subset of Euclidean space and to have a relatively weak differentiability criterion for M.

To bound the convergence of the confidence intervals for persistence diagrams, we define

$$\rho(x, t) = \frac{\Pr(B_{\frac{t}{2}}(x))}{t^d} \qquad \text{and} \qquad \rho(t) = \inf_{x \in M} \rho(x, t).$$

Then $\rho = \lim_{t \to 0} \rho(t)$ captures relevant information about the local variation in the probability measure on M.

We now fix our space $M \subset \mathbb{R}^k$ and let $\mathcal{P}$ denote the persistent homology of the sublevel sets of the function

$$\partial_M(z) = \inf_{y \in M} \partial_{\mathbb{R}^k}(y, z).$$

(Recall from Remark 2.3.5 that this is a version of the Čech complex.) For a sample of size n, let $\widehat{\mathcal{P}_n}$ denote the empirical persistent homology, i.e., the persistent homology of the sublevel sets of ∂_M restricted to $\mathcal{P}_n$. We have the following analogue of Theorem 2.2.1.

Proposition 3.5.5. *Under the hypotheses above,*

$$\Pr(d_B(\mathcal{P}, \widehat{\mathcal{P}_n}) > t) \le \frac{2^d}{\rho(\frac{t}{2})t^d} e^{-n\rho(t)t^d}.$$

The associated confidence region is the collection of boxes of side length t centered at the points of the persistence diagram $\mathcal{P}_n$.

In particular, setting

$$t_n = \left(\frac{4 \log n}{\rho n} \right)^{\frac{1}{d}},$$

we have that

$$\Pr(d_B(\mathcal{P}, \widehat{\mathcal{P}_n}) > t_n) < \frac{2^{d-1}}{n \log n}.$$

Making use of this result involves estimating ρ, which can be done using the plug-in estimator

$$\hat{\rho}_n = \min_i \frac{P_n(B_{\frac{r_n}{2}}(x_i))}{r_n^d},$$

where r_n is a sequence of numbers approaching 0 and P_n denotes the empirical measure for the sample $\{x_1, x_2, \ldots, x_n\}$.

There are a number of other methods of obtaining similar confidence interval estimates that are of broader interest; we turn to discussion of those in the remainder of the section.

3.5.1 Estimating Persistent Homology by Density Estimation

Another approach to computing the persistent homology from samples of a density in Euclidean space is to use standard techniques for density estimation to approximate the support of the density (e.g., see [429] for a modern theoretical analysis). Given a suitable probability density ρ on $\mathbb{R}^d$, the problem of estimating the superlevel sets

$$\Gamma_\rho(z) = \{x \in \mathbb{R}^d \mid \rho(x) > z\}$$

is a classical question in statistics. The path-connected components of $\Gamma_\rho(z)$ have long been studied in the context of unsupervised clustering and classification [229].

From the perspective of persistence, a natural question is to try to estimate the persistent homology of the level set filtration determined by the inclusions

$$\Gamma_\rho(z_2) \subseteq \Gamma_\rho(z_1)$$

for $z_1 < z_2$. A standard approach is to use a kernel density estimator; this is a smoothed version of the empirical density. The specific choice of kernel function employed is not important for our discussion, except for the following properties. We require a function $K \colon \mathbb{R} \to \mathbb{R}$ such that

1. $\int K = 1$,
2. the kernel has mean 0,
3. $\sup_x K(x) = K(0)$, and
4. K is Lipschitz for some constant ℓ.

Typically we will think of a smooth symmetric kernel, e.g., the Gaussian kernel $K(t) = \frac{1}{\sqrt{2\pi}} e^{-\frac{t^2}{2}}$.

For a bandwidth parameter h (this controls the amount of smoothing), define the measure

$$K_h(A) = h^{-d} \int_A K(h^{-1}t)dt.$$

Given the density ρ and associated measure P on $\mathbb{R}^d$, we want to study the convolution $P_h = K_h * P$, which we regard as a smoothed version of P. Denote the level set persistent homology of P_h by $\mathrm{PH}_k(P_h)$.

We can form an empirical approximation as follows. The density of the convolution is

$$p_h(x) = \int_M \frac{1}{h^d} K\left(\frac{\partial_{\mathbb{R}^d}(x, u)}{h}\right) dP(u),$$

and so the standard estimator given points $\{x_1, \dots, x_n\}$ is given by

$$\hat{p}_h(x) = \frac{1}{n} \sum_{i=1}^n \frac{1}{h^d} K\left(\frac{\partial_{\mathbb{R}^d}(x, x_i)}{h}\right). \tag{3.1}$$

We can now compute the persistence diagram associated to the level set filtration determined by the estimated density $\hat{p}_h$, which we will denote by $\mathrm{PH}_k(\hat{P}_h)$.

Remark 3.5.6. We note that this is estimating a somewhat different quantity than the persistent homology of the support of ρ; instead, we are in some sense directly estimating the homology of the support of ρ using the persistent homology of the level set filtration. This increases the robustness of the result, due to smoothing.

For simplicity, we assume that the support of the distribution P is contained in the Euclidean box $[-c, c]^d \subseteq \mathbb{R}^d$. Standard arguments show that $\hat{p}_h$ converges to p_h; this follows from Hoeffding's inequality, for example. (And stronger statements can be derived from tighter refinements of this sort of bound.) Translating this into a statement about persistence diagrams, we obtain the following result.

Theorem 3.5.7. *Under the hypotheses above, for fixed α and for any distribution P supported on the box $[-c, c]^d$*

$$\Pr(d_{W_p}(\mathrm{PH}_k(\hat{P}_h), \mathrm{PH}_k(P_h)) > \delta_n) \le \alpha$$

and where δ_n is a solution to the equation

$$2\left(\frac{4c\ell\sqrt{d}}{\delta_n h^{d+1}}\right)^d e^{-\frac{n\delta_n h^{2d}}{2K(0)^2}} = \alpha.$$

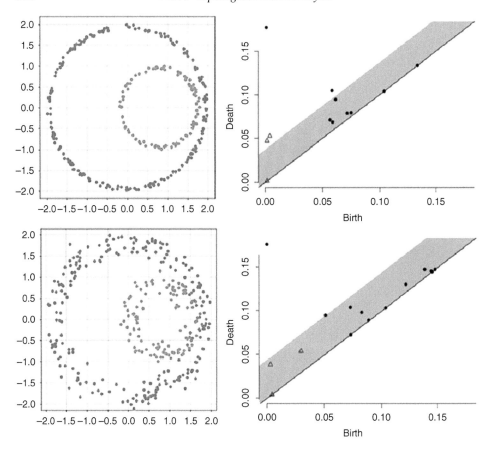

Figure 3.18 In the top panel, the 95% confidence interval contains the two bars for both H_0 and H_1 (dots represent H_0, triangles H_1); this correctly separates signal from noise. However, in the bottom panel, the 95% confidence interval suggests that all of the H_1 bars are noise.

As we can see in Figure 3.18, the confidence intervals computed in this fashion are fairly conservative.

For data embedded in Euclidean space, density estimation can also be used to eliminate outliers by smoothing to remove regions of low density. For example, this was performed manually in the famous example of the Klein bottle in visual image data [95], and is a standard data analysis tool [251]. Specifically, the persistent homology associated to the level set filtration of a density estimator is robust in the presence of outliers.

Let $X \subseteq \mathbb{R}^n$ denote the set of all points that might be returned by sampling, including both data points and noise points, i.e.,

$$X = X' \cup Z, \quad \text{where} \quad Z \cap X' = \emptyset,$$

where we regard X' as real data and Z as noise. Assume that the distribution on X we have experimental access to is

$$\Psi = \epsilon\theta + (1 - \epsilon)\mu,$$

for $0 \le \epsilon \le 1$, where μ is supported on X' and is the distribution we wish to estimate. We make no assumptions about θ.

Denote by $\mathcal{P}_\rho$ the persistence diagram associated to the level set filtration of the standard density estimator of equation (3.1), for fixed width parameter h, applied to empirical samples from a distribution ρ. The following lemma is now a simple calculation [170].

Lemma 3.5.8. *Let $X \subseteq \mathbb{R}^n$ be a subspace with probability density $\Psi = \epsilon\theta+(1-\epsilon)\mu$. Then*

$$d_B(\mathcal{P}_\Psi, \mathcal{P}_\mu) \le C\epsilon,$$

where C is a constant that depends on h.

This result implies that when ϵ is small and h is chosen appropriately, $\mathcal{P}_\Psi$ is a good approximation to $\mathcal{P}_\mu$ no matter what θ is, in particular, no matter how far away from X' the points of Z may be. Simple experiments in low dimensions validate this result [170].

Although this result is very encouraging, the general problems with density filtering remain – namely, choosing the width parameter requires either knowledge of the feature scale of the underlying data or a lot of experimentation, and density filtering is really only tractable for data embedded in Euclidean space or comparatively simple manifolds (see Figure 3.19). (Nearest neighbor density estimators do not perform well for realistic numbers of sample points.)

We believe that density filtering could be an ideal application of multidimensional persistence.

3.5.2 Estimating Persistent Homology by Resampling

Resampling is a standard technique for estimating confidence intervals around an empirical estimate of some quantity by generating many new finite subsamples from the given finite sample. Given n data points $X = \{x_1, x_2, \ldots, x_n\}$, there are two distinct possibilities for resampling estimators.

1. Subsampling involves estimating confidence intervals from empirical quantiles computed from subsamples $\{S_i\}$ of size $k < n$ generated by drawing *without* replacement from the empirical distribution on X (e.g., [412]).

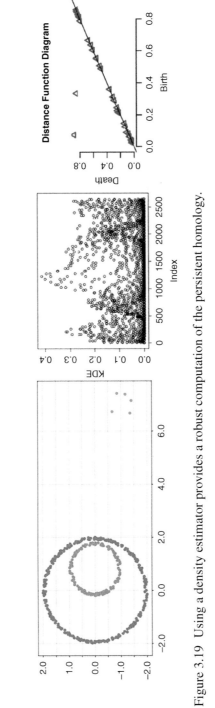

Figure 3.19 Using a density estimator provides a robust computation of the persistent homology.

2. The bootstrap involves estimating confidence intervals from empirical quantiles computed from subsamples $\{S_i\}$ of size $k < n$ generated by drawing *with* replacement from the empirical distribution on X (e.g., [54]).

We now discuss the use of these ideas to estimate persistent homology from finite samples. We start with the first case above, subsampling. Results in this regime are asymptotic and so stated in terms of the convergence of both n and k to ∞. We first work with the hypotheses of Proposition 3.5.5.

Remark 3.5.9. In the following discussion, to talk about asymptotic convergence we use "big-O" and "little-o" notation.

1. To say that a sequence $\{x_n\}$ is $o(f(n))$ means that for every $k \in \mathbb{R}$, there exists an $N \in \mathbb{N}$ such that for all $m > N$, $x_m < kf(m)$.
2. To say that a sequence $\{x_n\}$ is $O(f(n))$ means that there exists a constant $k \in \mathbb{R}$ and $N \in \mathbb{N}$ such that for all $m > N$, $x_m < kf(m)$.

Roughly speaking, the sequence is $o(f(n))$ if it grows strictly more slowly than the function f whereas the sequence is $O(f(n))$ if it grows at most as fast as a constant times $f(n)$.

Let b_n denote a sequence such that

$$b_n \to \infty \qquad \text{and} \qquad b_n = o\left(\frac{n}{\log n}\right).$$

Let $N = \binom{n}{b_n}$, and denote by $\{S_i\}$ the collection of all N subsamples of size b_n from the given sample $\{x_1, x_2, \ldots, x_n\}$. Set

$$L_n(t) = \frac{1}{N} \sum_{j=1}^{N} I(d_H(S_i, S) > t),$$

where I is the indicator function and d_H is the Hausdorff metric. For a given $\alpha \in (0, 1)$, let

$$c_n = 2L_n^{-1}(\alpha).$$

The arguments of [412] then imply convergence of the subsamples to the underlying metric space in Hausdorff measure and hence the following theorem providing confidence regions.

Theorem 3.5.10. *Under the hypotheses of Proposition 3.5.5, for large n (and $\rho > 0$), we have*

$$\Pr(d_B(\mathcal{P}, \widehat{\mathcal{P}_n}) > c_n) \leq \alpha + O\left(\left(\frac{b_n}{n}\right)^{\frac{1}{4}}\right).$$

We can also apply the bootstrap; in this situation, the best results come from considering the context of level set estimation from the density estimator. We work with the hypotheses of Section 3.5.1.

Then we have the following theorem.

Theorem 3.5.11.   *Under the hypotheses of Theorem 3.5.7, we have that*

$$\lim_{n \to \infty} \Pr\left(d_B(\mathcal{P}_h, \widehat{\mathcal{P}_h}) > \frac{q_\alpha}{\sqrt{n}}\right) \leq \alpha.$$

Here q_α is the $1 - \alpha$ quantile and is described below. The estimated confidence interval is then of width $\frac{2q_\alpha}{\sqrt{n}}$.

We can estimate the value q_α as

$$\hat{q}_\alpha = \inf_q \left(\frac{1}{N} \sum_{i=1}^{N} I(\sqrt{n}\|\widehat{p_h^i} - \widehat{p_h}\|_\infty \geq q) \leq \alpha \right),$$

where $\widehat{p_h^i}$ is the empirical probability density of the ith bootstrap subsample and $\|(-)\|_\infty$ denotes the L_∞ norm. Figure 3.20 has an example of confidence regions produced in this fashion; again, notice that these regions are quite conservative.

Remark 3.5.12.   It is also possible to show that resampling methods and the bootstrap can be applied directly in barcode space; this is more challenging technically due to the complexity of the metric geometry of $\mathcal{B}$. The issue is that establishing the asymptotic consistency of the bootstrap depends on obtaining control on the complexity of the class of functions used to describe empirical processes. For example, in $\mathbb{R}$, one uses the indicator functions supported on intervals $(-\infty, t]$. In barcode space, bounding the complexity of natural function classes is difficult and requires imposing further finiteness restrictions on the allowable barcodes.

3.6 Summarizing Persistence Diagrams

The results of Section 3.3 and Section 3.4 imply that it is possible in some circumstances to reliably estimate the persistent homology of a geometric object from samples. However, as we have emphasized, it remains difficult to directly apply the estimated barcode to inference. In view of this, a compelling approach is to study associated features produced by a choice of measurable map

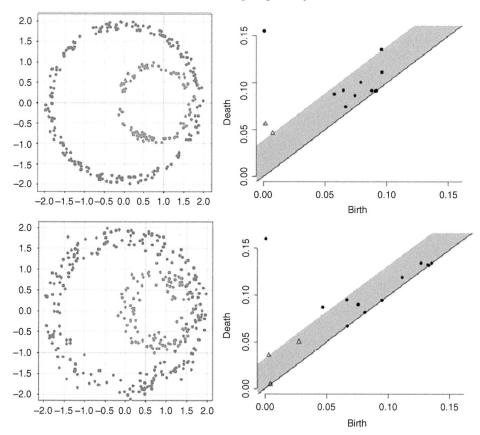

Figure 3.20 In the top panel, the 95% confidence interval clearly contains one bar and has a second at the edge for both H_0 and H_1; this correctly separates signal from noise. However, in the bottom panel, the 95% confidence interval suggests that all of the H_1 bars are noise whereas both the H_0 bars appear significant.

$$\theta\colon \mathcal{B} \to \mathbb{R}^n.$$

More generally, we might consider a measurable map

$$\theta\colon \mathcal{B} \to V,$$

for a vector space V which has a compatible topology (e.g., induced by the norm metric); V is regarded as equipped with the Borel σ-algebra. Then a distribution ρ on barcode space induces a pushforward distribution $\theta_*\rho$ on $\mathbb{R}^n$ or V.

This methodology has two substantial concrete benefits.

1. Many standard techniques in classical statistics apply essentially immediately to the distribution $\theta_*\rho$ on $\mathbb{R}^n$ or V. For example, summary statistics for $\theta_*\rho$, while not necessarily corresponding to any barcode, are now easy to compute

and work with. Consistency and convergence rates for empirical estimates can be quickly derived.

2. The resulting statistics can be used as input to visualization techniques or also as features for machine learning, e.g., classification and clustering algorithms. A particular advantage here is that such features can be combined with other sources of information or statistics produced from the raw data.

We can summarize the benefits of this simplification approach in terms of the following meta-theorem.

Theorem 3.6.1 (Meta-theorem of real projections from barcode space). *For any reasonable real-valued test statistic of barcodes, i.e., a suitable map $\mathcal{B} \to \mathbb{R}^n$, all the standard theorems and techniques of statistics and machine learning can be applied to the pushforward of any distribution on $\mathcal{B}$.*

There is infinite variety in the choice of feature maps to apply; in the remainder of this section, we discuss some representative examples.

3.6.1 Tractable Features from Persistence Diagrams

We begin by considering two simple and generic approaches for embedding arbitrary metric spaces in $\mathbb{R}^m$: the distance distribution and landmark embeddings. Both of these are easy to apply to distributions of barcodes, and yield distributions on Euclidean space. Then, for example, the mean of the pushforward distribution is a useful summary statistic.

The distance distribution is simply the induced distribution produced by computing distances between points; the next definition makes sense since the metric is always a measurable map on a metric measure space. See Figure 3.21 for a simple example.

Definition 3.6.2. Let (X, μ_X, ∂_X) be a metric measure space. The *distance distribution* on $\mathbb{R}$ is defined to be the pushforward $(\partial_X)_* \mu_X^{\otimes 2}$ of the distribution $\mu_X^{\otimes 2}$ on $X \times X$ along the function $\partial_X \colon X \times X \to \mathbb{R}$.

There are various elaborations of this example; for instance, one could consider distributions of $k \times k$ distance matrices induced by samples of size k. (Distance matrices as summaries of barcodes were studied in [97].)

Remark 3.6.3. In fact, a famous result of Gromov implies that a metric measure space is uniquely characterized by such distance matrix distributions for all k, in

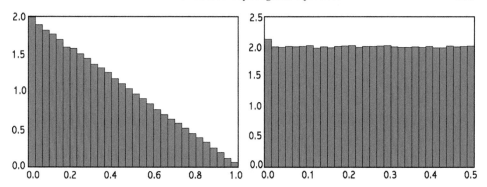

Figure 3.21 Left: The distance distribution for 1000 points sampled uniformly from [0, 1]. Right: The distance distribution for 1000 points sampled uniformly from S^1.

the sense that two metric measure spaces (X, ∂_X, μ_X) and (Y, ∂_Y, μ_Y) are isomorphic if and only if the distance distributions coincide as k goes to ∞ [212].

Another possibility is to consider distances to a fixed collection of points. Choose k landmark points $\{\ell_1, \ldots, \ell_k\}$; these can be selected arbitrarily, or as points of interest based on domain knowledge, or via a randomized algorithm biased to choose a point far from the existing landmarks, etc.

Definition 3.6.4. Let (X, μ_X, ∂_X) be a metric measure space and take a finite subset $\{\ell_1, \ldots, \ell_k\} \subset X$. Then the *landmark embedding distribution* on $\mathbb{R}^k$ is the pushforward of μ_X along the function $X \to \mathbb{R}^k$ specified by the formula

$$x \mapsto (\partial_X(x, \ell_1), \partial_X(x, \ell_2), \ldots, \partial_X(x, \ell_k)).$$

Remark 3.6.5. The selection of landmarks introduces many new statistical problems. For instance, the choice of k introduces a rough notion combining dimension and feature scale; the larger the dimension and the smaller the feature scale, the more landmark points one needs. Moreover, questions of stability of the results in the face of shifts in landmark points immediately arise. Currently, there are not many theoretical results in this regime (e.g., recall the discussion of the properties of the weak witness complex in Section 2.7). On the other hand, standard statistical tools (e.g., empirical confidence intervals for quantities computed from these distributions) can be applied to handle such issues.

The landmark distribution is a first guess at how to embed a metric space in Euclidean space. There is in fact an enormous literature on the problem of

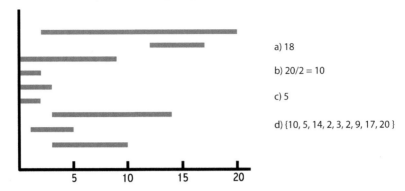

a) 18

b) 20/2 = 10

c) 5

d) {10, 5, 14, 2, 3, 2, 9, 17, 20 }

Figure 3.22 (a) The length of the longest bar, (b) the ratio of the endpoints of the longest bar, (c) the number of bars of length over 4, and (d) the righthand endpoints of each bar.

efficiently embedding a finite metric space in Euclidean space in a way which minimizes distortion (e.g., see [168] for a celebrated and essentially optimal result); although there has not been much investigation so far of these techniques in TDA (although see [455] for work that employs methods from this literature) we expect that this will be a useful avenue of research.

There are also many specific invariants of barcodes that provide values in $\mathbb{R}$ or $\mathbb{R}^n$; we provide some representative examples. (See Figure 3.22 for a specific example.) Note that a basic and important issue to consider for any such feature is whether it is stable with respect to perturbation in the barcode metric.

1. For a barcode B we can define

$$g_m(B) = |B(m)| - |B(m + 1)| \qquad \text{and} \qquad h_m(B) = \frac{B(m)}{B(m + 1)}$$

where $B(k)$ denotes the kth largest interval in B.

2. Given a barcode B, we can consider the set of birth-times $\{x_i\}$ or the set of death-times $\{x_i\}$ to provide a map to $\mathbb{R}^n$, where n is a bound on the size of the barcodes we consider.

3. Given a barcode B, we can consider a map to $\mathbb{Z}$ given by the number of non-zero bars or the number of non-zero bars greater than some minimum length ϵ.

4. Given a barcode B, we can consider a map to $\mathbb{R}^n$ given by the set of lengths $\{y_i - x_i\}$ or the set of size ratios $\left\{\frac{x_i}{y_i}\right\}$; to make sense of this, we must again bound the size of the barcodes and also sort the bars by length.

In [49], an explicit embedding of persistence diagrams in high-dimensional Euclidean space is considered; the idea is to take a grid on the persistence diagram and count barcode points within it. Unfortunately, this is not stable for certain kinds of perturbations of the barcodes that are small in the bottleneck distance.

3.6.2 Kernel Methods for Barcodes

The idea of *kernel methods* for machine learning involves embedding the data points in some kind of infinite-dimensional vector space where standard machine learning techniques apply. The trick is that rather than working with the embedding directly, it turns out to be sufficient to understand the inner product between two points in the embedded space; this is the kernel function. We say a bit more about the specifics of this in Section 4.3.4. Here, we focus on explaining the construction and definition of kernels for barcodes based on approximating a persistence diagram with a sum of Gaussian functions [4, 425]. These sorts of approaches yield kernels that are stable in the bottleneck and p-Wasserstein metrics on barcodes and provide sensible feature vectors for machine learning.

In [425], the kernel at scale σ for persistence diagrams D_1 and D_2 is computed by the formula

$$k_\sigma(D_1, D_2) = \frac{1}{8\pi\sigma} \sum_{\substack{p \in D_1 \\ q \in D_2}} e^{\frac{-\partial(p,q)^2}{8\sigma}} - e^{\frac{-\partial(p,\bar{q})^2}{8\sigma}},$$

where $\bar{q}$ denotes the reflection across the line $x = y$. Roughly speaking, we can think of this as a approximation by positive and negative Gaussians. The basic idea is that a persistence diagram can be approximated in function space as a sum of Dirac δ-functions centered at the points. However, the resulting metric on functions does not incorporate information about the proximity to the diagonal (i.e., bars of zero length). So instead, the δ-functions are used to specify a diffusion equation with the diagonal providing boundary constraints; the resulting solutions are Gaussians.

In contrast, in [4] a closely related approach was studied which uses weighted positive Gaussians; the difference in weights permits more flexibility in focusing on different features in the barcodes, and the use of positive Gaussians in some circumstances can provide computational efficiency. Although this is not phrased as a kernel method per se (but simply as a vector-space valued summary), it can be applied to produce a kernel just as in [425].

Remark 3.6.6. We can regard the grid counting method of [49] as a discretization of the Gaussian kernel description.

3.6.3 Persistence Landscapes

Another systematic approach to producing features from persistence diagrams is provided by Bubenik's *persistence landscapes* [76]. Suppose that we are given a barcode $\{[x_i, y_i)\}$, which we regard as a persistence diagram in $\mathbb{R}^2$. Changing coordinates via the transformation

$$[x, y) \mapsto \left[\frac{x+y}{2}, \frac{y-x}{2} \right),$$

we can equivalently represent a barcode as the multiset $\{[\frac{x_i+y_i}{2}, \frac{y_i-x_i}{2})\}$ in $\mathbb{R}^2$; we will assume that all persistence diagrams are represented in this format for the remainder of the section.

Next, define the piecewise-linear function

$$\Lambda_{(x,y)}(t) = \begin{cases} t - x, & t \in [x, \frac{x+y}{2}] \\ y - t, & t \in (\frac{x+y}{2}, y] \\ 0, & \text{otherwise.} \end{cases}$$

Definition 3.6.7. Let $B = \{[x_i, y_i)\}$ be a persistence diagram. The *persistence landscape* is the collection of functions $\lambda_B^k : \mathbb{R} \to \mathbb{R}$ for $k \in \mathbb{N}$, defined as

$$\lambda_B^k(t) = \lambda_B(k, t) = \text{kmax}_{[x_i, y_i) \in B} \Lambda_{[x_i, y_i)}(t),$$

where kmax denotes the kth largest value, defined to be 0 if the set in question contains fewer than k points. (We will often regard this collection as a single function $\Lambda : \mathbb{N} \times \mathbb{R} \to \mathbb{R}$.)

See Figure 3.23 for an example of a persistence landscape. One advantage of working with the persistence landscape is that for any fixed k this is a 1-Lipschitz function, and the set of all such functions is a $\mathbb{R}$-vector space with a metric induced by a norm that is complete and separable. As a consequence, one can easily define the mean landscape $\bar{\Lambda}$ for a collection of barcodes $\{B_i\}$, which is simply computed pointwise:

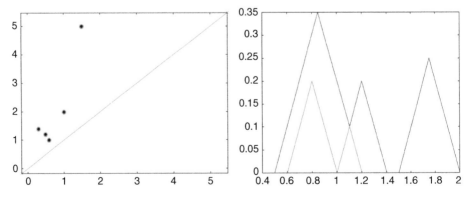

Figure 3.23 Left: A persistence diagram. Right: The associated persistence landscapes for $k = 1$ and $k = 2$.

$$\bar{\Lambda}_n = \frac{1}{n} \sum_{i=1}^{n} \lambda_{B_i}(k, t).$$

The mean landscape is the average value of the largest bar contained in k intervals. It is important to emphasize again that the mean landscape need not correspond to any particular barcode.

In this context, there is both a law of large numbers and a central limit theorem; these say that the mean of the landscapes of samples converges to the mean of the underlying distribution, and explain how fast this convergence occurs. Moreover, the average persistence landscape weakly converges to a Gaussian process (with a known rate of convergence) [76]. Specifically, we have the following result.

Theorem 3.6.8. *Provided that the expectation is finite,*

$$\bar{\Lambda} \to E(\Lambda),$$

where $\bar{\Lambda}_n$ is the empirical mean of the first n sample landscapes and $E(-)$ denotes the expected value.

Theorem 3.6.9. *Provided that the expectation and variance are both finite, then*

$$\sqrt{n}[\bar{\Lambda} - E(\Lambda)]$$

converges to a Gaussian random variable with the same covariance structure as Λ. (Here recall that the covariance structure determines the width of each Gaussian in the random variable.)

The following corollary allows us to perform inference.

Corollary 3.6.10. *The random variable produced by applying any functional (i.e., function from the space of landscapes to $\mathbb{R}$) also satisfies the central limit theorem.*

Of course, a choice of a useful and informative functional depends on the data and is not always evident. A simple approach is to use an indicator function for t in an interval $[-B, B]$ and k bounded by K.

Remark 3.6.11. In fact, we can prove a uniform version of the central limit theorem and bound the rate of convergence. This implies in particular that the bootstrap is asymptotically consistent and so can be used to estimate confidence intervals for persistence landscapes; see [109] for results of this form involving the multiplier bootstrap.

Furthermore, landscapes satisfy an evident analogue of the stability theorem: the L_∞ distance between landscapes is bounded by the Gromov-Hausdorff distance between point clouds.

A natural application of persistence landscapes to robust inference was studied in [110], where they used the average persistence landscape of the samples in Φ_k^n as a summary; this has the advantage of being easy to compute and study. In analogy with Theorem 3.4.2, one can show that the average persistence landscape is Wasserstein stable. Moreover, explicit estimates of the bias of this estimator as a function of the number of sample points can be obtained. (See Figures 3.24 and 3.25 for examples of this approach.)

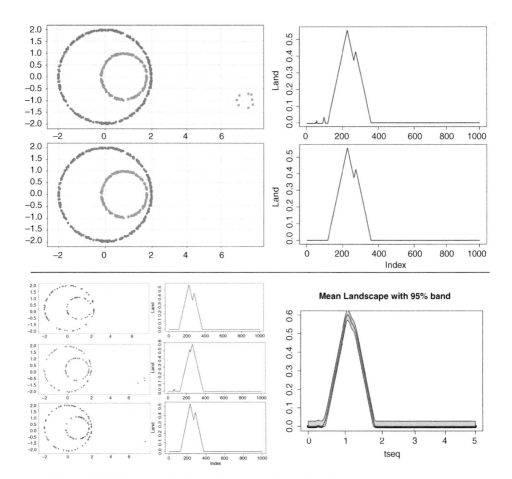

Figure 3.24 Since the landscape is a real-valued function, the pointwise average is easy to compute. The top two panels show the landscape for samples from two circles plus a noisy circle far away and two circles without the noisy circle. The bottom panels represent the effect of subsampling and averaging to remove the effect of the noisy circle.

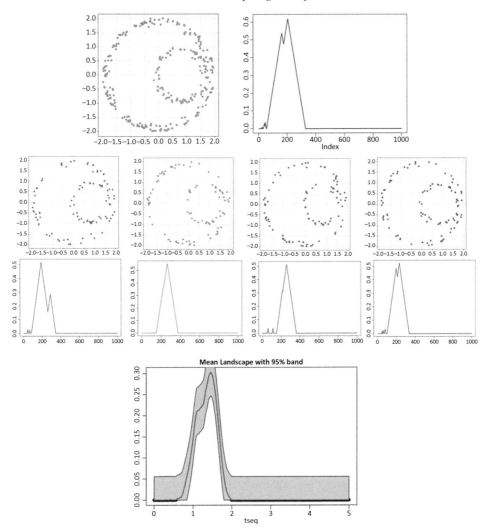

Figure 3.25 Subsampling and averaging is also effective with noisy data where the noise is concentrated around the underlying space.

3.6.4 Coordinates on Persistent Homology

A more principled source of real-valued invariants from barcodes comes from considerations from algebraic geometry. Adcock, Carlsson, and Carlsson introduced the idea of regarding subsets of barcode space as *algebraic varieties* and studying their coordinate rings [5]. *Coordinates* on a barcode just means a collection of functions from a space of barcodes to $\mathbb{R}$. In [5], the basic idea is to use symmetric polynomials in the start and endpoints of the bars, for barcodes with a fixed number of bars. (The symmetry of the polynomials is a consequence of the fact that we do not care about the ordering of the bars within the barcode.)

Remark 3.6.12.   This approach was extended to multidimensional persistence in [465].

   Unfortunately, these coordinates are not stable with respect to perturbation of the barcode in the bottleneck or p-Wasserstein metric; this is clear, as very short bars with large start and endpoints can affect these polynomials dramatically. To fix this problem, Verovšek [283] (building on [94]) introduced ideas from *tropical geometry* to build stable coordinates. Tropical geometry studies a semiring structure on $\mathbb{R}$ where addition of x and y is computed by $\max(x, y)$ or $\min(x, y)$ and multiplication of x and y by $x + y$ (ordinary addition on real numbers). This is a semiring in the sense that we do not require every number to have an additive inverse.

   The work of [283] showed that stable coordinates on barcode space could be obtained from rational functions (i.e., fractions) in "polynomial" expressions on the bar endpoints using the max-plus tropical structure. In [357], it is further shown that these coordinates provide sufficient statistics suitable for parametric inference; applications to reassortment in avian flu are discussed.

3.7 Stochastic Topology and the Expected Persistent Homology of Random Complexes

In the preceding sections, we have discussed techniques to produce stable persistent homology invariants of data despite the presence of noise. Another part of the statistical aspect of the story is to quantify the effect of idealized noise by describing the expected persistent homology of a "random complex." For example, such a description yields a family of strong null hypotheses. However, despite the mathematical interest and depth of theoretical work of this kind, in practice it is typically more suitable to use Monte Carlo simulation to find empirical estimates.

   As a consequence, our discussion is brief and we refer the interested reader to the primary sources for precise theorem statements (see also Kahle's survey [281] and the article [61]).

   In order to specify the problem, we need a model for generating random complexes. Recall that the Vietoris-Rips complex is completely determined by its 1-skeleton (see Definition 2.1.6), which is a graph. Therefore, processes that generate random graphs can also be regarded as producing random simplicial complexes.

   The most familiar model of a random graph is the Erdös-Renyi model, which connects vertices with some fixed probability. However, although there is a substantial literature on random simplicial complexes from this perspective (e.g., see [64] for a classic exposition), this is not a sensible model of random simplicial complexes in the geometric setting. The most relevant definition of a random complex from this perspective arises from the definition of a geometric random graph. (See [403] for an extensive treatment of the properties of geometric random graphs.)

Definition 3.7.1. Let (M, ∂_M, μ_M) be a metric measure space. Fix $\epsilon > 0$. A *geometric random graph* with k points is generated by sampling k points $\{x_i\}$ from M according to μ_M and forming the graph with k vertices and an edge (i, j) if $\partial_M(x_i, x_j) < \epsilon$.

Example 3.7.2. The most frequently studied example is the case when M is the unit cube $[0, 1]^n \subseteq \mathbb{R}^n$.

Definition 3.7.3. Let (M, ∂_M, μ_M) be a metric measure space. Fix $\epsilon > 0$. A *geometric random complex* with k points is generated by sampling k points $\{x_i\}$ from M according to μ_M and forming either the Vietoris-Rips or Čech complex associated to ϵ and the finite metric space $\{x_i\}$.

Although we have stated the definitions in full generality, most existing work studies distributions supported either on $\mathbb{R}^n$ or in a few cases on a smooth compact manifold embedded in $\mathbb{R}^n$ (e.g., see [62] for the latter).

Most current results (e.g., the work of Kahle) about geometric random complexes consider the expected ranks of the homology groups β_ℓ as simultaneously $\epsilon \to 0$ and $k \to \infty$. The results are controlled by $k\epsilon^n$:

1. in the *sub-critical* regime, $k\epsilon^n \to 0$,
2. in the *critical regime*, $k\epsilon^n$ goes to a constant, and
3. in the *super-critical regime*, $k\epsilon^n$ goes to ∞.

We now summarize what is known in these various settings.

1. **Sub-critical.** There are various results on the expected Betti numbers [282]. Here the situation is sometimes referred to as "dust," since there are many disconnected components and so the most important contribution is to H_0. This is the easiest non-trivial regime to analyze.
2. **Critical.** There is an enormous amount of non-trivial homology, and [543] provides detailed estimates on the expected rank of the homology for certain distributions on $\mathbb{R}^d$ and weak and strong laws of large numbers describing convergence.

3. **Super-critical.** The complex is asymptotically contractible and so there is no contribution to homology (and the analysis is basically trivial). This is analogous to the emergence of the "giant component" in the classical results on the behavior of random graphs.

A closely related but distinct perspective is provided by the work of Adler, Bobrowski, and Weinberger [7]. They consider distributions with infinite support on $\mathbb{R}^n$, and observe that sufficiently large samples separate into

- the "core," which is densely sampled and contractible, and
- the periphery, which "crackles" with homology.

This perspective is a variation on the results summarized above, insofar as the core and periphery correspond to super-critical and critical regimes simultaneously arising due to variation in the density.

The conceptual frameworks of "core" and "crackle" provide two kinds of indications of the limits of certain approaches to topological data analysis:

- a large core will obscure the signal, and
- the crackle will generate spurious homology classes.

All of the work discussed so far has focused on understanding homology for complexes with specific ϵ; only very recently has there been work extending this to persistent homology [63]. Here, there is more similarity between the regimes, but the scale of events differs. (See Figure 3.26 for a representative example.)

Notably, in the critical regime the longest bar in the barcode appears to satisfy a "law of the iterated logarithm" describing its length, for certain distributions on a cube (notably the Poisson distribution) and both the Čech and Vietoris-Rips complexes. Such a bound gives a precise estimate for how fast the length increases as the number n of sample points increases; roughly $\left(\frac{\log n}{\log\log n}\right)^{\frac{1}{k}}$ for kth homology. (This phenomenon is also mentioned in passing in the Adler-Bobrowski-Weinberger work.)

3.8 Euler Characteristics in Topological Data Analysis

A reasonable conclusion to draw from the discussion of this section is that it is advantageous to use the simplest possible topological invariants, e.g., low-dimensional persistent homology. This perspective suggests consideration of the Euler characteristic as a potentially interesting topological invariant which is robust and easy to compute and yet rich enough to capture topological properties of the underlying space.

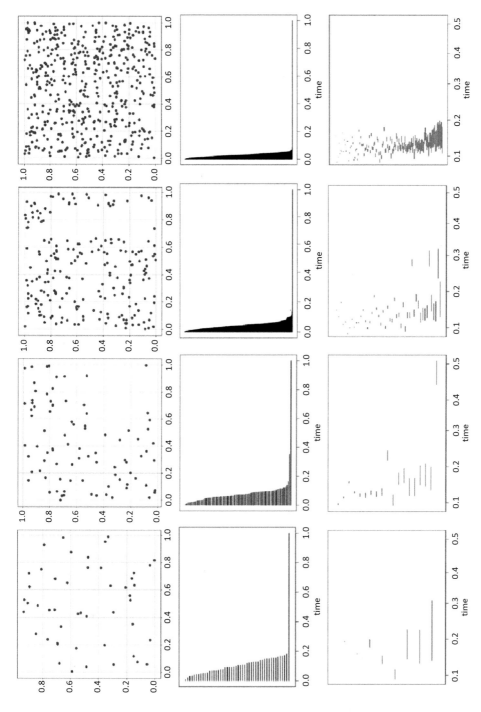

Figure 3.26 Persistent homology of points sampled uniformly from a unit square.

To further motivate this focus, Weinberger has pointed out that the Euler characteristic of a simplicial complex is *locally testable* [160]. Locally testable in this case means that the Euler characteristic can be computed from a small number of random samples from a simplicial complex, with high probability [204]. Specifically, fix $\epsilon > 0$. A tester for the Euler characteristic chooses $K(\epsilon)$ random vertices of the complex and has access to neighborhoods of size $D(\epsilon)$ around those vertices. The tester then returns a guess $\chi'(X)$ for the Euler characteristic such that

$$\Pr\left(\frac{\chi(X) - \chi'(X)}{|X_0|} \geq \epsilon\right) \leq \epsilon.$$

The existence of a tester is interesting because the functions K and D do not depend on the size of the complex but only on ϵ! Weinberger proposes that local testability is a good proxy for understanding when a topological invariant will be robust and reasonable to compute for small samples [533].

Remark 3.8.1.   Although more generally rational homology groups are known to be locally testable [160], no such results are known for other coefficients.

There has been a great deal of study of the special case of the Euler characteristic of Gaussian random fields. Let M be a smooth compact manifold and f a Gaussian random field on M; then Adler and Taylor provide formulas describing the expected Euler characteristic of the "excursion sets" $f^{-1}(u, \infty)$. See [8] for an overview of this work, and [9] for an interpretation in terms of persistent homology. These kinds of results have had numerous applications in situations where smooth processes of this sort arise, notably imaging. However, application of these techniques in genomics is in its infancy, although searching for ways to apply them seems like a productive endeavor.

A potentially promising direction for problems related to genomics comes from the *smooth Euler characteristic transform*, a generalization of the persistent homology transform [127]. We again assume we are working with a finite simplicial complex M embedded in Euclidean space $\mathbb{R}^d$. For a given direction v, let a_v and b_v denote the minimum and maximum values of $x \cdot v$ over the points of M. The Euler characteristic curve in the direction v is now defined to be the function

$$[a_v, b_v] \rightarrow \mathbb{Z}$$

defined by $t \mapsto \chi(M(v)_t)$. Let $\bar{\chi}(M(v))$ denote the average value of the Euler characteristic curve in the direction v.

Definition 3.8.2. The *smoothed Euler characteristic curve* for the direction v is defined to be the function

$$F_v^M(y) = \int_{-\infty}^y (\chi(M(v)_x) - \bar{\chi}(M(v))) \, dx.$$

Observe that by construction this is a smooth piecewise-linear function with compact support.

Definition 3.8.3. The *smooth Euler characteristic transform* is the function

$$\text{SECT} \colon S^{d-1} \to L_2(\mathbb{R})$$

specified by

$$v \mapsto F_v^M.$$

Interestingly, when $d \le 3$, the SECT can be shown to be injective; this is a sufficient statistic for describing the underlying distribution. Moreover, since the result is a function in L_2, just as in the case of the discussion of persistent landscapes, the SECT can be used as input to standard statistical models and resampling techniques can be used to obtain confidence intervals for predictors and summary statistics. This approach has been used to generate clinically meaningful conclusions from imaging data from glioblastoma tumors in [127].

3.9 Exploratory Data Analysis with Mapper

Because of the tremendous possible space of topological hypotheses, the framework of exploratory data analysis is very well suited for TDA. That is, rather than seeking to confirm specific hypotheses or test existing ideas about the data set, it is often much more sensible to simply attempt to find structure in the data.

The Mapper algorithm (as discussed in Section 2.8) is particularly well suited for this.

- The output of Mapper is a colored graph representing a multiscale clustering; it is often possible to visually interpret the results.
- As Mapper requires choices about bin sizes and filter functions, varying these allows us to explore structural properties of the data. For example, Mapper can account for the measure on the data by using a density estimator as the filter function.

Remark 3.9.1. Although Mapper output is not stable with regard to perturbation of these choices, in the exploratory paradigm this is not as substantial a problem as it might seem. One can use the same statistical tools normally used to assess the stability of the results of clustering, i.e., cross-validation. There are different ways to do this, but all of them boil down to either partitioning or subsampling the data and then comparing clustering results by counting pairs which end up changing depending on whether they are in the same or different clusters. But perhaps more importantly, there is a strong sense in which instability is not as big an issue in genomics as one might expect. Exploratory analysis will typically be validated by further experiment. That is, in this kind of usage, predictions from TDA are confirmed by a follow-up experiment before being regarded as a reliable discovery. As such, the consequence of errors in inference due to instability is a wasted experiment; this is in stark contrast to applications in machine learning such as, for example, self-driving cars or clinical recommendations.

A common experimental application of Mapper is to explore various choices of filter function and other parameters in order to find clusterings of the data such that the clusters correlate strongly with other known properties of the data (e.g., clinically significant variables). More precisely, we have the following setup.

1. In addition to the data (X, ∂_X), filter function, and cover, we have an additional function $\theta \colon X \to \mathbb{R}$.
2. We extend θ to a function with domain the Mapper complex by defining θ on a point in the complex to be the average or median of the values of f along the corresponding data points.
3. We want to identify regions in the Mapper complex where θ is unusually large.

Now we can apply permutation tests (i.e., randomly relabeling the points and computing the values of the function θ) to determine the significance of an observed value. To be precise, we carry out the following.

1. We generate a distribution on values of θ by randomly shuffling the values of θ on X and recomputing the values on the points of the Mapper complex.
2. We then regard an actual value as significant if it is larger than 99% of the values produced in this fashion, for example. (The specific cutoff for significance is a parameter choice as usual.)

This procedure has been used in applications, for instance in the cell differentiation example we described previously in Example 2.8.3. However, note that as is usual with permutation tests, it can be expensive computationally to obtain confidence intervals as opposed to simply p-values. Also, the stability of this procedure does not yet have sound theoretical foundations in general, although in practice it appears to be stable with respect to cross-validation.

3.10 Summary

- This chapter provides tools with which we may formally discuss sampling from geometric objects. We adopt the working hypothesis that we have data randomly sampled from an underlying metric measure space (X, ∂_X, μ_X) (see Definition 3.2.10).
- In order to state probabilistic stability theorems, we need distances between distributions and more generally metric measure spaces. Toward this goal, we use the Gromov-Prohorov distance (see Definition 3.2.33) and the Gromov-Wasserstein distance (see Definition 3.2.34).
- We can study probability measures on barcode space; Section 3.3 provides a formal approach to probability theory on barcodes.
- Using metrics on distributions, Theorem 3.4.2 provides an analogue of the stability theorem of persistent homology (Theorem 2.4.10) in the context of metric measure spaces. Another version of a probabilistic stability theorem is given by Theorem 3.4.5.
- Section 3.5 provides a rigorous approach to this chapter's overarching goal of estimating persistent homology by taking sufficiently many samples from a space in order to recover the persistent homology of the support of the probability distribution.
- Summarizing distributions of barcodes turns out to be a challenging problem. One possibility is to consider techniques that involve extracting real-valued features from persistence diagrams.
- We may also approach this problem via *kernel methods* (see Section 3.6.2), *persistence landscapes* (see Section 3.6.3) or *coordinates* on a barcode (see Section 3.6.4); all of these methods map barcodes to a vector space where traditional statistical methods can be applied.
- In addition to the study of techniques to produce reliable persistent homology invariants despite the presence of noise, we are interested in considering the effect of idealized noise itself through the persistent homology of random complexes.
- Adaptation of the Euler characteristic is an attractive idea due to the advantages of using simple topological invariants.
- The Mapper algorithm (see Section 2.8) is a useful tool for exploratory data analysis. Section 3.9 outlines a procedure for the use of Mapper in applications.

The integration of topological data analysis with statistical methods is still in its infancy. As the discussion in the next part of the book makes clear, the kinds of techniques presented in this chapter have not yet made it into practice. Some of this is due to the lack of consensus about the best way to handle some of the issues that arise. But the lack of power of some of the tests (e.g., techniques for estimating confidence intervals) combined with difficulties in producing topological summaries also provides a substantial impediment. We hope that the readers of this book will feel particularly motivated to work to develop standards for statistical practice in topological data analysis.

3.11 Suggestions for Further Reading

For background in probability theory, we recommend Billingsley's textbook [57]. For discussion of probability theory in non-positively curved metric measure spaces, Gromov's book [212] and Sturm's article [487] are very informative. However, in general, there are not yet any good survey articles or textbooks about probability theory in the context of topological data analysis; as an exception, Kahle's survey article on random complexes [281] is comprehensive. For a review of statistics, Wasserman's books [526, 527] provide good introductions, and Freedman's classic introduction to statistical modeling [184] teaches a healthy dose of skepticism about the power of statistical inference.

4

Dimensionality Reduction, Manifold Learning, and Metric Geometry

> A map is not the territory it represents, but, if correct, it has a similar
> structure to the territory, which accounts for its usefulness.
>
> *Alfred Korzybski*

Although topological data analysis is new, the idea of studying data by analyzing shape is classical. The original forms of this kind of analysis (regression, principal components analysis (PCA), and multidimensional scaling (MDS)) make the assumption that the data lies on a linear subspace in $\mathbb{R}^n$. In contrast, TDA makes minimal assumptions about the underlying metric measure space generating the data. On the one hand, this means that we can apply TDA to data sets where we have no reason to expect linear structure. On the other hand, strong geometric assumptions have many benefits. For example, assuming that the data lies on a k-dimensional subspace of $\mathbb{R}^n$ characterizes the problem as searching for a linear transformation $\theta \colon \mathbb{R}^n \to \mathbb{R}^k$ such that $\{\theta(x_i)\}$ retains something about the structure of $\{x_i\}$. Assuming linearity

1. provides coordinates for describing the data and predicting where new data points might lie,
2. allows the application of standard statistical inference methods, and
3. makes it straightforward to perform dimensionality reduction by constraining the value of k. For example, even if we believe that the data lies on a plane of dimension $\ell > 3$, it can be useful to project into $\mathbb{R}^2$ or $\mathbb{R}^3$ for visualization purposes.

Linear models are arguably the most frequently used tools in applied mathematics; however, the assumption of linearity is often unreasonable. As a result, there has been a lot of recent work generalizing these methods to algorithms that operate under the assumption that the data has been sampled from a compact manifold $M \subseteq \mathbb{R}^n$ of *much lower dimension* than n. These algorithms, loosely referred to as *dimensionality reduction* or *manifold learning*, then seek to infer a parameterized representation of the data in terms of a coordinate system for a manifold. It is interesting to point out that in many biological applications we do not expect the data to lie on a manifold. For instance, the intrinsic dimension of transcriptomic data is related to active transcription programs (see Chapter 7). The number of these programs, relative to the intrinsic dimension, is not expected to be constant.

Although the manifold assumption is usually unrealistic for genomic data, dimensionality reduction has been successfully applied in various ways to analyze real biological data. For example, most applications of clustering in genomic analysis use dimensionality reduction as a preprocessing step (e.g., the frequent application of *t*-SNE), which is becoming standard in single cell analysis, see Chapter 7. More interesting from our point of view is the fact that some of the most successful genomic applications of Mapper have used coordinates from PCA as filter functions.

In the first part of this chapter we give a rapid overview of manifold learning and dimensionality reduction, starting with the classical techniques and moving on to recent generalizations. There is a vast literature on this subject, and we cannot hope to do more than give a flavor of these techniques. Our goal is to convey the central ideas underlying these approaches to analyzing data. Roughly speaking, the basic strategy of most manifold learning techniques is to take the *k*-nearest neighbors of a point *x* and use the vectors specified by the line segments from *x* to its neighbors as an approximation for the tangent plane at *x*. Global optimization then sews these local approximations together to produce a low-dimensional representation of the data. In a sense that can be made precise, the efficacy of these approaches depends on the fact that the Laplace-Beltrami operator on the manifold (which describes heat flow) can be approximated from finite samples by a certain *graph Laplacian* matrix.

In Figure 4.1, we indicate the results of different manifold learning representations on data that lies on a plane in $\mathbb{R}^3$; all of them recover coordinates for the plane. In Figure 4.2, we show a plane that has been rolled up – although the plane is flat, the embedding is twisted and so cross-cutting connections are potentially a problem (recall Section 2.2). Here, there is a noticeable difference in performance between classical techniques that assume linearity and manifold learning algorithms that do not.

In contrast to these cases, we will explore our running example of nested circles (which are not linear at all), and also consider nested arcs. In this context, manifold learning algorithms do a much worse job at recovering meaningful parametrizations. These simple experiments highlight the ways that topological data analysis

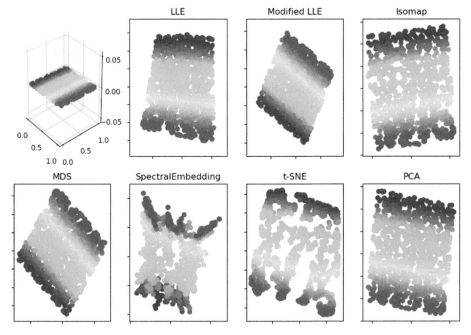

Figure 4.1 When the data lies on a plane in $\mathbb{R}^3$, all algorithms successfully recover a representation of the original data.

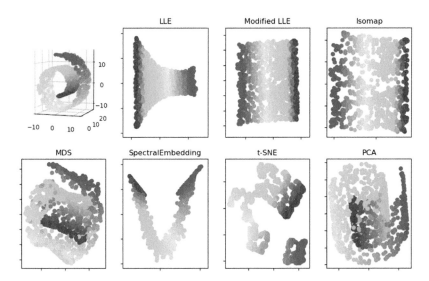

Figure 4.2 When the data lies on a rolled-up sheet, classical algorithms like PCA and MDS perform very poorly, whereas manifold learning techniques successfully capture the intrinsic shape of the data.

can be useful, even in situations where the data does lie on a low-dimensional manifold. Of course, in general, we do not necessarily expect such a hypothesis to hold.

However, in genomics there is an even more specialized geometric assumption
that is frequently warranted. When working with data generated by evolutionary
processes, it is standard to assume that the data can be organized into a phylogenetic
tree. A phylogenetic tree is typically represented as a metric tree; recall from Example 1.2.4 that this is a graph with no cycles and weighted edges, where the metric
is computed as the sum of the weights along the shortest path between two points.

In the second part of this chapter we give an overview of mathematical frameworks for dealing with phylogenetic trees. Again, this is a vast area of research with
many excellent books; Felsenstein's text is a classic exposition [172]. We begin by
giving a quick treatment of how to infer a phylogenetic tree from genomic data presented as a finite metric space; see Appendix C for a more detailed review. We then
explain celebrated work of Billera, Holmes, and Vogtmann [55] that shows that
phylogenetic trees can themselves be organized into a metric space; we will later
see in Chapter 5 that the associated metric geometry (see Section 4.7.3) supports
clinically significant analysis.

4.1 A Quick Refresher on Eigenvectors and Eigenvalues

Almost all of the dimensionality reduction techniques we will describe in this section involve computation of the eigenvectors of a matrix formed from the data
points. Although we have assumed that the reader has familiarity with basic linear
algebra, in this section we briefly review the relevant definitions. In the following,
we always work with real vector spaces.

Definition 4.1.1. Let A be an $n \times n$ matrix. An *eigenvector* v for A with *eigenvalue*
λ is a non-zero vector $v \in \mathbb{R}^n$ such that

$$Av = \lambda v.$$

The first key observation is that for symmetric matrices A (i.e., matrices such
that $A = A^T$), eigenvectors with distinct eigenvalues are orthogonal.

Proposition 4.1.2. *Let A be a symmetric $n \times n$ matrix and let $v_1, v_2 \in \mathbb{R}^n$ be
eigenvectors with distinct eigenvalues $\lambda_1 \neq \lambda_2$. Then v_1 is perpendicular to v_2.*

This suggests that we can think of eigenvectors for different eigenvalues as giving a preferred alternative set of coordinates for $\mathbb{R}^n$ which are adapted to the linear
transformation represented by A. We do not always have enough eigenvectors to
form a basis for all of $\mathbb{R}^n$, however.

Proposition 4.1.3. *An $n \times n$ matrix A has at most n distinct eigenvalues and at most n linearly independent eigenvectors. When there are exactly n independent eigenvectors, they form a basis.*

It is standard to sort the eigenvectors by the size of the associated eigenvalues; when we talk about the "top k" eigenvectors, we mean those with the k largest eigenvalues.

4.2 Background on PCA and MDS

A classical example of dimensionality reduction is principal component analysis (PCA). The idea here is, given a set of points $\{x_1, x_2, \ldots, x_m\}$ in $\mathbb{R}^n$ as data, to find an "optimal" linear projection $\theta: \mathbb{R}^n \to \mathbb{R}^k$, for $k < n$. Here is an outline of the algorithm.

1. We normalize to center the data and define

$$\widetilde{x}_i = x_i - \mu, \quad \text{where} \quad \mu = \frac{1}{n} \sum_i x_i.$$

2. We then form the covariance matrix

$$C = \frac{1}{n} \sum_i \widetilde{x}_i \widetilde{x}_i^T.$$

3. We compute the top k eigenvectors $\{v_1, \ldots, v_k\}$ of C to use as our basis.
4. These eigenvectors span a hyperplane (subspace) of $\mathbb{R}^n$ that is isomorphic to $\mathbb{R}^k$; the projection $\theta: \mathbb{R}^n \to \mathbb{R}^k$ of the data is precisely the orthogonal projection onto this plane followed by a choice of identification of the plane with $\mathbb{R}^k$.
5. We can also regard θ as producing vectors in $\mathbb{R}^n$; adding back μ yields approximations $y_i = \theta(\widetilde{x}_i) + \mu$ of each x_i.

This process chooses the basis which maximizes the variance captured by the representation; the eigenvector v_1 with the largest eigenvalue is the single direction which captures the maximal amount of information about the variance in the points, the plane spanned by $\{v_1, v_2\}$ is the plane with the most variance, and so forth. Interestingly, we can also characterize the output of PCA as the projection that minimizes the error function

$$E = \sum_{i=1}^{m} \partial_{\mathbb{R}^n}(x_i, y_i)^2.$$

That is, PCA produces the points $\{y_i\}$ which minimize the reconstruction error among all projections onto a k-dimensional subspace.

In fact, another classical approach to dimensionality reduction is to take minimization of E as a point of departure. Metric multidimensional scaling (metric MDS), takes as input a finite metric space (X, ∂_X) and computes an optimal embedding of X into a Euclidean space $\mathbb{R}^k$. Here the optimality criterion is to preserve the original metric data as much as possible, i.e., to minimize an analogue of E. Specifically, in MDS we search for a map $\theta \colon X \to \mathbb{R}^k$ that minimizes

$$\mathcal{E} = \sum_{x_i, x_j \in X} \left(\partial_X(x_i, x_j) - \partial_{\mathbb{R}^k}(\theta(x_i), \theta(x_j)) \right)^2.$$

We do this as follows.

1. Let D denote the matrix with entries $D_{ij} = \partial_X(x_i, x_j)$.
2. Set

$$H = I - \frac{1}{n} e e^T \qquad \text{and} \qquad Z = -\frac{1}{2} H D H,$$

   where as usual I denotes the identity matrix and e is the vector with all entries 1. (This step centers the results; since the minimizing embedding is not unique as distances are preserved by translation, we need to impose such a constraint to get a specific output.)
3. The embedding that minimizes $\mathcal{E}$ is then given by finding the eigenvectors $\{v_j\}$ of Z. Specifically, the embedding $\theta(x_i) \in \mathbb{R}^k$ is specified by normalizing so that $\|v_j\|^2 = \lambda_j$, making a matrix with the eigenvectors $\{v_j\}$ as columns, and taking the ith row.

When the metric space (X, ∂_X) arises as a subspace of $\mathbb{R}^n$, then it turns out that PCA and metric MDS coincide.

Theorem 4.2.1. *Given $\{x_1, x_2, \ldots, x_\ell\} \subset \mathbb{R}^n$ and $k < n$, the results of metric MDS and PCA embedding $\{x_i\}$ into $\mathbb{R}^k$ are isometric.*

However, metric MDS has the advantage that it can be applied to arbitrary metric spaces, i.e., metric spaces that are not subspaces of $\mathbb{R}^n$. Moreover, posing the problem as minimizing the embedding error function $\mathcal{E}$ allows us to consider variants which minimize different error functions. For example, work on "antigenic maps" describing genomic and phenotypic variability in the flu virus uses an MDS variant [467]. Of course, changing $\mathcal{E}$ can result in substantially more difficult optimization problems.

These procedures are very widely used in data analysis because they are in general easy to compute and (especially when k is chosen to be 2 or 3) result in convenient visualizations of the embedded data. However, these algorithms can be very unstable in response to perturbations of the data (although there is a growing literature on robust variants of MDS and PCA, e.g., [89]), especially when noise processes vary in different directions (e.g., see [411]). Under assumptions that the signal is low rank, variants known collectively as sparse PCA do a good job at recovering a sparse basis to describe the signal [552]. Sparsity of the data can also result in serious distortions; this is a particular problem in single-cell expression data. For a more extensive discussion of this problem and its relation to random matrix theory, see [15]. Another issue is that the optimal choice of k is a priori unknown. In practice, one often looks for an "eigenvalue gap," i.e., a natural splitting of the eigenvalues into a group of "large" eigenvalues and then a collection of much smaller eigenvalues. However, this procedure requires a threshold for deciding where the gap is, and is in general more of an art than a science.

A more serious issue from our perspective is the fact that when the data cannot be isometrically embedded as a Euclidean subspace of $\mathbb{R}^n$, PCA and MDS simply do not work particularly well to capture the intrinsic geometric structure. In Figure 4.3, we see that for a single curved ribbon in $\mathbb{R}^3$, PCA captures the intrinsic geometry with some distortion. But in Figures 4.4 and 4.5, for more complicated geometric objects (the union of two ribbons and a sphere in $\mathbb{R}^3$, respectively), PCA does not recover the intrinsic coordinates along the circle but rather just embeds a flattening of the circle in Euclidean space.

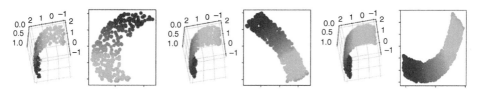

Figure 4.3 When the data lies on a single curved ribbon, the embedding into $\mathbb{R}^2$ exhibits distortion arising from the curvature.

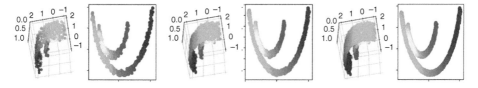

Figure 4.4 When the data lies on nested ribbons, the embedding is further distorted by the proximity of the two components.

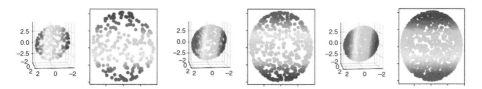

Figure 4.5 When the data lies on the standard sphere S^2, the embedding flattens the sphere and distorts the distances along an arbitrary axis.

Remark 4.2.2. When using the eigenvectors from PCA to describe the data, one possible source of difficulty in interpretation arises from the fact that the relevant linear combinations might include negative terms. In many applications, subtraction of basis vectors does not make sense. For example, many genomics applications of dimensionality reduction are interpretable only for positive combinations of terms. In this context, an algorithm called "non-negative matrix factorization" (NMF) is often used. In contrast to PCA, NMF is an iterative optimization procedure. See [319] for a classic rigorous discussion.

4.3 Manifold Learning

Suppose that we are given data points $\{x_1, \ldots, x_m\} \subseteq \mathbb{R}^n$, but we no longer assume that they admit a nearly isometric embedding as a hyperplane (i.e., an affine linear subspace). As we explained in Section 2.2, even when the points $\{x_i\}$ are produced by sampling from some embedding $\gamma \colon M \to \mathbb{R}^n$ of a compact Riemannian manifold M, the distance $\partial_{\mathbb{R}^n}(\gamma(x_i), \gamma(x_j))$ may not be very representative of the *intrinsic distance* $\partial_M(x_i, x_j)$. For example, as Figure 4.2 indicates, even when the manifold in question is homeomorphic to a plane, PCA and MDS can perform very poorly.

Consider the case of a line segment $\gamma \colon [0, 1] \to \mathbb{R}^2$ which is very twisted. Clearly, the distance along the curve $\gamma([0, 1])$ is poorly approximated by the Euclidean distance, especially near kinks. However, when γ is sufficiently smooth, there exists a feature scale at which Euclidean distances and intrinsic distances agree up to small error. In the work described in Section 2.2, this observation was leveraged to justify an algorithm for recovering the homology (and in fact homotopy type) of M. Here, we are interested in recovering coordinates on the manifold. This is a meaningful and potentially subtle question even in the case where M is contractible, and in fact most manifold learning algorithms focus on the case where M is contractible but the embedding $\gamma \colon M \to \mathbb{R}^n$ is twisted.

Manifold learning approaches ultimately rely on the fact that in favorable cases the manifold structure can be reconstructed by considering the "short distances" as reliable indicators of the intrinsic distance and ignoring the "long distances." One way to express this idea is to hypothesize that the basis determined by the k-nearest neighbors of a point z give a good approximation of the tangent plane to M at z.

4.3.1 Isomap

An early and prominent manifold learning algorithm is Isomap, which simply applies MDS to an empirical approximation of the intrinsic metric [495]. The procedure works as follows. We assume we are given data points $\{x_1, \ldots, x_n\} \in \mathbb{R}^n$. We fix a scale parameter ϵ and a target dimension parameter k.

1. Form the weighted graph G with
   - vertices the points $\{x_i\}$, and
   - edges (i, j) with weight $w_{ij} = \partial_{\mathbb{R}^n}(x_i, x_j)$ when $\partial_{\mathbb{R}^n}(x_i, x_j) \leq \epsilon$.
2. We now form a new metric space X' with points $\{x_1, \ldots, x_n\}$ but distance given by the graph metric on G. Recall from Example 1.2.4 that this means that the distance between two vertices is the length of the shortest path in the graph. The graph metric can be efficiently computed, for example via Dijkstra's algorithm (e.g., see [125, 24.3]).
3. Finally, we use MDS to embed this new metric space into $\mathbb{R}^k$ as above, producing points $y_i = \theta(x_i)$.

When the points $\{x_i\}$ are sampled from a convex subset $M \subseteq \mathbb{R}^m$ embedded isometrically into $\mathbb{R}^n$, $k \geq m$, and ϵ is in the right range, Isomap can recover almost exactly the coordinates for M. (Here recall that a subset A of $\mathbb{R}^n$ is convex if for $x_1, x_2 \in A$, the line between x_1 and x_2 is entirely contained in A. In particular, this implies that A is contractible.) The recovery guarantees follow from the fact that for sufficiently dense sampling from a Riemannian manifold and suitable ϵ, the graph metric computed in the second step of the procedure approximates the underlying distance [51].

In Figure 4.6, we see that for a single curved ribbon in $\mathbb{R}^3$, Isomap does recover the intrinsic distances, with a small amount of distortion. But in Figures 4.7 and 4.8,

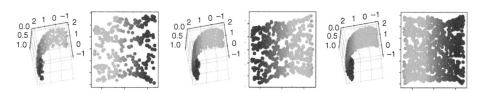

Figure 4.6 When the data lies on a single curved ribbon, Isomap does a good job of recovering the intrinsic coordinates.

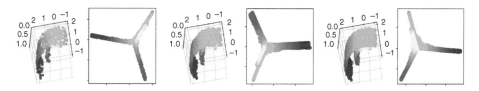

Figure 4.7 When the data lies on nested ribbons, Isomap collapses the two components to single lines.

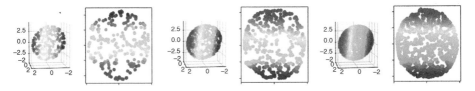

Figure 4.8 When the data lies on the standard sphere S^2 in $\mathbb{R}^3$, Isomap is not able to recover the intrinsic distances and embeds a flattening of the sphere in $\mathbb{R}^2$.

for more complicated geometric objects (the union of two ribbons and a sphere in $\mathbb{R}^3$, respectively), Isomap again fails to recover the intrinsic coordinates of the data. These examples illustrate some of the problems with Isomap.

1. An intrinsic issue is that MDS presumes that M can be isometrically embedded in Euclidean space; if it is not, the procedure seriously distorts the coordinates [335]. As a consequence, M must be flat in the sense of having zero curvature. Moreover, Isomap performs poorly on non-convex but contractible subspaces of Euclidean space, e.g., a space in the shape of the letter "Y."
2. Given new data points, the Isomap embedding has to be recomputed; there is no way to adapt an existing embedding.
3. A further issue in practice is that the algorithm is not robust to outliers and is very sensitive to differences in density or the precise value of ϵ. (For example, see [373] for discussion of these points.)
4. Finally, efficiency can also be a problem, especially for large numbers of samples. These issues arise both from the substantial costs of computing the graph metric and from the size of the resulting MDS problem. Some efforts to approach this by subsampling have been studied, e.g., see [460] for a sparse version of Isomap.

4.3.2 Local Linear Embedding (LLE)

A closely related approach is the *local linear embedding* (LLE) algorithm [439]. Once again, we assume we have data points $\{x_1, \ldots, x_n\} \subset \mathbb{R}^n$ and we fix a target dimension parameter k and a neighborhood size K.

1. For each point x_i, we solve for weights w_{ij} which minimize the expression

$$\mathcal{E}(x_i) = \sum_i \left(x_i - \sum_j w_{ij} x_j \right)^2,$$

subject to the constraints

$$
\begin{cases}
w_{ij} = 0 & x_j \text{ not a } K\text{-nearest neighbor of } x_i \\
\sum_j w_{ij} = 1.
\end{cases}
$$

Roughly speaking, we are solving for weights that optimally reconstruct each point x_i from its K-nearest neighbors. The weights can efficiently computed via least squares.

2. Embedding points $\{y_i = \theta(x_i)\} \subseteq \mathbb{R}^k$ are computed so that

$$
\mathcal{E} = \sum_i \left(y_i - \sum_j w_{ij} y_j \right)^2
$$

is minimized. This problem can be solved by computing the top k eigenvectors of the matrix corresponding to the associated quadratic form, subject to some nondegeneracy constraints.

Broadly speaking, LLE has fairly similar qualitative properties as Isomap; this is illustrated in Figures 4.9, 4.10, and 4.11. In practice, it turns out to work somewhat better than Isomap on samples of non-convex contractible subsets $M \subseteq \mathbb{R}^k$ (e.g., regions with dents in them), and also has the advantage that the eigenvector problem involves a matrix that is always sparse, and hence it can be run on substantially larger data sets than Isomap.

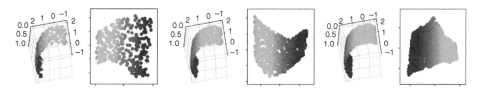

Figure 4.9 When the data lies on a curved ribbon, LLE does a good job of recovering the intrinsic coordinates and unfolding the ribbon.

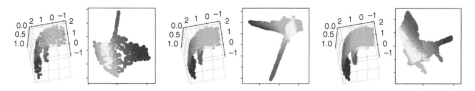

Figure 4.10 When the data lies on nested ribbons, LLE does a better job than PCA or Isomap but still engages in serious distortion of the intrinsic metric.

Figure 4.11 When the data lies on the standard S^2 in $\mathbb{R}^3$, LLE does not do a good job of capturing the intrinsic distances and simply embeds a flattening of the sphere.

4.3.3 Laplacian Eigenmaps

Isomap and LLE implicitly use the tangent plane of a manifold to perform local reconstruction of points. A more explicit use of the manifold structure is to try to exploit the existence of the Laplace-Beltrami operator, a map from functions on M to functions on M which is computed as the divergence of the gradient; on $\mathbb{R}^n$, this takes the classical form

$$\Delta f = \sum_{i=1}^{n} \frac{\partial^2 f}{\partial x_i^2}.$$

The first technique to take this approach is the *Laplacian eigenmaps* algorithm due to Belkin and Niyogi [45].

Once again, we assume we are given data points $\{x_1, \ldots, x_k\} \subseteq \mathbb{R}^n$ and we form a neighborhood graph that captures the "small" distances between points. Precisely, we fix a width parameter σ and proceed as follows.

1. Form the weighted graph G with
   - vertices in bijection with the points $\{x_i\}$, and
   - edges (i, j) with weight $w_{ij} = e^{-\frac{\partial(x_i, x_j)^2}{\sigma}}$ when $\partial(x_i, x_j) \leq \epsilon$.
2. We let D denote the diagonal matrix specified by $D_{ii} = \sum_j w_{ij}$ and define the *graph Laplacian* as $L = D - W$, where W is the matrix of edge weights from G.
3. We solve $Lf = \lambda Df$ for the top k eigenvectors, which determine the embedding; we form the matrix with columns these eigenvectors, and the rows are the embedded points $\{y_i\}$. This procedure can be viewed as finding a solution to the optimization problem of determining $\{y_i\}$ that minimize

$$\mathcal{E} = \sum_{i,j} (y_i - y_j)^2 W_{ij},$$

i.e., finding an embedding that penalizes nearby points x_i and x_j being sent to distant points y_i and y_j.

Here the basic technical underpinning is one of the fundamental insights of spectral graph theory, namely that the graph Laplacian we describe above shares many interesting properties with the Laplacian of a manifold [116]. Moreover, as a basic consistency check, when the points $\{x_i\}$ are sampled from a compact Riemannian manifold, as the number of points increase and σ decreases, the graph Laplacian converges in a precise sense to the Laplace-Beltrami operator on the manifold [46].

As with Isomap and LLE, Laplacian eigenmaps is expected to work best on convex subsets of $\mathbb{R}^n$; like LLE, the eigenvector problems involved tend to be sparse and so Laplacian eigenmaps can handle comparatively larger data sets. However, as Figures 4.12, 4.13, and 4.14 indicate, Laplacian eigenmaps has distinctly worse performance than either Isomap or LLE. We discuss the method due to its historical importance and conceptual clarity.

Remark 4.3.1. A refinement of the Laplacian eigenmaps algorithm, called Hessian eigenmaps [146], uses a discretized version of the Hessian matrix of second

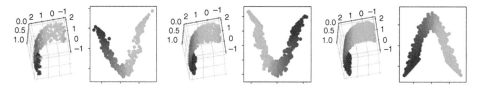

Figure 4.12 When the data lies on a single curved ribbon, Laplacian eigenmaps does not unfold the ribbon properly.

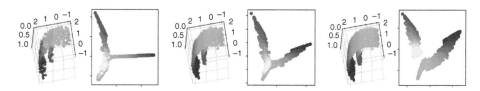

Figure 4.13 When the data lies on nested ribbons, Laplacian eigenmaps does in fact manage to recover some aspects of the relationship between the ribbons, although each one is compressed and distorted.

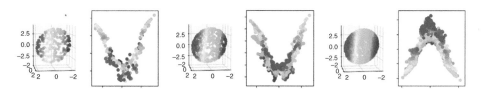

Figure 4.14 When the data lies on the standard S^2 in $\mathbb{R}^3$, Laplacian eigenmaps does not do a good job of capturing the intrinsic distances.

partial derivatives. The advantage of using the Hessian is improved theoretical guarantees: the Hessian eigenmaps algorithm can be shown to be asymptotically correct for arbitrary connected subsets of $\mathbb{R}^n$. Although in idealized situations Hessian eigenmaps outperforms other manifold learning algorithms, in practice it does not work well – estimating second derivatives is well known to be numerically unstable.

4.3.4 Manifold Learning and Kernel Methods

There is a basic resemblance between all of the manifold learning techniques described in the preceding subsections; at a high level, it appears that they are relying on similar geometric ideas. It turns out that this connection can be made precise using a standard body of techniques for handling nonlinearity in data analysis, *kernel methods*.

The basic idea is to choose a nonlinear embedding Ψ of the points $\{x_i\}$ into an infinite-dimensional inner-product space H, and use the inner product and norm on H to analyze the points. The idea is that Ψ will *unfold* the data so that linear techniques applied to $\Psi(x_i\}$ will reveal nonlinear structure in $\{x_i\}$. For example, *kernel PCA* performs PCA on the embedded points and is frequently applied in conjunction with clustering algorithms.

Although it is easy to see that for a suitable nonlinear embedding, such techniques would be very effective, a number of questions about how to implement this procedure arise.

1. It is not clear how easy it will be to produce a suitable map Ψ without possessing a priori knowledge of the data, and
2. it is not clear that working directly in the infinite-dimensional space H is algorithmically tractable.

The key insight that makes kernel methods effective is the observation that a wide variety of algorithms (including clustering and PCA) can be computed without explicit knowledge of Ψ or H provided one has access to a *kernel K*, which is a map

$$K : X \times X \rightarrow \mathbb{R}$$

such that

$$K(x_i, x_j) = \langle \Psi(x_i), \Psi(x_j) \rangle_H,$$

where the expression on the right denotes the inner product in H. This formulation reduces the question to producing a kernel K which is algorithmically tractable and also encodes geometric information about X. The construction of such kernels turns out to be much more tractable than producing Ψ.

Example 4.3.2.

1. A standard kernel is the *radial basis function* or Gaussian kernel

$$K(x_i, x_j) = e^{-\frac{\partial_X(x_i, x_j)^2}{2\sigma^2}}.$$

   This kernel is standard in classification and clustering applications.
2. On the set of trees (acyclic connected graphs), an interesting kernel is the *subtree kernel* which is defined as

$$K(t_i, t_j) = \#\{\text{isomorphic subtrees of } t_i \text{ and } t_j\}.$$

   This kernel is frequently used in analysis of linguistic and phylogenetic data. (See [121] for an early paper on kernel methods in natural language processing.)
3. On the set of strings, the (k, m) mismatch kernel is defined as

$$K(w_i, w_j) = \#\{\text{substrings } s_1, s_2 \text{ of } w_i \text{ and } w_j \text{ such that}$$
$$|s_1| = |s_2| = k \text{ and } s_1, s_2 \text{ agree up to } m \text{ mismatches}\}.$$

   This kernel has seen notable applications to protein matching [322].

It now turns out that many manifold learning approaches can be interpreted as kernel PCA; notably, we can represent Isomap, LLE, and Laplacian eigenmaps in this fashion [225]. For example, for a data set $\{x_i\}$, Isomap is (up to scaling) identical to kernel PCA for the kernel

$$K(x_i, x_j) = \left(-\frac{1}{2}(I - ee^T)D_2(I - ee^T)\right)_{ij},$$

where D_2 denotes the matrix of squared distances, and e denotes the vector with all entries 1.

4.3.5 Discrete Harmonic Analysis

As we have seen, many manifold learning algorithms essentially involve an approximation to the Laplace-Beltrami operator on a manifold via the graph Laplacian. That is, one way to think about the underlying mathematics of manifold learning is in terms of discrete approximations of heat flow on the underlying manifold. In classical physics, the Laplace operator Δ arises in the heat equation, which in its simplest form can be written

$$\frac{\partial f}{\partial u} = \Delta f.$$

The heat equation describes how the geometry of a manifold interacts with local temperature information to control the diffusion of heat over time.

Elaborating on this perspective, work of Jones, Coifman, Maggioni and collaborators has produced a large body of work on *discrete harmonic analysis* in terms of the heat kernel, the function that describes the infinitesimal flow. Two basic observations in the mathematical setting are that:

• the heat flow on the manifold describes the geometry of the manifold, and
• harmonic analysis (a generalization of Fourier analysis) in terms of the basis of powers of the heat kernel gives a good description of functions on the manifold.

The idea of discrete harmonic analysis is to analyze data using discretized approximations to the heat flow. Given the data represented as a finite metric space (X, ∂_X), we fix a rapidly decaying function $K(x_i, x_j): X \times X \to \mathbb{R}$. For example, a standard similarity measure is given by

$$K(x_i, x_j) = e^{-\frac{\partial_X(x_i, x_j)}{\sigma}},$$

where σ is a width parameter. We now proceed as follows.

1. Form the weighted graph G with vertex set in bijection with the points $x \in X$ and an edge (x_i, x_j) of weight $w_{ij} = K(x_i, x_j)$ provided that $K(x_i, x_j) > 0$.
2. Writing $D_{ii} = \sum_j w_{ij}$, define the (normalized) graph Laplacian to be

$$\mathcal{L} = D^{-\frac{1}{2}}(D - W)D^{-\frac{1}{2}},$$

where $W_{ij} = w_{ij}$. (Notice that there is a slight difference from the graph Laplacian used in Section 4.3.3.)

The discrete version of the heat flow is given by the random walk on the graph, sometimes referred to as the diffusion walk in this situation; this is a Markov process on the vertices where the vertex at time t is selected from the neighbors of the vertex at time $t - 1$ according to the edge weights. Precisely, the probability of moving to vertex k from vertex i is

$$p_{ki} = \frac{w_{ki}}{\sum_j w_{kj}}.$$

The random walk on the graph G above has transition probabilities determined by $D^{-1}W$. It is sometimes useful to make this walk symmetric (i.e., to ensure that $p_{ki} = p_{ik}$); in this case, the transition probabilities for the symmetrized walk are given by

$$T = I - \mathcal{L} = D^{-\frac{1}{2}}WD^{-\frac{1}{2}}.$$

Roughly speaking, the diffusion walk on the graph describes how a point mass at a given point in the graph spreads out over time as it diffuses according to the edge weights. In the limit as the number of points increases and σ decreases, the diffusion walk converges to the actual heat flow on the manifold described by the heat equation.

The key observation is now that harmonic analysis of T gives rise to geometric descriptions of the data set. To explain, the eigenfunctions of the powers T^t determine a metric on X; this is the so-called "diffusion distance" at scale t. Specifically, if we denote the eigenvectors of T (regarded as an L^2 operator) by $\{\Psi_k\}$ and the associated eigenvalues $\{\lambda_k\}$, the diffusion distance at scale t is given by

$$D_t(x,y)^2 = \sum_{k \geq 0} \lambda_k^{2t}(\Psi_k(x) - \Psi_k(y))^2.$$

Since T and its powers describe an ergodic Markov process, the eigenvalues λ_i satisfy $|\lambda_0| = 1$ and

$$|\lambda_0| \geq |\lambda_1| \geq |\lambda_2| \geq \ldots .$$

Truncating the expression for D_t by removing eigenfunctions corresponding to eigenvalues smaller than some threshold ϵ provides a tractable approximation. Roughly speaking, the diffusion distance between two points is a measure of how connected the points are; i.e., it reflects the probability of moving from x to y in the diffusion walk.

Moreover, we can use eigenvectors of T to embed the data $\{x_i\}$ in $\mathbb{R}^k$ so as to optimally preserve the diffusion distance; just as in the manifold learning algorithms described above, we put the scaled eigenvectors $\lambda_t^i \Psi_i$ as the columns of a matrix and take the rows to compute the embedding. This embedding is such that the Euclidean distance between the embedded points is close to the diffusion distance on G.

Furthermore, wavelet bases for functions on the data can be constructed using powers of T; this gives a geometric basis for representing functions. The diffusion process can also be used to smooth data before applying machine learning algorithms. See [119] for the original paper; more generally, Maggioni's research group has an extensive bibliography.

4.3.6 Other Manifold Learning Techniques

We have chosen to highlight manifold learning techniques that are conceptually significant and of historical importance; however, this has subsequently become a very active area of research. There are now many other techniques, each with slightly different virtues, insights, and limitations. As a few examples, some interesting methods include the following.

1. *Local tangent space alignment* [519], which uses the nearest neighbors of a point to estimate the tangent space locally and then performs a global optimization to align these compatibly.
2. *Maximum variance unfolding* [532], which forms a neighborhood graph that maximizes distance between far-away points by solving a semidefinite programming (SDP) problem and then computes eigenvectors, and uses these as the basis for an embedding.
3. *Manifold charting* [69], which solves for local neighborhood coordinate patches for each point and then sews them together using a global optimization process.

4.3.7 Manifolds of Differing Dimension

An obvious extension of the setup for manifold learning is the case where the data is generated from the union of manifolds of differing dimension. This situation is the simplest case of the general problem of "stratified space learning" (see Example 1.11.9). Of course, ad hoc adaptations of manifold learning techniques could be used: for example, cluster points by some estimate of local dimensionality (so that points in a cluster come from a subset of roughly constant dimension) and then apply manifold learning techniques to each cluster separately.

However, it is reasonable to expect that more systematic approaches would be superior. So far, there have been two main settings studied.

1. The data is assumed to lie on the union of hyperplanes of different dimensions; i.e., $M = \cup_i \mathbb{R}^{n_i}$, where each $\mathbb{R}^{n_i}$ is presented as embedded in an ambient space $\mathbb{R}^m$ via a linear map $\gamma_i \colon \mathbb{R}^{n_i} \to \mathbb{R}^m$ [481].
2. The data is assumed to lie on a metric graph. Recall from Example 1.11.9 that these are stratified spaces with a zero dimensional stratum for the vertices and a one dimensional stratum for the edges. We discuss the special case of trees further below in Section 4.7.1. See for example [106] for an approach to general graphs using Reeb graphs.

(Although note that [48] studied the problem of clustering points into different strata, using estimates of local homology.)

4.4 Neighbor Embedding Algorithms

In this section, we discuss a different approach to dimensionality reduction, *stochastic neighbor embedding* (SNE) [242] and its more popular descendant *t*-distributed stochastic neighbor embedding (*t*-SNE) [336]. One issue with many of the manifold learning algorithms we have discussed so far is that they do not work well when the data points are of non-uniform density. The stochastic neighbor embedding algorithms address this by constructing a similarity measure between points that reflects the local density around each point. They are much more explicitly probabilistic (and less geometric) in their design.

4.4.1 Stochastic neighbor Embedding (SNE)

Let $\{x_1, \ldots, x_m\} \subset \mathbb{R}^n$ denote the data and $\{y_1, \ldots, y_m\} \subset \mathbb{R}^k$ denote a candidate set of corresponding image points, for $k \leq n$. Then we define

$$p_{j|i} = \frac{e^{-\frac{\partial_{\mathbb{R}^n}(x_i, x_j)^2}{2\sigma_i^2}}}{\sum_{k \neq i} e^{-\frac{\partial_{\mathbb{R}^n}(x_i, x_k)^2}{2\sigma_i^2}}}$$

and

$$q_{j|i} = \frac{e^{-\partial_{\mathbb{R}^n}(y_i, y_j)^2}}{\sum_{k \neq i} e^{-\partial_{\mathbb{R}^n}(y_i, y_k)^2}},$$

where the variances σ_i are obtained by an optimization process we will describe shortly and in the second equation we are fixing all of the variances to be identically $\frac{\sqrt{2}}{2}$. We set $p_{i|i} = q_{i|i} = 0$.

The idea behind SNE [242] is that good image points $\{y_1, \ldots, y_n\}$ have the property that the difference between $p_{j|i}$ and $q_{j|i}$ is minimized, in the sense that we minimize the summed Kullback-Leibler divergences via the cost function

$$C = \sum_i \sum_j p_{j|i} \log \frac{p_{j|i}}{q_{j|i}}.$$

(See Remark 3.2.32 for discussion of the Kullback-Leibler divergence as a dissimilarity measure on probability distributions.) Notice that ensuring these local distributions are similar means that SNE is sensitive to variation in density; the density around a point is explicitly represented in the cost function.

Remark 4.4.1. Recall that the Kullback-Leibler divergence is a dissimilarity measure but not a metric: it is not symmetric. In the context of SNE, this asymmetry is a regarded as a feature – it serves to enforce a preference for preserving local distances.

In practice, the SNE algorithm proceeds by solving for minimizing points $\{y_i\}$ via a gradient descent procedure (often with diminishing amounts of noise added as a form of simulated annealing) in order to find a good local minimum for C. Convergence is often slow and depends critically on good choices of the variances σ_i, which we now explain how to obtain. In principle, differing choices of σ_i

amount to enforcing variable numbers of neighbors used to do the local estimation of the coordinates; this is expressed here via the use of the "perplexity," which is computed as

$$P = 2^{-\sum_j p_{j|i} \log_2 p_{j|i}}.$$

Roughly speaking, P controls the number of effective neighbors that are used; recall that this is a loose estimate of the local dimension. The desired perplexity (typically in the interval $[10, 100]$) is a parameter, and we solve for values σ_i which achieve the perplexity. The output of the algorithm can be fairly sensitive to the choice of perplexity value.

4.4.2 t-*Distributed Stochastic Neighbor Embedding (t-SNE)*

In practice, a refinement of stochastic neighbor embedding is commonly used. One of the problems with the original SNE procedure is that the gradient descent optimization procedure is slow and it can be difficult to get it to converge. Another problem afflicts the standard application of SNE to visualization. Specifically, in order to use SNE to visualize data, one solves for embedded points $\{y_i\}$ in $\mathbb{R}^2$ or $\mathbb{R}^3$. When there are even a moderately large number of points, the cost function can cause compression of the points so that they all lie very close to the center of mass, which makes the visualization hard to use. (This compression also defeats clustering algorithms.)

To resolve these problems, van der Maaten and Hinton [336] proposed the variant algorithm t-SNE. This is quite similar to SNE, with the following modifications.

1. We symmetrize $p_{j|i}$ as follows:

$$p_{ij} = \frac{p_{j|i} + p_{i|j}}{2}.$$

2. We define a symmetrized variant of $q_{j|i}$ as follows:

$$q_{ij} = \frac{\left(1 + \partial_{\mathbb{R}^m}(y_i, y_j)^2\right)^{-1}}{\sum_{k \neq \ell}\left(1 + \partial_{\mathbb{R}^m}(y_k, y_\ell)^2\right)^{-1}}.$$

In the original definition, the $q_{j|i}$ was defined using a Gaussian; this expression replaces that with the Student t-distribution with one degree of freedom (i.e., a Cauchy distribution), which has more weight in the tails.

3. Finally, the cost function is replaced with the alternative expression

$$C = \sum_i \sum_j p_{ij} \log \frac{p_{ij}}{q_{ij}}.$$

Once again, the cost function is optimized via a gradient descent procedure. These modifications have a number of interesting consequences.

1. The use of a heavier tailed distribution in the embedding space means that outliers have less impact on the overall results and the compression effects around the center of mass are alleviated to some degree (although there are still upper bounds on the number of points that can reasonably be embedded before "clumping" occurs).
2. The adjusted formula for q_{ij} also has the effect of substantially improving the efficiency and quality of the gradient descent procedure.

Figures 4.15, 4.16, and 4.17 show that the *t*-SNE procedure can produce very reasonable embeddings recovering local geometry; in particular, *t*-SNE arguably performs best of the methods we have examined on the nested ribbons.

However, caution is required when interpreting the results of *t*-SNE. In contrast to other dimensionality reduction methods, *t*-SNE does not directly depend on a discretization of the Laplace-Beltrami operator or approximation of the local tangent

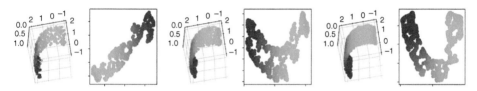

Figure 4.15 Especially at lower densities, *t*-SNE unfolds the ribbon to recover its intrinsic coordinates.

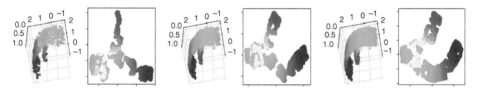

Figure 4.16 When the data lies on the nested arcs, *t*-SNE actually does a reasonable job at recovering the intrinsic coordinates.

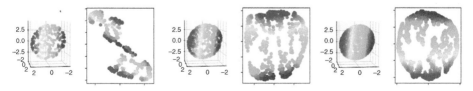

Figure 4.17 When the data lies on the standard S^2 in $\mathbb{R}^3$, *t*-SNE does not do a good job of capturing the intrinsic distances and instead flattens the sphere.

planes. Put another way, it is more sensitive to different properties of the underlying geometric object than other manifold learning methods. On the one hand, this flexibility can be a real asset when working with data that does satisfy the manifold hypothesis. On the other hand, the geometric properties of $\{y_i\}$ can be difficult to relate to the geometric properties of $\{x_i\}$. For example, when clustering points after computing the t-SNE embedding in $\mathbb{R}^2$, inter-cluster distances do not reflect global properties faithfully, and relative sizes of clusters are usually meaningless. Moreover, there are no geometric theoretical guarantees about the ideal behavior of t-SNE. We now turn to discuss an application highlighting best practice in using t-SNE.

4.4.3 Reliable Use of t-SNE

A common usage pattern for t-SNE is to project into $\mathbb{R}^2$ and then apply a standard clustering method. This application of the algorithm has had some impressive successes, as a method which adjusts for local density can reveal clusterings which would not be evident using methods which impose a global constraint. A celebrated application is the viSNE procedure [13], a tailored use of t-SNE, which has been used for visualizing and classifying single-cell expression data, notably to distinguish healthy and cancerous bone marrow samples. We highlight the protocol here as it provides an exemplary case study of how to robustly apply dimensionality reduction.

The basic approach of viSNE applies t-SNE to embed gene expression data collected from single-cell bone marrow samples, regarded as vectors in $\mathbb{R}^n$ with the correlation metric, into $\mathbb{R}^2$ and then performs clustering on the embedded data. The overall conclusion is that in this embedding, healthy bone marrow cells are close together across samples and quite far from cancerous samples. This conclusion was carefully validated.

1. The stability of each clustering was tested via standard cross-validation; some data points were removed and deviation in the clusters was measured.
2. In order to handle the limits on the total numbers of points for embeddings in $\mathbb{R}^2$, the algorithm was run repeatedly on subsamples from the data. The clusters were compared to ensure that the analysis was robust to this subsampling.
3. To demonstrate more global stability, samples from different normal patients were compared (and observed to be extremely similar in terms of the resulting clusterings).
4. To ensure that the results were not artifacts of experimental procedure, different experimental methods for obtaining the expression data were compared by contrasting the resulting clusters.

The procedures outlined above give very good confidence that the results of the viSNE procedure are capturing real geometric information about the expression data of bone marrow. In general, this example is a good model for an analytical protocol for topological data analysis.

4.5 Mapper and Manifold Learning

In principle, one could use the coordinate charts provided by manifold learning algorithms for all sorts of geometric inference about the data. In practice, dimensionality reduction procedures are most commonly used as a preprocessing step before applying some kind of clustering algorithm. From the perspective of TDA, a very interesting extension of this approach is to use the output of manifold learning algorithms as filters for Mapper.

Recall from Example 2.8.3 that using PCA coordinates of expression data as a filter function for Mapper captured cell differentiation trajectories [431]. A reasonable question to ask is what Mapper adds over standard manifold learning; to answer this, we can directly compare the output of Mapper to the output of various manifold learning procedures. In Figure 4.18, we represent both the output of Mapper and the raw outputs of PCA, MDS, and t-SNE for the differentiation process.

The results are informative.

1. Mapper is only very slightly better than MDS and t-SNE for estimating a cell's position along the differentiation trajectory.
2. However, the graphical representation of Mapper contains additional informative structure; the loops and flares in the resulting Mapper graph are biologically relevant.

In general, we expect that this kind of fusion of topological data analysis and dimensionality reduction will provide a useful technique for describing the structure of genomic data.

4.6 Dimensionality Estimation

Recall that the basic operating assumption in dimensionality reduction is that the data points $\{x_i\} \subseteq \mathbb{R}^n$ lie on a geometric object that has much lower intrinsic dimension k than the ambient space. As such, a natural problem to consider is whether we can directly determine the dimension of $\{x_i\}$ without actually computing a description of the lower dimensional object.

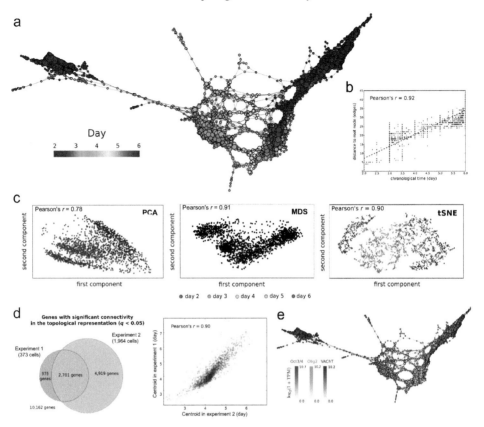

Figure 4.18 The differentiation timeline can be extracted from various manifold learning techniques. Source: [431]. From Abbas H. Rizvi et al., *Nature Biotechnology* 35, 551-560 (270). © 2017 Nature. Reprinted with Permission from Springer Nature.

There are several reasons why we might want to have efficient approaches to this problem. Although the very idea of intrinsic dimension presupposes the data has enough geometric structure to give rise to a notion of dimension, there are many classes of objects that have a good definition of dimension which are not manifolds (e.g., fractals). Moreover, estimating the intrinsic dimension can provide a sense of how good low-dimensional summaries can be; for example, projecting a data set with intrinsic dimension 4 into $\mathbb{R}^2$ will typically result in much less distortion than projecting a data set with intrinsic dimension 50 into $\mathbb{R}^2$. Finally, understanding the intrinsic dimension gives us a sense of the number of sample points required to accurately estimate geometric features of the underlying object.

There are a number of ways to try to directly estimate the dimension of the points $\{x_i\}$; almost all of them consider the rate of growth of the number of points within an ϵ-ball as ϵ increases. For example, the correlation dimension [205] of a set of points $\{x_1, \ldots, x_n\}$ is computed by considering the curve produced by plotting

$$(x, y) = \left(\log \epsilon, \frac{2}{n(n-1)}\theta_\epsilon\right),$$

where θ_ϵ is the count of the number of pairs (x_i, x_j) such that $\partial(x_i, x_j) \le \epsilon$, and performing a regression to estimate the slope of this curve. The correlation dimension has proven useful in handling estimation of dimension for geometric objects like fractals that are not well described by more classical measures of dimension.

When working under the manifold learning assumptions, i.e., that the data is given as points $\{x_i\} \subset \mathbb{R}^n$ sampled from a k-dimensional manifold, a more geometric version of this idea can be applied. The basic observation is that when points are sampled from a density ρ in $\mathbb{R}^k$, the number of points expected in a ball of radius ϵ centered around z is approximately $\rho(z)$ times the volume of the ball. Therefore, empirical estimates of the rate of growth of the count of sample points in Euclidean balls of expanding radius can be used to estimate dimension. A very clean form of this approach is given by the maximum-likelihood estimator of Levina and Bickel [327], which assumes a Poisson distribution for the data and is given at the point x by the formula

$$\left(\frac{1}{N(R, x)} \sum_{j=1}^{N(R,x)} \log \frac{R}{T_j(x)}\right)^{-1},$$

where $N(R, x)$ is the number of points in the ball of radius R around x and $T_j(x)$ is the distance from x to the jth point in this ball. To compute a global estimate, we can average the likelihood estimators.

Remark 4.6.1. MacKay and Ghahramani observe that the estimator above has substantial bias (even for low dimensions) which can be corrected by a slightly different approach to combining the points, namely again using maximum likelihood estimation. This amounts to computing the following:

$$m^{-1} = \frac{\sum_{i=1}^{n} \sum_{j=1}^{N(R,x_i)} \log \frac{R}{T_j(x_i)}}{\sum_{i=1}^{n} N(R, x_i)}.$$

Notice that this is very similar to the original approach; here we are just averaging the inverses. Numerical experiments suggest this correction reduces both bias and variance.

Another interesting approach in the manifold setting, due to Little et al. [332], combines local PCA estimates of the dimension at various scales. The idea is around each data point z to choose k-nearest neighbors and perform PCA on the vectors determined by the pairs of z and a nearest neighbor to obtain a local estimation of the tangent plane. More precisely, we perform the following algorithm.

1. For each point z, we compute the eigenvalues $\lambda_1(r), \ldots, \lambda_K(r)$ of the covariance matrix

$$C = \frac{1}{n} \sum_i x_i x_i^T,$$

restricted to the data points $x_i \in B_r(z)$, as r varies. (We assume that the data has been centered.)

2. Ideally, we will see the magnitudes of the eigenvalues cluster into two groups, i.e., there will be a substantial eigenvalue gap over a broad range of values of r. We regard the smallest ones as noise. We then restrict attention to the region of the eigenvalue curves (i.e., plots of the values of the kth eigenvalue as a function of r) where the rate of growth of the noise eigenvalues is flat.

3. Of the remaining eigenvalues, we separate those having linear growth as a function of r from those having quadratic growth via a regression. The linear growth eigenvalues are regarded as corresponding to eigenvectors in the direction of the local tangent plane, whereas the quadratic growth eigenvalues are coming from eigenvectors in the direction of the local curvature of the manifold. The number of linear growth eigenvalues provides a local dimension estimate at z.

4. Finally, we average the dimension estimates over all points z.

4.7 Metric Trees and Spaces of Phylogenetic Trees

In this section we explain an approach to the analysis of phylogenetic trees based on endowing sets of trees with geometric structure. The importance of phylogenetic tree structures in biological sciences cannot be overstated; their use begins with Darwin's proposal of the tree as a metaphor for the process of species generation through branching of ancestral lineages [133]. Since then, tree structures have become ubiquitous in biology for describing evolutionary relations: notably, clonal evolution events that start from asexual reproduction of a single organism (primordial clone), which mutates and differentiates into a large progeny (see the left panel of Figure 4.19) [293]. Examples of these processes include single gene phylogeny in non-recombinant viruses, bacteria that are not involved in horizontal gene transfer events, and metazoon development from a single germ cell.

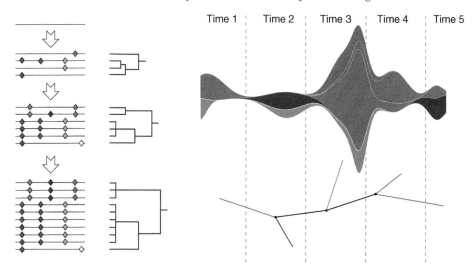

Figure 4.19 Clonal evolution of an asexually reproducing organism. Left: Through acquiring mutations and differentiation, the primordial clonal organism gives rise to a large heterogeneous population, whose evolution can be described with tree-like structures; here horizontal lines represent organisms, and symbols on the line represent mutations. Right: Longitudinal sampling of a clonal population permits the construction of phylogenetic trees that describe its evolutionary history. Here, subpopulations are represented by different colors; subsampling of a particular subpopulation is illustrated by the color of the branch in the tree, one of the many trees that can be reconstructed from this population. Source: [545]. From Zairis et al., Genomic data analysis in tree spaces, arXiv: 1607.07503 [q-bio.GN].

There are a number of basic mathematical and algorithmic questions that arise in this context.

1. Given genomic data, how can we fit a "best" phylogenetic tree to this data that optimally encodes the evolutionary relationships in the data?
2. Given two trees, how can we assess quantitatively how different they are? Given a collection of trees, can we compute a "summary" or average tree?
3. More generally, how can we describe probability distributions on trees? (For example, more sophisticated output of algorithms to answer the first question might provide a distribution of trees.)

There is a tremendous body of work on the first question; in the first part of this section, we focus on a purely metric method (neighbor-joining) that takes as input a finite metric space and produces a corresponding metric tree. We give a rapid but more comprehensive treatment of tree inference algorithms in Appendix C.

In the remainder of the section, we assume that our raw data has been turned into phylogenetic trees, and describe an approach to the second and third questions based on using the *metric geometry* (Section 4.7.3) associated to a specific metric on the set of phylogenetic trees.

4.7.1 Inferring Trees from Metric Data

One way to formulate the problem of inferring a phylogenetic tree structure from genomic data is to regard the data as a finite metric space (X, ∂_X) and postulate that the metric ∂_X is a *tree metric*, i.e., the points correspond to the leaves of a tree and the distance corresponds to the length of the shortest path in the graph. (Recall the discussion of graph metrics from Example 1.2.4.)

Definition 4.7.1. A *phylogenetic tree* with m leaves is a weighted, connected graph with no circuits, having m distinguished vertices of degree 1 labeled $\{1, \dots, m\}$ (referred to as *leaves*), and all other vertices of degree ≥ 3.

We refer to edges that terminate in leaves as *external* edges and the remaining edges as *internal*. We will use the term tree metric to refer to the metric induced on the leaves from the graph metric of the phylogenetic tree.

Of course, not every metric arises from a tree metric. Specifically, given a metric space (X, ∂_X), the metric ∂_X is a tree metric if and only if it satisfies the *four point condition* [82].

Lemma 4.7.2. *A metric space (X, ∂_X) is isometric to a tree metric space if and only if for any $u, v, w, x \in X$, two of the three sums*

$$\partial_X(u, v) + \partial_X(w, x), \qquad \partial_X(u, w) + \partial_X(v, x), \qquad \partial_X(u, x) + \partial_X(v, w)$$

are equal and greater than the third.

But although this can be used as a test, it does not provide an algorithm for producing a tree. A good solution to the tree inference problem then ideally has the following properties.

1. When ∂_X really is the metric corresponding to a tree metric, the algorithm recovers a tree such that the associated metric is isometric to the input.
2. When ∂_X is "close" to a tree metric in a suitable sense, the algorithm recovers a tree T such that the associated tree metric is close to ∂_X.

An influential method to do this is *neighbor-joining* [442], which recursively constructs the output tree by selecting a pair of points, joining them as leaves coming out from an internal vertex, and then repeating the process with the new vertex regarded as a leaf and the joined points removed, until all of the points are part of the tree. More precisely, the algorithm works as follows. We assume we are given as input a finite metric space (X, ∂_X) such that $|X| = n$.

1. We initialize the output tree T to have a vertex for each point of X, each connected to a central root and with no other edges.
2. Calculate the function

$$Q(x_i, x_j) = \partial_X(x_i, x_j) - \frac{1}{n-2} \left(\sum_{k \neq i} \partial_X(x_i, x_k) + \sum_{k \neq j} \partial_X(x_j, x_k) \right).$$

3. Find the points x_i and x_j that minimize $Q(x_i, x_j)$.
4. Define a new point z, and form T' from T by adding edges from x_i and x_j to z, deleting the edges from x_i and x_j to the root, and connecting z to the root. Define the edge weights as

$$w_{x_i, z} = \frac{1}{2} \left(\partial_X(x_i, x_j) + \frac{1}{n-2} \left(\sum_{k \neq i} \partial_X(x_i, x_k) - \sum_{k \neq j} \partial_X(x_j, x_k) \right) \right)$$

and

$$w_{x_j, z} = \frac{1}{2} \left(\partial_X(x_i, x_j) + \frac{1}{n-2} \left(\sum_{k \neq j} \partial_X(x_j, x_k) - \sum_{k \neq i} \partial_X(x_i, x_k) \right) \right).$$

5. Form the discrete metric space $X' = X - \{x_i, x_j\} \cup \{z\}$, with

$$\partial_{X'}(x_k, z) = \frac{1}{2} (\partial_X(x_i, x_k) + \partial_X(x_j, x_k) - \partial_X(x_i, x_j)).$$

6. If X' consists only of $\{z\}$, terminate and return T'. Otherwise, return to step 2 with T' and X' in place of T and X.

First, the algorithm is sound, in the sense that when the metric ∂_X actually is a tree metric, the neighbor-joining algorithm recovers the tree. More interestingly, it is fairly robust to noise; notice that neighbor-joining does not really require a metric space as input, as the triangle inequality is never used. We have the following consistency result [23].

Theorem 4.7.3. *Let (X, ∂_X) be a tree metric space and $D \colon X \times X \to \mathbb{R}$ a function satisfying*

$$|D(x, y) - \partial_X(x, y)| \leq \frac{1}{2} \min_{\substack{x_1, x_2 \in X \\ x_1 \neq x_2}} \partial_X(x_1, x_2)$$

for all $x, y \in X$. Then neighbor-joining applied to D returns a metric space isometric to (X, ∂_X).

Neighbor-joining turns out to work surprisingly well in practice; a theoretical justification for this is given in [351]. Nonetheless, there are no global guarantees about the behavior of the procedure for metric spaces far from trees; for instance, in some cases neighbor-joining can produce negative edge lengths or exhibit other perverse behavior.

Remark 4.7.4. A natural question to ask is how to determine whether a metric space is far from being tree-like. One measure of the divergence from being a tree metric is given by Gromov's δ-*hyperbolicity*, which is a relaxation of the four-point condition.

Persistent homology also gives an interesting approach to detecting whether a metric space is a tree: metric trees are contractible and should have no homology. Therefore computing PH_k for any $k > 0$ yields information about divergence from being tree-like. We discuss applications of this idea to population genetics in Section 5.2.

4.7.2 The Billera-Holmes-Vogtmann Metric Spaces of Phylogenetic Trees

For many applications, it would be very desirable to have a metric on the set of phylogenetic trees. A distance function would permit quantitative comparisons. It would also allow one to apply clustering algorithms to collections of trees (e.g., produced from samples from distinct patients). A metric would also provide some of the foundations for dealing with probability distributions on phylogenetic trees as well as summary statistics.

Billera-Holmes-Vogtmann (BHV) constructed a metric on weighted phylogenetic trees using the tools of metric geometry [55]. They defined a metric space BHV_m of isometry classes of rooted phylogenetic trees with m labeled leaves where the non-zero weights are on the internal branches. The space BHV_m is constructed by gluing together (recall Section 1.5) $(2m-3)!!$ positive orthants $\mathbb{R}^n_{\geq 0}$ of dimension $n = m - 2$, where

$$\mathbb{R}^n_{\geq 0} = \{(x_1, x_2, \ldots, x_n) \in \mathbb{R}^n \mid x_1 \geq 0, x_2 \geq 0, \ldots, x_n \geq 0\};$$

each orthant corresponds to a particular tree shape, with the coordinates specifying the lengths of the internal edges. A point in the interior of an orthant represents a binary tree; if any of the coordinates are 0, the tree is obtained from a binary tree by collapsing the internal edges with length 0. We glue orthants together such that a (non-binary) tree is on the boundary between two orthants when it can be obtained by collapsing edges from either tree geometry. Put another way, orthants corresponding to two tree topologies are adjacent when they are connected by a *rotation*, i.e., one topology can be generated from the other by collapsing an edge to length 0 and then expanding out another edge from the incident vertex.

The metric on BHV_m is induced from the standard Euclidean distance on each of the orthants.

1. For two trees t_1 and t_2 which are both in a given orthant, the distance $d_{\mathrm{BHV}_m}(t_1, t_2)$ is defined to be the Euclidean distance between the points specified by the weights on the edges.
2. For two trees which are in different orthants, there exist (many) paths connecting them which consist of a finite number of straight lines in each orthant. The length of such a path is the sum of the lengths of these lines, and the distance $d_{\mathrm{BHV}_m}(t_1, t_2)$ is then the minimum length over all such paths.

For many points, the shortest path goes through the "cone point," the star tree in which all internal edges are zero. See Figure 4.20 for a picture of tree space.

As explained in [55, §4.2], efficiently computing the metric on BHV_m is a non-trivial problem. However, there exists a polynomial-time algorithm based on successive approximation of geodesic paths [394].

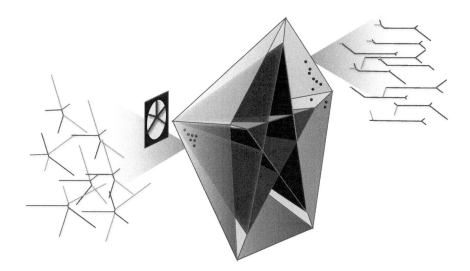

Figure 4.20 Moduli space of phylogenetic trees describing clonal evolution. Collections of trees can be mapped onto a geometric space, forming a point cloud. Trees with the same topology will live in the same orthant, and crossing into an adjacent orthant corresponds to a tree rotation (collapsing an edge to 0 and expanding out a new edge). Points closer to the vertex of the cone have relatively little internal branch length, while points near the base of the cone have little weight in the external branches. Source: [545]. From Zairis et al., Genomic data analysis in tree spaces, arXiv: 1607.07503 [q-bio.GN].

The main result of Billera, Holmes, and Vogtmann is that the length metric on BHV_n endows this space with a (global) CAT(0) structure (see Definition 4.7.9 below). The fact that BHV_n is a CAT(0) space means that points are connected by unique geodesics (which realize the distance between them) and there are unique centroids. As a consequence, it is reasonable to consider geometric inference in this setting. Moreover, BHV_n is clearly a complete metric space and is separable, which means that it contains a countable dense subset; any tree can be approximated by a sequence of trees in the same orthant that have rational edge lengths. That is, BHV_n is a Polish space and so as discussed in Section 3.3 is a reasonable space on which to apply the standard machinery of probability theory (see [247, 248] for work in this direction). In some applications it is also useful to consider a *projectivized* variant of the tree space where the internal edges are constrained to have lengths that sum to 1. We denote this subspace of Σ_n by $\mathbb{P}\Sigma_n$ and refer to it as the projective tree space.

Remark 4.7.5. The space of phylogenetic trees turns out to appear in various other contexts in mathematics; for instance, it is closely related to the moduli space of algebraic curves [141]. Perhaps more relevantly, it appears in the context of Diaconis and Sturmfels' algebraic statistics [145] as a tropical Grassmannian [479] (see also [395]).

4.7.3 Metric Geometry

Although metric spaces often arise in contexts in which there is no evident notion of geometry, it turns out that under very mild hypotheses a metric space (X, ∂_X) can be endowed with structures analogous to those arising on Riemannian manifolds. See [73, 83] for a comprehensive treatment of metric geometry. The basic approach to this involves the notion of *length* of a path in a metric space.

Definition 4.7.6. Let (X, ∂_X) be a metric space. Let $I \subset \mathbb{R}$ be an interval $[a, b]$. The *length* of a path $\gamma: I \to X$ is

$$L(\gamma) = \sup \sum_{i=1}^{n} \partial_X(\gamma(x_i), \gamma(x_{i-1})),$$

where the sup is taken over all collections $a = x_0 < x_1 < \ldots < x_{n-1} < x_n = b$.

We can define a potential distance function on (X, ∂_X) by defining the distance between x and y to be

$$\inf_{\gamma} L(\gamma),$$

where the infimum is taken over all $\gamma : [0, 1] \rightarrow X$ such that $\gamma(0) = x$ and $\gamma(1) = y$. A metric space is a *length space* if this distance agrees with the metric. When the infimum can be achieved, we have the notion of a geodesic metric space.

Definition 4.7.7. A metric space (M, ∂_M) is a *geodesic metric space* if any two points x and y can be joined by a path with length precisely $\partial_M(x, y)$.

Any Riemannian manifold is a geodesic metric space. But more generally, a good notion of *curvature* makes sense in any geodesic metric space [11]. The idea is that the curvature of a space can be detected by considering the behavior of the area of triangles, and triangles can be defined in any geodesic metric spaces. Specifically, given points p, q, r, we have the triangle $T = [p, q, r]$ with edges the paths that realize the distances $\partial_M(p, q)$, $\partial_M(p, r)$, and $\partial_M(q, r)$. The connection between curvature and area of triangles is revealed by the observation that given side lengths $(\ell_1, \ell_2, \ell_3) \subset \mathbb{R}^3$, a triangle with these side lengths on the surface of the Earth is "fatter" than the corresponding triangle on a Euclidean plane. To be precise, we consider the distance from a vertex of the triangle to a point p on the opposite side – in a fat triangle, this distance will be larger than in the corresponding Euclidean triangle. (Thin triangles are defined analogously.) See Figure 4.21 for examples of thin and fat triangles.

Given a triangle $T = [p, q, r]$ in (M, ∂_M), we can find a corresponding triangle $\tilde{T}$ in Euclidean space with the same edge lengths. Given a point z on the edge $[p, q]$, a comparison point in $\tilde{T}$ is a point $\tilde{z}$ on the corresponding edge $[\tilde{p}, \tilde{q}]$ such that $\partial_{\mathbb{R}^2}(\tilde{z}, \tilde{p}) = \partial_M(p, z)$.

Definition 4.7.8. Let (M, ∂_M) be a metric space. We say that a triangle T in M satisfies the CAT(0) inequality if for every pair of points x and y in T and comparison points $\tilde{x}$ and $\tilde{y}$ on $\tilde{T}$, we have $\partial_M(x, y) \leq \partial_{\mathbb{R}^2}(\tilde{x}, \tilde{y})$.

Definition 4.7.9. If every triangle in M satisfies the CAT(0) inequality then we say that M is a *CAT(0) space*.

Figure 4.21 Thin triangles have angles that add up to less than 180 degrees; fat triangles have angles that add up to more. We can detect the curvature of the Earth by observing that big triangles are fat.

More generally, let M_κ denote a complete and simply connected two dimensional Riemannian manifold with curvature κ; the classification results discussed above show that there is a unique such manifold up to homeomorphism. The diameter of M_κ will be denoted D_κ. A D_κ-geodesic metric space is one in which all pairs of points p and q such that $\partial_M(p,q) < D_\kappa$ are connected by a geodesic.

Definition 4.7.10.   A D_κ-geodesic metric space M is a *CAT(κ) space* if every triangle in M with perimeter $\leq 2D_\kappa$ satisfies the inequality above for the corresponding comparison triangle in M_κ.

Clearly, if $\kappa' \leq \kappa$, any CAT(κ') space is also CAT(κ). More importantly, this notion coincides with standard ideas about curvature in geometric examples: An n-dimensional Riemannian manifold M that is sufficiently smooth has sectional curvature $\leq \kappa$ if and only if M is CAT(κ). For instance, Euclidean spaces are CAT(0), unit spheres are CAT(1), and hyperbolic spaces are CAT(-1).
  As described, CAT(κ) is a global condition.

Definition 4.7.11.   A metric space (X, ∂_X) is *locally CAT(κ)* if for every x there exists a radius r_x such that $B_{r_x}(x) \subseteq X$ is CAT(κ).

Example 4.7.12.   For example, the flat torus (formed by taking the box $[0, 1] \times [0, 1]$ and gluing together the edges $\{0\} \times [0, 1]$ to $\{1\} \times [0, 1]$ to make a cylinder and the edges $[0, 1] \times \{0\}$ to $[0, 1] \times \{1\}$ to make a torus) is locally CAT(0) but not globally CAT(0).

Theorem 4.7.13 (Cartan-Hadamard).   *A simply connected metric space that is locally CAT(0) is also globally CAT(0).*

A remarkably productive observation of Gromov is that many geometric properties of Riemannian manifolds are shared by CAT(κ) spaces. In particular, CAT(κ) spaces with $\kappa \leq 0$ (referred to as *non-positively curved metric spaces*):

1. admit unique shortest paths joining each pair of points x and y,
2. have the property that all balls $B_\epsilon(x)$ are convex and contractible for all x and $\epsilon \geq 0$, and
3. have stable midpoints of shortest paths.

It is in general very difficult to determine for an arbitrary metric space whether it is CAT(κ) for any given κ. Even for finite simplicial complexes where the metric is induced from the Euclidean metric on each face, this problem does not have a general solution.

4.8 Summary

- A standard approach in data analysis is to search for low-dimensional structure in high-dimensional data points using the geometry encoded in the interpoint distances.
- Principal component analysis (PCA) takes a finite set of points in $\mathbb{R}^n$ (with the assumption that these points admit an isometric embedding as a plane) and seeks to find an optimal linear projection of the data into $\mathbb{R}^k$ for $k < n$.
- Metric dimensionality scaling (MDS) is another classical method which determines an optimal embedding of a finite metric space (X, ∂_X) into a Euclidean space. MDS is similar to PCA, but can be applied to arbitrary metric spaces.
- Isomap and local linear embedding (LLE) are two related algorithms that apply MDS to empirical approximations of the intrinsic metric of a manifold. Isomap and LLE differ slightly in their procedures. LLE is slightly more successful in practice on non-convex contractible subsets of Euclidean space.
- Almost all manifold learning algorithms depend on approximating the local tangent structure of the manifold from the data, typically by studying the spectrum of the graph Laplacian. Heat flow provides a conceptual framework for describing these approximations.
- Neighbor embedding algorithms like stochastic neighbor embedding (SNE) make different geometric assumptions and are effective in working with data points of non-uniform density. A descendant of SNE, *t*-distributed stochastic neighbor embedding (*t*-SNE), is an extremely popular choice in applications.
- As synthetic examples illustrate, classical dimensionality reduction and manifold learning techniques work best under restrictive hypotheses about the geometry of the data.
- The coordinates provided by manifold learning algorithms can be used as filters for Mapper. This is an interesting avenue for combining TDA and geometric dimensionality to provide a more flexible description of the underlying structure of the data.
- Metric geometry is the study of geometric structures on metric spaces that are similar to those that arise in Riemannian geometry.
- The space of phylogenetic trees may be endowed with geometric structure. This structure provides the foundations for dealing with probability distributions and summary statistics.

4.9 Suggestions for Further Reading

There is of course a tremendous literature on PCA and many different variants of MDS; we particularly recommend Hastie, Tibshirani, and Friedman's classic text [233] for a wonderful exposition in the context of classification and learning. The area of manifold learning and the problem of working with non-Euclidean manifolds embedded in Euclidean space is substantially more recent. A nice survey of manifold learning techniques (discussed in a broader machine learning context) is available in [50].

Part II
Biological Applications

5

Evolution, Trees, and Beyond

The affinities of all the beings of the same class have sometimes been represented by a great tree. I believe this simile largely speaks the truth. The green and budding twigs may represent existing species; and those produced during each former year may represent the long succession of extinct species.

Charles Darwin

Any living cell carries with it the experience of a billion years of experimentation by its ancestors.

Max Delbruck

5.1 Introduction

It is impossible not to marvel at the richness of life on Earth: from the large mammals in the sea and on the plains, to the hardy plants of the high mountains, to the microbes living in hydrothermal vents and under the Antarctic ice. The adaptability and sheer quantity of cellular life on this planet is staggering. The challenge of classifying its diversity was recognized as far back as the fourth century BC, when Aristotle (384–322 BC) introduced one of the first systematic taxonomies of living organisms. He began his work by separating the plants from animals. Then, he split the animals into those that walked, swam or flew, and the plants into those small, medium or large in size. He further subdivided these groups based on other criteria. Beyond his taxonomy, Aristotle also proposed a hierarchy of animals, known as the "Ladder of Life," with simple organisms on its lower rungs and humans at the top.

Modern taxonomy was founded in the eighteenth century by the Swedish scientist Carolus Linnaeus, who undertook the colossal task of classifying all known animals, plants and even minerals. In *Systema Naturae* (1735), Linnaeus proposed a hierarchical structure where similar organisms were first grouped into species, similar species would be grouped into one genus, and similar genera would be grouped into one family, and so on, generating a phylogeny built of six ranks of taxa: kingdom, class, order, family, genus, species. Thus the highest rank taxa in the Linnaean taxonomy were the three kingdoms (plants, animals, and minerals). The animal kingdom, for instance, was divided into six classes: Mammalia, Aves, Amphibia, Pisces, Insecta, and Vermes. The Linnaean system is the forerunner of most modern classifications of living and extinct organisms. Each lower rank taxon belongs to a higher rank taxon, ultimately generating a tree structure (Figure 5.1). Despite the elegance of his taxonomy, Linnaeus was troubled by the possibility that a given organism could present characteristics common to several taxa of the same rank. To deal with this contingency, he annotated some *animalia paradoxa*, or contradictory animals, that resisted hierarchical classification. Amongst these were the dragon – which looks like a snake but has wings like a bird – and the legendary Borometz or Scythian Lamb, a tree that grew lambs (Figure 5.2). Based on their failure to fit within the hierarchy, it was eventually concluded that some of these animals were mythical and did not exist beyond the medieval bestiaries and the human imagination.

In 1859, Charles Darwin published *On the Origin of Species* [133], in which he introduced his landmark theory of evolution by natural selection. Evolution arises when parent organisms reproduce, generating progeny that resemble their parents but have additional variation that allows them to adapt to different environmental pressures. Some of this variation in the progeny is inheritable and is passed on to future generations. The accumulation of inherited variation over time eventually leads to the formation of new species. *On the Origin of Species* contained a single figure, depicting the ancestry of species as a phylogenetic tree (Figure 5.3). The idea of variation and its inheritance provided a beautiful explanation of the generation of species through a branching process: different organisms in the same taxa resemble each other because they share a common ancestor. Darwin revealed that a taxonomy is fundamentally a historical document – a record of the development of life on Earth.

Since then, the tree structure has become a dominant framework for representing evolutionary processes. Prior to the advent of sequencing technologies, most comparisons between organisms were based on phenotypic traits, the set of an organism's observable characteristics. The development of technologies to decode genomic material has provided a means to track the source of inherited variation and has enabled the comparison of organisms at the most fundamental level. Inferring evolutionary trees from molecular data, the practice known as molecular phylogenetics, has become a standard process in the study of evolution.

A species tree can only be inferred from genomic data if different regions of the genome provide similar trees. In humans, however, this is not the case. We

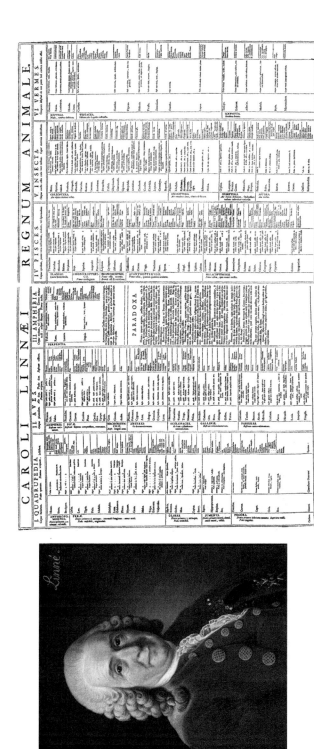

Figure 5.1 In 1735, the Swedish scientist Carolus Linnaeus published the *Systema Naturae*, a hierarchical classification of all known animals, plants, and minerals. These three groups formed his kingdoms, each of which was further divided into classes. For instance, except for a few exceptions (*animalia paradoxa*), all known animals were segregated as mammals, birds, amphibians, fish, insects or worms. Interestingly, the *animalia paradoxa* presented features from several classes and could not be unequivocally classified. Source: (1) Portrait of Carl Linnaeus, 1707–1778, Painted by Alexander Roslin in 1775, NMGrh 1053, Nationalmuseum, Stockholm, public domain. (2) Table of the Animal Kingdom (Regnum Animale) from Carolus Linnaeus's first edition (1735) of *Systema Naturae*.

Figure 5.2 The "animalia paradoxa" were animals that challenged Linnaean tax-onomy because they possessed similarities with organisms belonging to at least two different higher taxa. The dragon, for instance, had a body similar to that of a reptile but also wings like birds (illustration from the *Liber Floridus*, or *Book of Flowers*, circa 1100AD, public domain). The Borometz, or Scythian Lamb, was a plant that grew lambs. Source: Lee, H. 1887. The Vegetable Lamb of Tartary: a Curious Fable of the Cotton Plant, to Which Is Added a Sketch of the History of Cotton and the Cotton Trade. S. Low, Marston, Searle & Rivington, London.

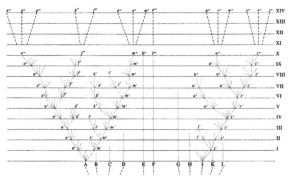

Figure 5.3 This tree appeared in Darwin's *On the Origin of Species* as a means of capturing the divergence of species. In this figure, time advances moving up the tree. The roots of the tree represent the original species that diversified according to a branching process through progeny variation and selection. The top branches constitute the modern species, and the branches that do not persist to the top represent extinct species. Source: Left: Library of Congress, Prints & Photographs Division, reproduction number, LC-DIG-ggbain-03485. Right: Darwin, C. R. 1859. *On the origin of species by means of natural selection, or the preservation of favoured races in the struggle for life*. London: John Murray.

know that some genomic material, like mitochondrial DNA, is inherited through the maternal line, while other material, like the Y chromosome in men, comes through the paternal line. Thus trees inferred from mitochondrial DNA will not agree with those inferred from the Y chromosome, as the evolutionary stories of our fathers and our mothers are different. The problem becomes more complex when different regions across chromosomes give rise to different potential trees. Genomic data has increasingly challenged the single-tree picture, as biological phenomena like species hybridization, bacterial gene transfer, and meiotic recombination have complicated the lineage of inheritance. Despite their smaller genomic size, viruses are also found to contain incompatible genomic tree histories. Frequent recombination events in HIV, for instance, have confounded attempts to reconstruct an early history of the epidemic.

In 1990, using molecular comparison, Carl Woese et al. proposed the organization of all cellular life forms into three large domains: the Bacteria, the Archaea, and the Eukarya [538]. This study showed the power of genetic information to elucidate deep phylogenetic relations that were hidden to other methods. Woese's tree, however, was based only on a small fragment of 1500 nucleotides in the 16S ribosomal RNA of prokaryotes, a tiny fraction of any organism's genome. One then wonders if the tree reconstructed from this small part of the genome can be extended to other parts of the genome, or if there exist other genes that could generate vastly different trees. Indeed, with the accumulation of genomic information, an increasingly complex picture of the relations between species is emerging, with different genes providing different incompatible tree phylogenies (see Figure 5.4), highlighting the need for new representations [147].

There are, broadly, two ways in which organisms acquire genomic material. The first, which we call here clonal evolution, is the consequence of direct transmission of genes from a single parent to the offspring. Clonal evolution is a type of vertical evolution, the direct transmission of genetic information from parents to offspring. Changes in genomic material are mediated by random mutations over multiple generations. The genomic material is inherited from a single parent, and mutations will lead to differences between a clone and its parents. This type of vertical evolution is best represented by a mathematical structure called a phylogenetic tree. The left of Figure 5.5 depicts a rooted tree where the root node at the apex represents an ancestor that propagates and diversifies over time, creating new lineages, called clades, in a branching pattern.

As has become increasingly apparent with the advent of sequencing technologies, genomic material may also be acquired through a second means: horizontal or reticulate evolutionary events. These events occur when distinct clades merge to form a new hybrid lineage. This phenomenon may be effected in a number of ways and occurs across all domains of life. This phenomenon is pervasively

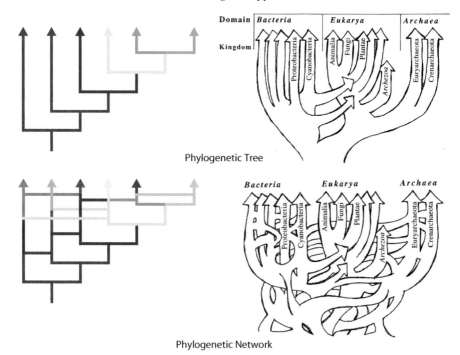

Domain | Bacteria | Eukarya | Archaea
Kingdom | Proteobacteria Cyanobacteria | Animalia Fungi Plantae Archezoa | Euryarchaeota Crenarchaeota

Phylogenetic Tree

Bacteria | Eukarya | Archaea
Proteobacteria Cyanobacteria | Animalia Fungi Plantae Archezoa | Euryarchaeota Crenarchaeota

Phylogenetic Network

Figure 5.4 Idealized, simplistic phylogenetic trees contrast with more realistic, complex reticulate networks. On the top right is the Doolittle representation of the Tree of Life, made before the advent of sequencing technologies. It was thought that most evolution occurred through branching processes, with the notable exceptions of mitochondria and chloroplasts – believed to be symbiotic bacteria that fused part of their genome to their host's. This picture is changing as the significance of horizontal exchange of genomic information is becoming more evident. Source: [147]. From Doolittle, W. F., Phylogenetic Classification and the Universal Tree, *Science*, 1999, 284 (5423): 2124–2128. © 1999 Reprinted with permission from AAAS.

found in viruses, for instance. As we will see later in detail, viral influenza undergoes horizontal evolution through reassortment and HIV undergoes horizontal evolution through recombination. Phylogenetic trees, however, are not able to capture these horizontal evolutionary events. Representing these events graphically requires a new structure called a *reticulate network*, in which branches are allowed to both join and split. Places in the network where branches merge are known as cycles and correspond to individual reticulate events (Figure 5.5, right). The resulting network is the result of merging many different trees with different topologies.

To detect reticulate events by phylogenetic means, one must first construct a tree for each gene in the genome and then cross-reference each pair of trees for conflicts in lineal history. The simple example of the Network of Life, depicted

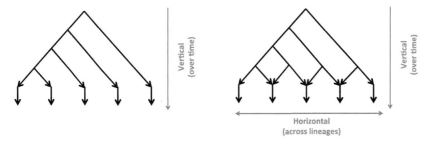

Figure 5.5 Examples of a phylogenetic tree (left) and a reticulate network (right) capturing clonal and horizontal evolution, respectively. Source: [100]. From Joseph Minhow Chan, Gunnar Carlsson, and Raúl Rabadán 'Topology of viral evolution', *Proceedings of the National Academy of Sciences* 110.46 (2013): 18566–18571. Reprinted with Permission from Proceedings of the National Academy of Sciences.

in Figure 5.4, illustrates the complexity of inferring the properties of phylogenetic networks summarizing complex data sets (from Doolittle [147]).

Some of the processes that lead to non-tree-like structures are shown in the table below.

Organism	Reticulate process	Description
Viruses	Homologous recombination	Intragenomic homologous crossover
	Reassortment	New sets of different segments in segmented viruses
Bacteria	Transformation	Acquisition of foreign DNA from environment
	Transduction	Viral-mediated exchange
	Conjugation	Exchange through cell-to-cell contact
Eukaryotes	Meiotic recombination	Crossover and gene conversion during meiosis
	Hybrid speciation	Hybridization between different species
	Endosymbiosis	Fusion of genomes of symbionts

5.2 Evolution and Topology

We now explain how to use topological data analysis to determine when evolutionary processes violate tree-like assumptions, i.e., to detect reticulate events, based on observed genomic data. First, to understand how reticulate events can be observed in genomic data, we will consider a very simple model.

Assume that we have a simplified genome with only two nucleotides, or basic informational units, 0 and 1. We will further assume that this genome is quite large

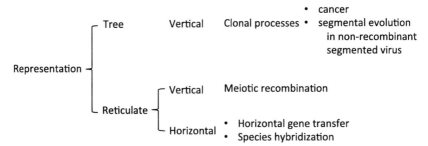

Figure 5.6 Summary of concepts used in this chapter. Clonal processes, such as tumor development or bacterial evolution without gene transfer, are well captured and represented by trees. Many processes, however, cannot be represented by a tree and require the reticulate representation. These include certain vertical processes where the genetic information is inherited from more than one parent, like in meiosis in eukaryotes. Other processes involve the transfer of genetic information between species, like in horizontal gene transfer in bacteria or species hybridizations.

and that mutations exchanging 0s and 1s occur at uniformly random positions along the genome. If the total number of bases is very large compared to the number of mutations, then, assuming that mutation sites are chosen at random, the probability that any particular site will be mutated twice is very small. In particular, as the genome length approaches infinity and the number of mutations is held constant, the probability of any site being mutated twice approaches zero. We can formalize this for genomes of finite length by imposing the constraint that any given site only mutates once; this is called the infinite-sites assumption. For this discussion, let us adopt the infinite-sites assumption and assume that an organism evolves only through a clonal process.

We now observe that certain mutational patterns are not possible given these assumptions. Suppose that we have a genome of length 2. Then we can have four possible alleles: 00, 01, 10, and 11. Consider an organism with ancestor 00. A mutation in the ancestor's first site generates 10 and a subsequent mutation in its second site generates 11. How can we generate 01 after these two mutations? The ancestral genome would have to mutate back at the first site; but this second mutation at the first site would violate the infinite-sites assumption and thus this mutational pattern would not be allowed in our model. Similarly, if the ancestor's second site mutated first, we would not be able to generate the allele 10. Therefore the presence of four alleles in the observed population is incompatible with a solely clonal evolutionary process from a single ancestor in this setup.

This observation can be turned into a test for reticulate events: checking for these four alleles is referred to as the *four gamete test*. In practice, no genome is of infinite length and so the infinite-sites hypothesis is not quite right; thus, violations of the four gamete test are possible even for strictly clonal evolution. However, if the violation is identified at multiple sites, chance becomes an unlikely explanation. For instance, in order to generate four genomes 0000, 1100, 0011, 1111, the infinite-sites model would need to be violated twice. The more violations we have, the less likely it is that our assumptions of clonal evolution are correct. So if we are confident that the infinite-sites model is a reasonable approximation, then a large number of incompatibilities casts doubt on the assumption of clonal evolution.

This raises the question of how to quantify what a "large number" of incompatibilities should be. One method is the Hudson-Kaplan test, which counts the minimum k such that there exists a partition of the data into k subsets such that within each subset all sites are compatible with the four gamete test [254]. For example, in the case of the genomes 0000, 1100, 0011, 1111, if we split the genome down the middle, and consider each half independently (00, 11, 00, 11 and 00, 00, 11, 11), then the four gamete hypothesis is no longer violated in each partition.

Besides the Hudson-Kaplan method and variations based on the four gamete test, there have been significant efforts to identify recombinants, their ancestors, and specific genomic break points, i.e., the points in the genome where recombination has occurred. Several major strategies have been developed to detect recombinants.

- Distance methods rely on differences between the genetic pairwise distances along the genome, usually using a sliding window technique. Based on some underlying model one can then evaluate the likelihood of a recombination [530].
- Phylogenetic methods are based on the idea that if a recombination has occurred, the trees inferred from different parts of the genome could have distinct topologies [432]. The same type of techniques can be used to identify genes that have been transferred when orthologous genes from different species are more similar to each other than expected, given the species' evolutionary relationship [27].
- Compatibility methods search for phylogenetic incongruence in a site-by-site basis, and, in general, do not require the phylogeny of the sequences to be known [414, 470].
- Substitution methods search for clustering of substitutions along the genome using some summary statistics in different phylogenetic partitions [414, 485].
- Linkage disequilibrium methods are one of the most popular techniques for studying large genomes, e.g., the human genome. The main idea in such methods is that if in a section of a genome there has not been recombination, the presence of a substitution is informative (linked) to nearby regions. Recombination breaks this linkage. We will describe in Section 5.8 how linkage is used to study recombination in the study of large numbers of human genomes.

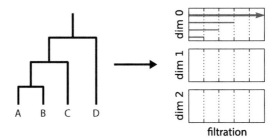

Figure 5.7 Tree topologies are contractible. When computing persistent homology, one can observe that there are no bars in barcodes of dimension bigger than zero. Source: [162]. Reprinted with permission: © EAI European Alliance for Innovation 2016.

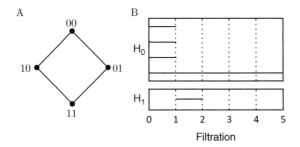

Figure 5.8 A simple reticulation event involving four genomes with only two sites and two bases 0 and 1. (A) If the four possible states 00, 01, 10 and 11 are present (four gamete test) one can suspect that a site has mutated twice or there has been a recombination between these sites. In the case of large genomes where mutations in the same site are considered highly improbable (infinite site models), the four gamete test is used in many statistical tests for the identification of recombination events and specific recombination sites. (B) When we apply persistent homology to the Hamming distance between these different small genomes, one clearly identifies an interval $[1, 2)$ in the first homology persistent diagram. The non-trivial homology classes in dimension one and higher can be used as indicators of the presence of recombination or multiple mutations in the same site. Source: [162]. Reprinted with permission: © EAI European Alliance for Innovation 2016.

A summary of these methods and accompanying software can be found at the end of the chapter in Section 5.12.

The methods used to identify recombinant sequences can suffer from prohibitive computational costs in large data sets. These methods are designed for the specific task of identifying recombinants and breakpoints. They use quantified measures of the violation of the tree assumption to infer these events. A natural question that then arises is whether there are better descriptors of the data in the event of recombination. The first hint that topological data analysis might be useful comes from the observation that trees are contractible (see Figure 5.7).

The four genomes $\{00, 01, 10, 11\}$ can be considered the fundamental and simplest model of recombination. Topologically the set forms a loop, as shown in Figure 5.8, i.e., the Vietoris-Rips complex (recall Definition 2.1.6) contains a complex with a non-trivial loop. Four of the six pairs of gametes are separated by a Hamming distance of 1, while the other two are separated by a distance of 2. (Here recall from Example 1.2.5 that the Hamming distance counts the number of positions at which the strings are different.) At a filtration distance of 1, the four pairs become connected, forming a loop. At a filtration distance of 2, the remaining two pairs become connected, destroying the loop. Thus, we have non-vanishing H_1 homology on the filtration interval $[1, 2)$. This simple example suggests that persistent homology provides a method for counting the number of incompatibilities and, at the same time, determining the scale of each incompatibility in terms of the distance between the alleles.

Each interval in the barcode can be interpreted as a sign of a recombination event involving a set of sequences including the common ancestor, parental, and recombinant strains. The interpretation and identification of the recombinant and parental strains could be complicated or impossible unless given further information. This is analogous to the problem of finding a root of a phylogenetic tree if no information about ancestral states is provided. Persistent homology can provide a simple way to estimate the number of incompatibilities.

For our purposes, we can assume the genomes of organisms in an evolving population forms a metric space (X, ∂_X), which we never directly observe. Instead, we observe a sample of data points (i.e., sequenced genomes of cells) that lie in X. Restricting the metric ∂_X on X to our sample gives us a distance between points. Considering genomes as a string of characters makes it easy to define distances, e.g., the Hamming distance. Metrics currently used in biological applications are based on different models of how mutations can occur. For instance, one can modify Hamming distances to account for the possibility of back mutations after some time (Jukes-Cantor distances [279]), account for the fact that different substitutions can occur with different probability (Kimura models [298]), allow different bases to occur at different frequencies [173, 231], among many possible refinements. Thus, we have a finite metric space generated by genomic sequences separated from each other by some genetic distance, to which we can apply the techniques of topological data analysis.

Recall from Section 4.7.1 that some finite metric spaces can be derived from a weighted tree, with the distances between two leaves calculated by adding up the weights associated to the edges connecting them (see Figure 5.9). Tree-like spaces can be used to represent clonal processes, with internal nodes representing unsampled ancestors.

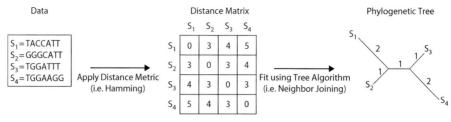

Figure 5.9 Pipeline for analyzing genomic data using persistent homology. Starting from a sample of sequences, one can compute distances reflecting the similarity between organisms. These distances provide a finite metric space, that, in some cases, can be summarized by a phylogenetic tree whose leaves correspond to points in the metric space. Distances between branches can be estimated by the addition of weights along the shortest path.

Obviously, not every metric space has this tree-like metric property. In general, finite metric spaces that can be represented by weighted trees are only a small subspace of all finite metric spaces. Indeed, Lemma 4.7.2 described the required four point condition satisfied by metric spaces generated by trees.

But when there is no underlying tree explaining the data, we can capture and represent evolutionary processes beyond trees using topology. A phylogenetic tree can be continuously deformed into a single point. The same action cannot be performed for a reticulate network without destroying the loops or cycles in the structure. The active hypothesis then is that the presence of these holes results directly from horizontal evolutionary events. This idea can be formalized into the following theorem [100].

Theorem 5.2.1. *Let (M, ∂_M) be any tree-like finite metric space, i.e., a space satisfying the four point condition, and let $\epsilon \geq 0$. Then the Vietoris-Rips complex $VR_\epsilon(M, \partial_M)$ is a disjoint union of acyclic complexes. In particular, $H_i(VR_\epsilon(M, \partial_M)) = \{0\}$ for $i \geq 1$.*

In other words, the presence of homology above dimension zero indicates that the metric space does not satisfy tree-like metric properties. Identifying the genomes that are the generators of these homology classes selects subsets of genomes whose derived distances do not satisfy the tree condition, indicating that non-tree-like evolutionary processes have occurred within these subsets (Figure 5.10).

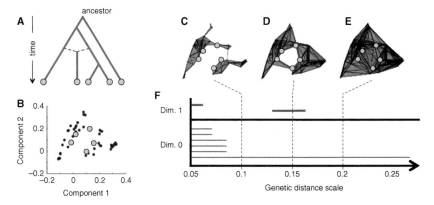

Figure 5.10 Persistent homology detects historical recombination events from population genetic data. Consider the reticulate phylogeny in panel A. Five genetic sequences sampled today (the yellow circles) developed from a single common ancestor through clonal evolution (solid blue lines) and recombinant evolution (dotted red lines). Panel B illustrates this sample within a larger sample of the population. Persistent homology is applied to this larger sample and three filtrations are shown in panels C, D and E. Panel F shows the resulting barcode. Note that these two dimensional plots (panels C, D, E), created by principal component projection, are used merely to visualize the sequences; projection is not part of the algorithm. The dimension 1 bar near the center of panel F identifies a recombination event involving the five highlighted sequences. The scale over which this bar persists captures the genetic difference between the parents of the recombinant [323]. Source: [100].

The persistent homology approach suggests a general strategy to study the space of genomes. Instead of considering trees and reticulate networks in the phylogenetic sense, we consider these structures in the context of simplicial complexes and compute their persistent homology. An additive tree is a single connected component without any loops which displays only zero dimensional topology. Reticulate structures, on the other hand, contain loops and therefore may contain non-trivial higher dimensional topology.

Recall from Section 2.3 that persistent homology can be displayed in a barcode plot where for a given filtration and dimension k, different bars represent independent k-dimensional cycles that generate non-trivial homology classes. As we have observed, the presence of non-zero homology above dimension zero indicates deviation from a tree metric. The next step is to define a quantity that captures the extent of deviation from a tree. In order to do this, we consider the distribution B_k of bar lengths of k-dimensional cycles for some $k > 0$.

Specifically, a natural measure of the deviation from a tree metric is some kind of count of the number of bars in PH$_1$. We define the *topological obstruction to phylogeny* (TOP) to be the L^∞ norm, or maximum, of the lengths of the bars. The work of [100] established that a filtration with non-zero TOP implies that the finite metric space is not tree-like. Another possible measure is the L^1 norm, which is equivalent to the sum of the bar lengths. In simulations of evolutionary data where we initially set a rate r of horizontal evolution, we find that of all L^p norms, the L^1 norm best correlates with r. Finally, we could also consider the L^0 norm, simply the count of the number of bars, which is also proportional to r. To approximate r, we consider either the L^1 or L^0 norm normalized by time; we define the irreducible cycle rate (ICR) to be precisely this normalization.

As we will see at the end of this chapter, the relationship between the recombination rate and the persistent homology of a sample of genetic sequences can be probed using coalescent simulations of evolution. Figure 5.11 shows how the number of persistent dimension 1 cycles, b_1, grows with the number of recombination events that occur in a simulation.

In [164], it was demonstrated how b_1, together with the birth and death scales of each cycle, can be used to estimate the population-scaled mutation rate ρ. The accuracy and precision of this estimator increases with sample size; we discuss this in Section 5.7.3.

These results suggest a map between algebraic topology invariants, such as Betti numbers and generators of homology classes, and different types of genomic exchange events (Figure 5.12). Persistent homology provides information about the obstructions to tree-like metrics due to homoplasies (shared mutations in different genomes that are not shared by their common ancestors), recombination, reassortment, or other modes of horizontal exchange of genomic material. By studying the cycles that generate higher dimensional classes (the witnesses to the violation of the tree-like assumption), we can infer what type of biological process occurred that violated the tree-like assumption. Later in this chapter, we will

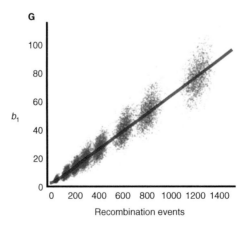

Figure 5.11 The number of one dimensional persistent homology classes, b_1, scales with number of recombination events in a coalescent simulation.

Persistent Homology	Viral Evolution
Filtration value ε	Genetic distance (evolutionary) scale
0-dimensional Betti number at filtration value ε	Number of clusters at scale ε
Generators of 0-dimensional homology	A representative element of the cluster
Hierarchical relationship among generators of 0-0-dimensional homology	Hierarchical clustering
1-dimensional Betti number	Number of irreducible recombination/reassortment events
Generators of 1-dimensional homology	Recombinant/reassortant events
Generators of 2-dimensional homology	Complex horizontal genomic exchange
Number of higher dimensional generators in time frame	Lower bound on recombination/ reassortment rate
Non-zero high dimensional homology (topological obstruction to phylogeny)	No phylogenetic representation

Figure 5.12 Rough dictionary between TDA notions and evolutionary concepts. Source: [100]. From Joseph Minhow Chan, Gunnar Carlsson, and Raúl Rabadán, 'Topology of viral evolution', *Proceedings of the National Academy of Sciences* 110.46 (2013): 18566–18571. Reprinted with Permission from Proceedings of the National Academy of Sciences.

explore this relationship in detail through a series of examples in the viral, bacterial, and eukaryotic worlds.

Remark 5.2.2. In some cases the map between homological invariants and evolutionary phenomena can be made more explicit [323]. This is the case for "galled trees," directed acyclic graphs that differ from trees by a few isolated recombinations. In that case, the homology in dimensions bigger than one vanishes, generalizing Theorem 5.2.2. These "galled trees" can be constructed by pasting tree-like and isolated recombination events that correspond to operations in the associated finite metric spaces [323].

5.3 Viral Evolution: Influenza A

5.3.1 Influenza A

Influenza A is a segmented single-stranded RNA orthomyxovirus that infects different hosts of many species. Indeed, the highest genetic diversity of these viruses is found in birds, mostly waterfowl, of the order of Anseriformes (ducks, swans and geese), Passeriformes, and Charadriiformes (including gulls). Waterfowl are the virus's natural reservoir, perpetuating the vast biodiversity of influenza, including all different subtypes. But influenza A has also been found in pigs, seals, and other mammals including, of course, humans. Classification of influenza viruses is traditionally made by the antigenic properties of the proteins displayed in the viral envelope hemagglutinin (HA), ranging from H1 to H16, and neuraminidase (NA), ranging from N1 to N9. Recently, new types of influenza viruses have been identified in bats in Central America [500], leading to two new hemagglutinin types (H17 and H18) and two new neuraminidase types (N10, N11). Of course, it is possible that, as surveillance programs get more extensive, new related viruses will be found in other hosts (Figure 5.13).

Infection in humans and other mammals usually occurs in the upper respiratory tract, lasts a couple of weeks, and is associated with symptoms that vary from fever, sore throat and other cold-like symptoms to more serious complications that can result in death. It has been estimated that near half a million deaths are associated to influenza infections every year around the world. Transmission of human influenza occurs mostly through the air, in the form of droplets of water released from coughs or sneezes, and through fomites, surfaces that carry infectious particles. These modes of transmission seem to be more effective at low temperatures and low humidity, factors that are probably relevant for the seasonal pattern observed in human influenza. Illness associated with influenza infection is most common in winter – from November to April in the Northern Hemisphere, and from May to October in the Southern Hemisphere.

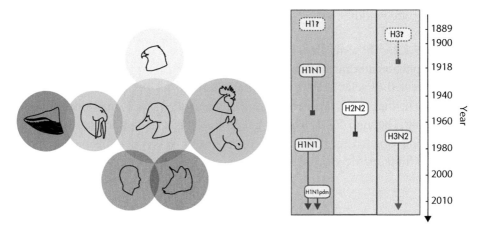

Figure 5.13 Left: Influenza A infects many different species, mostly birds and mammals. The greatest diversity of the virus can be found in waterfowl. Occasionally, viruses can jump species and infect other hosts. Influenza A has been reported in humans, swine, horses, seals, camels, bats and even whales. Right: Twentieth century influenza pandemics. Pandemics are caused by viruses containing genes from other species. Although there is some speculation about pandemics in the nineteenth century, the first well-characterized influenza pandemic was the so-called Spanish flu in 1918. Since then influenza pandemics have occurred every 30 years, with the last pandemic originating in swine in 2009. The Influenza A virus infects different species and generates pandemics.

In contrast to mammals, most birds show no clinical signs of infection by the virus, which replicates in their gut and sheds into the water through feces [529]. However, mutations occasionally occur that increase its pathogenicity, resulting in a highly pathogenic avian influenza (HPAI), which causes a multi-organ systemic disease that can kill birds. Large surveillance programs are dedicated to detecting HPAI outbreaks; HPAI transmission to humans is a chief concern.

What factors allow a virus to be transmitted between individuals or species? In both humans and waterfowl, the virus must recognize specific molecules on the surface of the cell in order to fuse with it and infect it. These molecules vary in different cells and hosts; however, the recognition of monosaccharide residues on epithelial cells by viral hemagglutinin is a common pathway. Avian influenza interacts with an α-2,3-sialic acid, prevalent in the intestinal tract of birds. In contrast, human influenza binds the α-2,6-sialic acid predominant in the human upper respiratory tract, begetting the flu-like symptoms of cough, sore throat, and rhinorrhea. Pig trachea contains both types of sialic acids. This unique feature of swine supports the mixing vessel theory that pigs provide a bridge for influenza from avian host to human, allowing the virus to adapt to recognize α-2,6-sialic acid through reassortment [269]. Host switching from waterfowl to human, however, does not require a swine intermediary. The HPAI H5N1 virus, for instance, infected 18 people and killed six in 1997 after first appearing in Guangdong in 1996 then spreading rapidly to poultry in Hong Kong. That year, Hong Kong culled more than 1 million poultry.

Since 2003, a number of sporadic H5N1 outbreaks with suspected poultry intermediaries have taken place among humans and other mammals, causing 860 human infections and 454 deaths as of February 2019 – a staggering mortality rate of nearly 60%. The fulminant progression of H5N1 infection most likely results from its specificity for α-2,3-sialic acids, which are present at a low concentrations in the human lower respiratory tract. Infection in the lower respiratory tract leads to the more flagrant symptoms of viral pneumonia. As such, avian H5N1 demonstrates high pathogenicity and productive infectivity in humans, but an inability for sustained transmission between humans. Given the high mortality rate of infection, it is a matter of the utmost importance to determine whether these viruses could become transmissible among humans like seasonal influenzas.

Recently, in the laboratory setting, teams led by Kawaoka [266] and Fouchier [240] demonstrated the pandemic potential of non-seasonal strains. They engineered H5N1 by mutating specific sites (site-directed mutagenesis) and passing the virus along ferrets (which share similar sialic acid distributions to humans) until they generated strains capable of transmission. Similarly, Zhu et al. showed that the 2013 H7N9 strain, which infected 131 humans and caused 32 deaths in two months in the Jiangsu province of China [191], infected and was transmitted between ferrets, suggesting that human to human transmission of H7N9 has most likely already occurred [548]. These outbreaks underscore the need for further investigation into the mechanisms of viral evolution and the adaptation of animal viruses to humans.

Influenza viruses are enveloped and nearly 100 nm in diameter. Their genome is 13,000 bases long and is composed of eight segments of single-stranded antisense RNA (Figure 5.14). Each segment encodes one or two viral genes. Antisense RNA is the complement of the RNA that codes for proteins; thus it cannot be directly translated into functional protein. In order for the influenza genome to express protein, positive-sense strands must be produced from the template of the antisense strands. Further complexity arises when the virus attempts to make new *virions*, the infectious particles that allow the virus to be transmitted outside of the host cell. The replicating virus must duplicate its original antisense RNA and, in order to do so, it must polymerize new strands of ribonucleotides complementary to the template of the positive-sense RNA. Influenza carries its own polymerase complex, which it uses for all of its RNA replication; in fact, the three longest genes of influenza (PB2, PB1, PA) code for the three proteins directly involved in replicating genomic material. The polymerase complex interacts directly with viral RNA and the nucleoproteins (NPs) that attach to it. An RNA segment, together with a copy

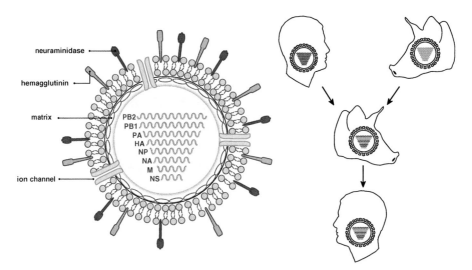

Figure 5.14 Influenza A is an antisense single-stranded RNA virus whose genome is composed of eight different segments containing one or two genes per segment. This virus contains an envelope borrowed from the infected cell that expressed two viral proteins, hemagglutinin and neuraminidase. When circulating viruses co-infect the same cell, new viruses can be created that contain segments from both parents. This phenomenon, called reassortment, can lead to dramatic adaptations to novel environments, and it is thought to be one of the contributing factors to human influenza pandemics.

of the polymerase complex and several NP proteins, forms the ribonucleoprotein (RNP) particle that is released into the cell cytoplasm and packaged in the virion or viral particle. The virion consists of a shell (capsid) of membrane proteins (MPs) and M2 proteins that form tetramers with ion channel activity. These ion channels help modulate pH within the virions and regulate the release of viral RNA into infected cells. Two other proteins, non-structural N1 and NS2, are found in infected cells but are absent, or have low expression, in virions. The existence of other proteins in alternative reading frames has been proposed; however, these proteins do not have a well-characterized role in the life cycle of the virus [507, 547].

Influenza evolves by accumulating mutations at a high rate. Estimates of evolutionary rates, or changes per unit time, indicate that influenza, like many other RNA viruses, evolves at a rate of $\sim 10^{-3}$ per nucleotide per year. This brisk evolutionary rate poses a significant challenge in the development of vaccines. Current vaccines for influenza rely on leveraging the antigenic response to epitopes (the sections of proteins recognized by antibodies) in hemagglutinin. However, these epitopes change as the virus accumulates mutations. The World Health Organization updates the composition of the vaccine with the hope that the updated vaccine will more faithfully resemble circulating strains. To help the WHO, national and international organizations put significant effort into collecting genomic and antigenic data from circulating strains of influenza. These large collections – more than 100,000 genomes, currently – constitute excellent material on which to test the mathematical and computational methods described in this book.

Substitutions (i.e., point mutations) accrued by influenza can be viewed as small changes in the nearly continuous evolution of its genome. However, point mutations are not the only way that influenza evolves; more dramatic change can occur. As discussed, influenza genomes consist of eight different segments. These segments are the viral analogue of chromosomes. When two viruses of different strains co-infect the same host cell, they can generate a progeny containing novel combinations of segments taken from both parental strains [416, 417]. This phenomenon, called reassortment, shuffles the genomic material of different strains and constitutes the underlying mechanism behind influenza pandemics.

A pandemic influenza is a viral strain that was initially endemic to animal hosts like waterfowl and swine that obtained the requisite mutations to infect and adapt to human hosts, thereby spreading on a global scale. Mutations necessary for human adaptation can be easily acquired by incorporating segments from viruses already adapted to human hosts through reassortment. Mutations and reassortments can introduce changes in the antigenic properties of the strain, which, in turn, can render antibodies raised against previously circulating viruses ineffective. The mutational change of seasonal influenza, referred to as *antigenic drift*, contrasts with the more dramatic reassortment in pandemic strains that creates entirely new viral genomes, referred to as *antigenic shift*.

In modern history, the most calamitous example of an influenza pandemic was the H1N1 (Spanish flu) epidemic of 1918. H1N1 claimed the lives of 50 to 100 million people worldwide [276]. As it disseminated throughout post-war Europe, it justified drastic public health measures including the widespread shuttering of theaters, schools and churches. A physician working at Camp Devens, a military base west of Boston, related the dramatic effects of the pandemic strain to a friend in a letter on September 29th, 1918 [210]:

This epidemic started about four weeks ago, and has developed so rapidly that the camp is demoralized and all ordinary work is held up till it has passed... These men start with what appears to be an attack of la grippe or influenza, and when brought to the hospital they very rapidly develop the most vicious type of pneumonia that has ever been seen. Two hours after admission they have the mahogany spots over the cheek bones, and a few hours later you can begin to see the cyanosis extending from their ears and spreading all over the face, until it is hard to distinguish the coloured men from the white. It is only a matter of a few hours then until death comes, and it is simply a struggle for air until they suffocate. It is horrible... We have been averaging about 100 deaths per day, and still keeping it up... We have lost an outrageous number of nurses and doctors ... It takes special trains to carry away the dead. For several days there were no coffins and the bodies piled up something fierce, we used to go down to the morgue (which is just back of my ward) and look at the boys laid out in long rows. It beats any sight they ever had in France after a battle. Good-by old Pal, God be with you till we meet again.

The genome and the virus itself were isolated from bodies buried in a mass grave in the permafrost of a remote Inuit village in Brevig Mission (called Teller Mission in 1918) on the Seward Peninsula of Alaska [423]. 85% of the adults that were buried in the mass grave died within the span of five days in November, 1918. In 1997, several of the bodies were exhumed. The viral sequence of this strain was recovered and can be found online under the name A/Brevig Mission/1/18 (H1N1). Despite knowledge of the sequence, many questions about the 1918 pandemic strain remain:

What was its original host?

Where and when did it first infect humans?

And why was it so pathogenic?

After a couple of waves of worldwide infection, the pandemic-causing strain became a seasonal influenza virus.

The next human pandemic, the so-called Asian flu, occurred in 1957. A descendant of the 1918 H1N1 pandemic strain, still circulating in humans, acquired three segments of avian origin (PB1, HA, NA), forming the H2N2 strain and causing a pandemic (Figure 5.13). The H2N2 virus circulated in humans, replacing the H1N1 virus, until the next pandemic. In 1968, a new reassortant, H3N2, which contained H2N2 and avian segments (PB1, HA), was identified in South Asia and rapidly spread across the world. H3N2 still circulates in humans today and is a major cause of morbidity associated with influenza. Interestingly, H1N1, which had not been found circulating in humans since the pandemic of 1957, reemerged in 1977 and co-circulated with H3N2.

In 2009, a swine-origin novel H1N1 virus marked the first pandemic of the twenty-first century (Figure 5.15). In mid March 2009, reports came from Mexico regarding an outbreak of respiratory illness. In April, two cases were documented in the United States in children from Southern California [200]. The CDC was alerted to the first case on April 13th: a ten-year-old boy who lived in San Diego County. The patient had fallen ill with fever, cough and vomiting on March 30th. None of his family members shared his symptoms. In the other case, a nine-year-old girl developed a respiratory illness in Imperial County. The CDC identified a new strain of influenza related to viruses circulating in swine and character-ized and published its genome. Since neither of the children had been in contact with pigs or each other – they lived 130 miles apart – the CDC suggested that the virus was already circulating in humans. A few days later, cases emerged in Texas. Within the following month, infection had struck every continent. The World Health Organization declared the strain a pandemic on June 11th, 2009.

The 2009 pandemic resulted from a reassortment between different influenza viruses circulating in swine [474, 506]. The pandemic virus showed relation to viruses isolated in swine more than a decade ago in North America and Asia. It is still unclear how, where and when these viruses developed into a human pandemic, and where the virus was circulating in the year before the pandemic. The most widely accepted conjecture is the *hidden pig herd hypothesis*, which proposes that incomplete surveillance missed strains in untested swine herds, and recent reports suggest that these viruses circulated in pigs in Mexico [241, 348].

The recent ancestors of a pandemic virus provide invaluable information about the set of minimal genomic alterations that can transform a zoonotic agent into a human pandemic. Understanding the origins of infectious strains can help us define scientifically based rules for the risk assessment of new strains and for the implementation of public health measures that might help avoid or mitigate future pandemics.

5.3.2 Reassortments in Influenza through TDA

We have seen that dramatic changes in the genetic makeup of an influenza virus can occur through reassortments, i.e., when two or more diverse viruses co-infect the same cell and create new viruses containing genomic material from the parental strains. Figure 5.16 shows three parental viruses with genomes comprising three

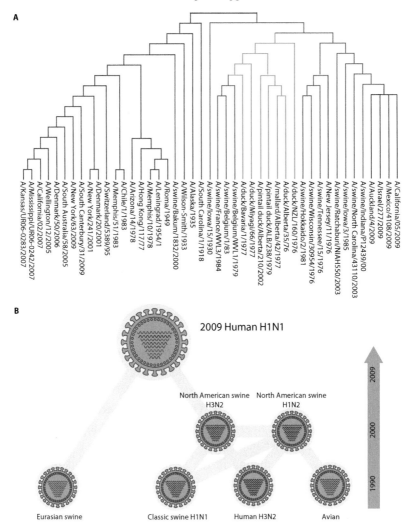

Figure 5.15 Origins of H1N1 2009 pandemic virus. Using phylogenetic trees, the history of the HA gene of the 2009 H1N1 pandemic virus was reconstructed. It was related to viruses that circulated in pigs potentially since the 1918 H1N1 pandemic. These viruses had diverged since that date into various independent strains, infecting humans and swine. Major reassortments between strains led to new sets of segments from different sources. In 1998, triple reassortant viruses were found infecting pigs in North America. These triple reassortant viruses contained segments that were circulating in swine, humans and birds. Further reassortment of these viruses with other swine viruses created the ancestors of this pandemic. Until this day, it is unclear how, where or when these reassortments happened. Source: [506]. From *New England Journal of Medicine*, Vladimir Trifonov, Hossein Khiabanian, and Raúl Rabadán, Geographic dependence, surveillance, and origins of the 2009 influenza A (H1N1) virus, 361.2, 115–119. © 2009 Massachusetts Medical Society. Reprinted with permission from Massachusetts Medical Society.

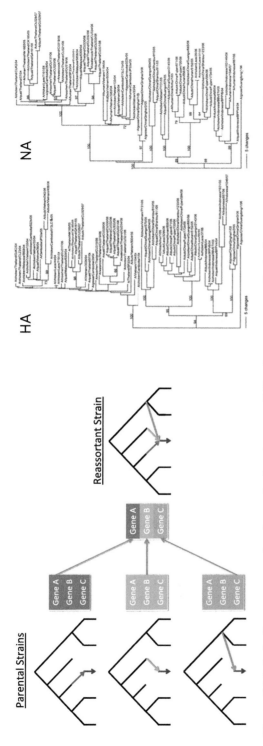

Figure 5.16 Left: Reassortments in viruses lead to incompatibility between trees. Reticulate network representing the reassortment of three parental strains. The reticulate network results from merging the three parental phylogenetic trees. Source: [100]. Right: Indeed, incompatibility between tree topologies inferred from different genes is a criterion used for the identification of events of genomic material exchange. Here we represent two genes of influenza A virus with different topologies using phylogenetic networks. From Joseph Minhow Chan, Gunnar Carlsson, and Raúl Rabadán, 'Topology of viral evolution', *Proceedings of the National Academy of Sciences* 110.46 (2013): 18566–18571. Reprinted with Permission from Proceedings of the National Academy of Sciences.

different genes and unique phylogenetic histories. All three can undergo a reassortment in which each parent donates a different gene. No single tree can capture the whole history. As such, incompatibilities between tree topologies derived from different genes may provide evidence of reassortment.

There are many interesting questions pertaining to reassortment. Imagine two different viruses infecting a cell. In principle, if each virus has eight segments, one could generate 2^8 different segmental combinations. Are these combinations all realized in nature? Is there any preference for certain combinations? Several reports have suggested that reassortments do not occur at random, but demonstrate clear preferences [206, 292, 416]. These apparent preferences may have multiple causes. The process of generating new viruses involves the packaging of eight different segments into the same virion and, although the packaging process is not completely understood, it is possible that different segments physically interact [385]. Cosegregation could also be due to selection. Given that different segments code for different proteins that work in conjunction, it is conceivable that two proteins that are co-adapted to work together will lead to offspring with higher fitness. Knowledge of these patterns may help reduce the number of potential viruses that we must consider in future pandemics.

We can study reassortments using the persistent homology framework described previously in this chapter. Let us start with a single segment: hemagglutinin [100]. To leverage persistent homology, we align our sequences, compute pairwise distances between them, and generate a finite metric space with points representing different sequences. The distance metric captures the genetic diversity present in the collection of sequences. We observe that most of the information in this metric space is contained in its zero dimensional homology with a few short bars in dimension one (see Figure 5.18 below). At this point, we can infer that a tree is a good representation of the evolution of one segment. The zero dimensional homology provides useful information about the clustering structure of different isolates. Looking at the generators of the zero dimensional classes, we can reconstruct a hierarchical clustering structure that resembles a phylogenetic tree. For example, when studying different subtypes of influenza A circulating in aquatic birds, we clearly see that the hierarchical structure derived from the zero dimensional homology correctly captures the splits between major subtypes. This phylogenetic information can be obtained easily by classical techniques that do not use persistent homology (Figure 5.17). Similarly, with our HA data, the sequences that generate zero dimensional homology can be assembled into a tree that closely resembles the unrooted phylogenetic tree created on the viral subtypes. This same analysis can be repeated for each of the eight segments of influenza (Figure 5.18). In each case, we do not recover large bars in the barcode diagram for non-zero dimensions. The few small bars at dimension one are associated with homoplasies. In cases of vanishing

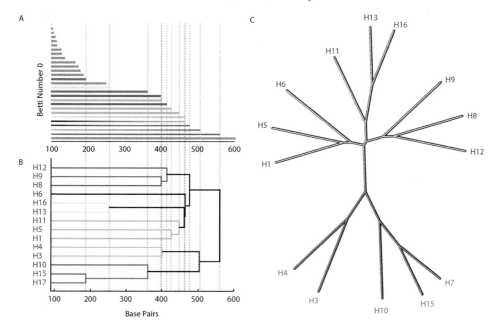

Figure 5.17 In case of vanishing higher dimensional homology, zero dimensional homology generates trees. When applied to only one gene of influenza A, in this case hemagglutinin, the only significant homology occurs in dimension zero (panel A). The barcode represents a summary of a clustering procedure (panel B), that recapitulates the known phylogenetic relation between different hemagglutinin types (panel C). Source: [100]. From Joseph Minhow Chan, Gunnar Carlsson, and Raúl Rabadán, 'Topology of viral evolution', *Proceedings of the National Academy of Sciences* 110.46 (2013): 18566–18571. Reprinted with Permission from Proceedings of the National Academy of Sciences.

higher homology, the zero dimensional homology closely follows the traditional tree structure.

However, when studying the persistent homology for several genes at the same time, large numbers of homology classes appear at dimensions one and higher, indicating pervasive reassortments. By looking in detail at the cycles in higher dimensional homologies, we can attribute these cycles to different biological processes that violate tree-like assumptions: homoplasies, recombinations or reassortments. If several sequences generate a large non-trivial class, a reassortment event likely took place among the ancestors of these isolates [100]. We can generate useful statistics based on barcode information; for instance, we can estimate how often different combinations of the eight segments cosegregate in an effort to identify preferences among the potential combinations. As an example, we rarely see cycles form with the segments that interact to form the polymerase complex PA, PB1, PB2, NP, indicating that these segments tend to cosegregate [100]. This

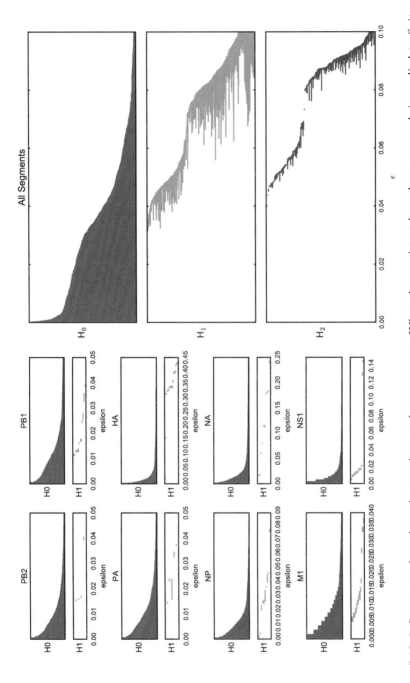

Figure 5.18 Influenza evolves through mutations and reassortment. When the persistent homology approach is applied to finite metric spaces derived from only one segment, up to small noise, the homology is zero dimensional suggesting a tree-like process (left). However, when different segments are put together, the structure is more complex revealing non-trivial homology at different dimensions (right). 3105 influenza whole genomes were analyzed. Data from isolates collected between 1956 to 2012; all influenza A subtypes.

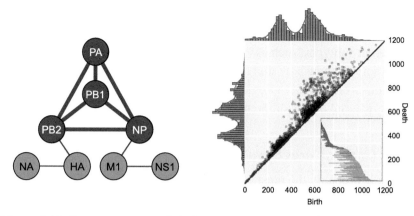

Figure 5.19 Co-reassortment of viral segments as structure in persistent homology diagrams. Left: The non-random cosegregation of influenza segments was
measured by testing a null model of equal reassortment. Significant cosegregation
was identified within PA, PB1, PB2, NP, consistent with the cooperative function of the polymerase complex. Source: [100]. Right: The persistence diagram
for whole-genome avian flu sequences revealed bimodal topological structure.
Annotating each interval as intra- or inter-subtype clarified a genetic barrier to
reassortment at intermediate scales. From Joseph Minhow Chan, Gunnar Carlsson, and Raúl Rabadán, 'Topology of viral evolution', *Proceedings of the National
Academy of Sciences* 110.46 (2013): 18566–18571. Reprinted with Permission
from Proceedings of the National Academy of Sciences.

finding is consistent with the cooperative functioning of these proteins, which
engenders negative selection against new combinations that do not cooperate as
effectively (Figure 5.19).

In addition, each of the sequenced viruses (isolates) comes with information
of where and when the virus was isolated, together with the hemagglutinin and
neuraminidase subtype. Under the assumption that smaller cycles in the non-trivial
homology classes are in some way closer genetically, one can also infer when and
where the event took place and what the types of the parental strains were. Other
relevant information is provided by the birth and death times of the class which
provide information about how genetically distant parental viruses were. Numbers
associated to one and higher dimensional classes (birth, death and size of bars in
the barcode diagram) provide a useful way to summarize the type of event. The size
of the bars associated to non-zero homology classes is also indicative of the type of
reassortment events that could occur. The persistence diagram for whole genomes
of avian flu sequences reveals bimodal topological structure (Figure 5.19, right).
In other words, there are smaller bars and larger bars. Inspection of generators
of different bars immediately reveals two types of reassortment processes. Small
bars are generated by mixing of viruses that are closely related, belonging to the

same subtype, such as two strains of H5N1 for example. Large bars, meanwhile, are generated by the mixing between the genomic material of distant viruses belonging to two different subtypes, such as H5N1 and H7N2, for example.

These examples show how studying finite metric spaces derived from large numbers of genomes can reveal biologically interesting phenomena and assess the flow of genomic material across different scales.

5.3.3 Influenza Virus Evolution and the Space of Phylogenetic Trees

Vaccination is probably the most effective method of reducing the morbidity associated with influenza infection. Administering a vaccine introduces a peptide with similar antigenic properties to circulating strains, causing the body to form protective antibodies against those strains. Every year, the World Health Organization selects strains for the Northern and Southern Hemispheres. Historically, it selected three different strains: two representing influenza A subtypes (H3N2 and H1N1) and one representing an influenza B subtype. Recently, a second influenza B subtype was added to make a quadrivalent vaccine containing peptides related to two influenza A and two influenza B strains. As viral genomes evolve, so does their antigenic presentation. This creates a continuous challenge to engineer new peptides that accurately represent circulating strains for use in vaccines. Ideally, one would like to have a universal vaccine able to target a wide spectrum of different strains and also future emerging strains. Interesting ideas in this vein have been put forward, but no such vaccine exists yet.

Hemagglutinin (HA) causes most of the body's antigenic response to influenza and it is the protein used in vaccines. The relation between the different isolates of the HA gene can be represented by a phylogenetic tree. Currently, more than 100,000 HA sequences can be found in public databases. With such a large sample of genomes, corresponding phylogenetic trees can become too complex to visualize or analyze. For instance, we would like to study these trees in terms of the geometry of the Billera-Holmes-Vogtmann metric space of phylogenetic trees (see Section 4.7.2). However, these spaces become increasingly complex as the number of leaves increases.

In Zairis et al., an approach involving reducing complicated trees to lower-dimensional structures by a process referred to as *tree dimensionality reduction* was proposed [545]. The idea behind tree dimensionality reduction is simple: instead of studying the properties of large trees like the one in Figure 5.20, one decomposes the large tree into a cloud of smaller trees by repeatedly subsampling the leaves of the large tree and taking the subtree determined by these leaves. In this way, one obtains a distribution of smaller trees that can capture a range of complex structural properties. This procedure has two advantages: first, it is far easier to visualize, extract, perform statistical analysis, and interpret different types of

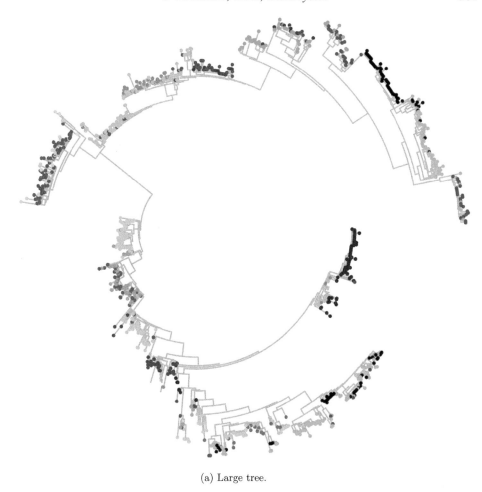

(a) Large tree.

Figure 5.20 Evolution of influenza A virus presenting clear seasonal variation. Identifying statistical patterns in large trees is often difficult. This phylogenetic tree of the hemagglutinin (HA) segment from selected 1089 H3N2 influenza viruses across 15 seasons can be subsampled for statistical analysis in lower dimensional projections. Source: [545]. Adapted from Zairis et al., Genomic data analysis in tree spaces, arXiv: 1607.07503 [q-bio.GN].

evolutionary relationships on these smaller trees; and, second, it avoids the poor scalability of phylogenetic algorithms.

As an illustration, we describe an analysis from [545] relating HA sequences from certain seasons to those of later seasons. Zairis et al. picked random strains from five consecutive seasons from a data set of 1,089 sequences of H3N2 HA collected in the United States between 1993 and 2015. Unrooted trees were generated using neighbor-joining based on Hamming distance (a visualization of the position of these trees in tree space is shown in Figure 5.21).

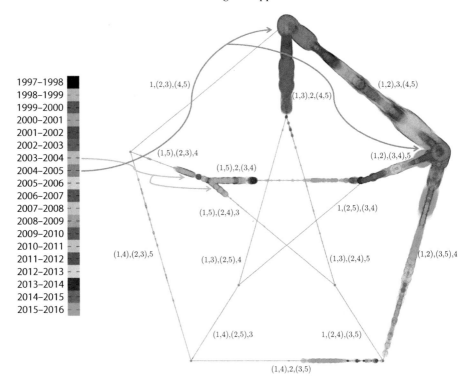

1997–1998
1998–1999
1999–2000
2000–2001
2001–2002
2002–2003
2003–2004
2004–2005
2005–2006
2006–2007
2007–2008
2008–2009
2009–2010
2010–2011
2011–2012
2012–2013
2013–2014
2014–2015
2015–2016

1,(2,3),(4,5)   (1,3),2,(4,5)   (1,2),3,(4,5)

(1,5),(2,3),4   (1,2),(3,4),5

(1,5),2,(3,4)

(1,5),(2,4),3   1,(2,5),(3,4)

(1,4),(2,3),5   (1,3),(2,5),4   (1,3),(2,4),5   (1,2),(3,5),4

(1,4),(2,5),3   1,(2,4),(3,5)

(1,4),2,(3,5)

Figure 5.21 Temporally windowed subtrees in the projectivized tree metric space $\mathbb{P}\Sigma_5$. The distribution of trees derived from five-consecutive-season windows in time are superimposed on a common set of axes for projective tree space. 1089 full-length HA segments from H3N2 were collected in New York state from 1993 to 2016. Two consecutive seasons of poor vaccine effectiveness in 2003–2004 and 2004–2005 are highlighted with green and gray arrows respectively. The green distribution strongly pairs the 1999–2000 and 2003–2004 strains, hinting at a reemergence. Source: [545]. Adapted from Zairis et al., Genomic data analysis in tree spaces, arXiv: 1607.07503 [q-bio.GN].

Most of the trees showed linear evolution between seasons (the topology of the trees follows a time ordered pattern, with ancestor of strains in a season directly related to strains in the immediate previous season), indicating genetic drift as the virus's dominant evolutionary process; however, there are distinct clusters of trees in other regions of the space that indicate reemergence of strains in the 2002–2003 season genetically similar to those circulating in the 1999–2000 season.

The data was analyzed to test the hypothesis that elevated HA genetic diversity in circulating influenza predicts poor vaccine performance in the subsequent season. This amounts to staggering the seasons sampled for distributions of trees from the season of the vaccine effectiveness label, to yield an honest prediction task. Distribution features that may intuitively predict future vaccine performance

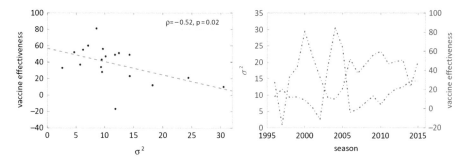

Figure 5.22 Diversity in recent circulating HA predicts vaccine failure. Negative correlation observed between vaccine efficacy in season $(t, t + 1)$ and the variance in trees generated from seasons $(t - 1, t), (t - 2, t - 1)$, and $(t - 3, t - 2)$. Source: [545]. Adapted from Zairis et al., Genomic data analysis in tree spaces, arXiv: 1607.07503 [q-bio.GN].

include the variance and the number of clusters in the point cloud. Given the limited number of temporal windows, too rich a feature space may lead to overfitting the vaccine efficacy. In Figure 5.22 we illustrate the predictions of the variance of a lagging length-3 window on vaccine effectiveness. Our notation is such that a window labeled year y would include the flu season of $(y - 1, y)$ and preceding years. The vaccine effectiveness figures represent season $(y, y + 1)$. It is clear, from both the left and right panels, that lower variance in a temporal window predicts increased future vaccine effectiveness, with a Spearman correlation of -0.52 and p-value of 0.02. The lone outlier season came in 1997–1998 [218], when the vaccine efficacy was lower than expected. In that season the dominant circulating strain was A/Sydney/5/97 while the vaccine strain was A/Wuhan/359/95.

5.4 Viral Evolution: HIV

5.4.1 Human Immunodeficiency Virus

Human Immunodeficiency Virus, or HIV, is one of the most devastating infectious diseases in modern history. Current estimates suggest 36.7 million people live with HIV today and more than 1 million die each year [244]. HIV mostly infects and destroys helper T-cells. These T-cells, also known as CD4$^+$ cells, play an essential role in the body's response to infection: they coordinate the immune response by promoting B-cells to produce antibodies and recruiting and activating neutrophils, macrophages, natural killer cells, and CD8$^+$ killer T-cells – a host of cells which neutralize invading pathogens. When CD4$^+$ T-cells die, the body's immune response is severely impaired. Pathogens that can normally be controlled by the immune system are able to infect HIV-positive patients. These "opportunistic infections" can result in the death of the infected individual. The process of CD4$^+$ T-cell depletion typically takes years and symptoms do not become evident until the cell population declines sufficiently. This clinical latent period of infection contributes to the spread of the virus through apparently healthy hosts.

HIV is a *retrovirus*. Retroviruses encode their genome in single-stranded and positive-sense RNA. When a retrovirus infects a cell, it converts its genome to double-stranded DNA in the cell's cytoplasm by first creating an antisense strand of DNA complementary to its RNA genome (cDNA) and then forming a positive-sense DNA strand complementary to the cDNA. The conversion of RNA to DNA is the opposite of the usual process in human cells, in which RNA is generated from a DNA template. It is termed reverse transcription and is facilitated by the viral enzyme reverse transcriptase (RT). After the creation of double-stranded DNA in the cytoplasm, the DNA is transported to the nucleus, where it is incorporated into the human genome. By this means, the virus gains access to the host cell's genomic machinery and its abilities to transcribe mRNA and thus the ability to translate viral proteins and replicate the viral genome (Figure 5.23).

Retroviruses are classified into two subfamilies (Orthoretroviridae and Spumaretroviridae) that include some oncoviruses, such as Rous sarcoma virus, which we will briefly describe when talking about cancer. HIV belongs to the Lentivirus genus, a taxon of retroviruses with long incubation periods before they become symptomatic and acquire the capability to infect non-replicating cells. The virions, or viral particles, of retroviruses have capsids, which surround and protect their genome, and envelopes (lipid bilayer surrounding the capsids) borrowed from the host-cell plasma membrane (Figure 5.23); specifically, HIV has a conical capsid of about 100 nm. Retroviruses contain two identical copies of the RNA genome, each around 10,000 bases in size. There are three major genes present in all retroviruses.

- The *gag* gene codes for the proteins that generate the capsid.
- The *pol* gene carries information about the enzymes necessary for replication and reverse transcription (i.e., reverse transcriptase), for integrating viral DNA into the host genome (i.e., integrase) and for cleaving viral polyproteins to activate them (i.e., protease).
- The *env* gene codes for the glycoproteins that bind to the T-cell's receptors and allow the virus to invade the host cell. Env translates directly to the polyprotein gp160, which is cleaved into two smaller proteins: gp120, which binds to the CD4 receptor and the co-receptors (CCR5 or CXCR4), and gp41, which promotes fusion of the cell membrane and viral envelope.

In addition to these three long genes, HIV has at least six smaller proteins that are involved in genomic regulation and interaction with host machinery; multiple roles

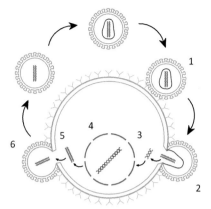

1. Virion attaches to receptor and co-receptors.
2. Viral RNA diploid genome is released into cytoplasm.
3. Reverse transcription and integration of provirus into cell genome.
4. Transcription and translation of viral proteins.
5. Viral RNA replication.
6. Budding from cell.

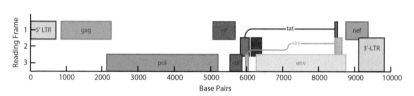

Figure 5.23 Life cycle and genomic structure of the HIV virus. Top: Life cycle of HIV. The virion attaches to CD4 receptors and co-receptors on the membrane of the CD4$^+$ T-cell, allowing for the fusion of the viral envelope with the T-cell membrane and the release of the viral RNA into the cell's cytoplasm. The viral reverse transcriptase reverse transcribes the viral genome into double-stranded DNA, which is transported into the cell's nucleus and integrated into the host's DNA. After integration, the host cell's genomic machinery treats the integrated virus, or provirus, as part of the host genome, generating mRNA and viral protein, and copies of the RNA genome. Two copies of the HIV genome are packaged in each virion and the virions bud from the host cell. In the final process of matura-tion, cell-free virions assemble conical capsids that stabilize their genomes. These mature virions are now able to infect other cells. Bottom: The genome of HIV con-sists of three large genes, gag, pol and env, common to most retroviruses, and six small genes that arise from subsequent splicing events.

have been reported for each of these proteins. The Trans-Activator of Transcription (Tat) is a small protein of around 100 amino acids that binds to cellular factors in order to increase transcription of all HIV genes, including itself, thus creating a positive-feedback loop of transcription. The regulator of the expression of virion proteins (Rev) is necessary for the synthesis, stability and transport of several viral mRNAs. The Viral protein R (Vpr) has about 100 amino acids and among other functions, transports the pre-integrated viral genome into the host's nucleus. The Viral infectivity factor (Vif) inhibits the cellular protein APOBEC3G. APOBEC proteins are cytidine deaminases, proteins that induce mutations in cytidines, that catalyze the deamination of cytidine to uridine, introducing a large number of C-to-U or C-to-T mutations in RNA or DNA respectively in localized settings.

APOBEC3G enters the virion and mutates the viral genome, resulting in hyper-mutated genomes causing defective viruses. Vif prevents APOBEC3G activity by targeting it for proteasomal degradation [454, 544]. Beyond blocking APOBEC3G activity, it has also been associated with the infectivity of virions. Finally, the Viral protein Unique (Vpu) has been implicated in the degradation of host-cell CD4 receptors and the release of virions.

It remains unclear exactly when, where, and how HIV became a human pathogen [453]. The disease associated to the virus, the Acquired Immunodeficiency Syndrome, or AIDS, is caused by two related retroviruses, HIV-1 and HIV-2. In the developed world, AIDS was identified through a sudden increase in rates of opportunistic infections and very rare tumors in injection drug users and men who have sex with men. The opportunistic infections included *Pneumocystis jirovecii* pneumonias, previously reported to occur in individuals with highly compromised immune systems, and the tumors included Kaposi sarcoma, later shown to be itself caused by an infection [185]. In 1983, two groups in the United States and France reported a new retrovirus associated with this immunodeficient state [36, 190]. For this work, Françoise Barré-Sinnousi and Luc Montagnier won the Nobel Prize in Physiology or Medicine in 2008 (Figure 5.24). A second HIV virus, named HIV-2, was reported in West Africa in 1986 with a similar, although not identical, genomic structure to HIV-1.

The virus was then identified in the general population living in Africa [410]. Infection rates indicated that the virus was already circulating in African

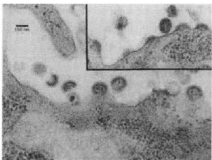

Figure 5.24 Identification of HIV as a cause of AIDS. Left: Electron microscopy of sections of HIV virus producing cells. Source: [36]. From F. Barré-Sinnoussi et al., Isolation of a T-Lymphotropic Retrovirus from a Patient at Risk for Acquired Immune Deficiency Syndrome (AIDS), *Science*, New Series, Vol. 220, No. 4599, pp. 868–871, 1983. © 1983 American Association for the Advancement of Science. Reprinted with permission from AAAS. Right: Françoise Barré-Sinnousi and Luc Montagnier, who won the Nobel Prize in Physiology or Medicine in 2008 for the discovery of the virus. Source: © The Nobel Foundation. Photo: Ulla Montan.

populations before it was identified in the Western world. More recently, sampling of HIV viruses in Central Africa has shown a higher genetic diversity compared with other viruses collected all around the world, suggesting an older African origin [516]. That was supported by retrospective studies that identified the virus in blood samples from patients in Kinshasa at the end of the 1950s [540]. A Norwegian sailor, Arvid Darre Noe, was reported to be infected with HIV-1 group O, most likely in 1961 or 1962 when working in Cameroon [186]. The closest relatives of these viruses infecting other species can be found in African primates. It is now believed that there were multiple transmission events leading to the major subclades of the virus. Some of these transmission events, such as that of group M from chimpanzees in Central Africa, led to rapid spread throughout the human population. A recent study using HIV-1 env sequence data from different countries in the Congo River basin suggests that the most recent common ancestor of all group M strains dates back to 1920 in the Democratic Republic of Congo [169]. Several societal changes occurring at that time, including the growth of African cities and the mobility of workers, have been discussed as potential factors contributing to the spread of the virus.

5.4.2 Viral Recombination in HIV

HIV is notorious for its high diversity, created and maintained not only by its high mutation rate but also by frequent recombination. Using data from patients, mutation rates of HIV have been estimated to be $(4.1 \pm 1.7) \times 10^{-3}$ per base per cell [129]. Many of these mutations, however, are lethal to the virus and only a small fraction can make functional viruses. The major causes of mutations in vivo are the reverse transcriptase and cytidine deaminases (in the process of retrotranscription), although human DNA-dependent RNA polymerase can also contribute when generating viruses from the integrated provirus. On average, mutations caused by RT only constitute 2% of all mutations; but this statistic varies to a large degree across patients. Patients that rapidly progress to the symptomatic stage experience fewer hypermutations (accumulation of a large number of mutations in a virus), suggesting that cytidine deaminases play an important role in HIV pathogenesis.

Because the genome of HIV is not segmented, reassortment does not occur. Instead, recombination is the major driver of horizontal evolution. RT's polymerase can use either genomic RNA strand as a template for reverse transcription and it can switch between strands during the process (Figure 5.25). If the two RNA strands packaged in a virion include distinct mutations or come from different parent viruses, template switching by RT can create a mosaic genome.

These recombinations can occur commonly, and recombinants can become the dominant forms circulating in large fractions of host populations. Circulating

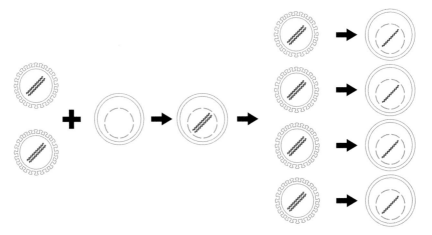

Figure 5.25 Recombination in HIV. The genome of HIV is diploid, containing two more-or-less identical copies of the RNA genome. Virions, however, can be packaged with two very different copies if two distinct HIV viruses co-infect the same cell. When reverse transcribing the RNA from these virions into a single copy of DNA, the polymerase can jump between the two strands, generating a mosaic virus containing fragments of both parental strands.

Recombinant Forms, or CRFs, are common recombinants deriving from recombination between viruses of different subtypes. The notation and naming of CRFs is complex because different "pure" parent subtypes can generate many different mosaic viruses. The breakpoint of recombination can occur anywhere along the genome and multiple breakpoints are common. Barred by frequent recombination, drawing an evolutionary tree from a single gene is virtually impossible. As expected, and in contrast to influenza, when applying persistent homology to HIV, individual genes reveal large numbers of one and higher dimensional homology classes, indicating a history of reticulate events, most likely recombination (Figure 5.26). When concatenating the large genes of the virus, large recombination events are uncovered, relating multiple parental strains of subtypes A. An example of a long bar observed in two dimensional homology is shown in Figure 5.27, revealing a complex recombination event between major HIV subtypes, B, C, D, F, and 13cpx, a complex recombinant strain.

5.4.3 Viral Recombination in Late-Stage HIV Infection

We have seen that untreated HIV can lead to an impaired immune system. However, there are other symptoms that occur in patients with long-term infections. HIV-associated dementia (HAD) is a condition associated with long-term viral progression and low CD4$^+$ T-cell counts. This condition is the most severe of

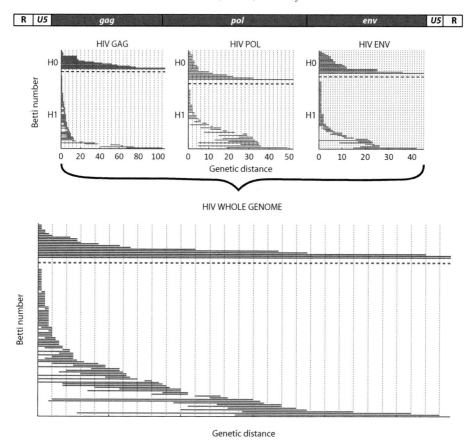

Figure 5.26 Persistent homology reveals recombination within genes and across the genome. Unlike in influenza, persistent homology barcodes of HIV reveal intragenic recombination in the three major HIV genes gag, pol and env. When concatenated and run through the persistent homology pipeline, the multi-gene fragments have homology classes in dimensions one and higher. Source: [100]. From Joseph Minhow Chan, Gunnar Carlsson, and Raúl Rabadán, 'Topology of viral evolution', *Proceedings of the National Academy of Sciences* 110.46 (2013): 18566–18571. Reprinted with Permission from Proceedings of the National Academy of Sciences.

the HIV-associated neurocognitive disorders, which are believed to result from exposure of the brain to high levels of HIV-1 following breach of the blood-brain barrier by HIV-infected monocytes [287]. While instituting combination antiretroviral therapy early in infection may prevent neurocognitive decline, later initiation of therapy does not appear to reverse pre-existing symptoms [503]. Understanding the nature of the viral population in the brain is therefore of ongoing interest. Virus sampled from the cerebrospinal fluid (CSF) or brain of HAD-affected individuals is often genetically distinct from that of the peripheral blood, suggesting

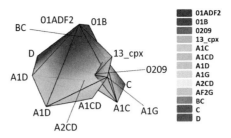

Figure 5.27 Here is a polytope representing complex recombination events with multiple parent strains. This polytope represents a two-dimensional class in persistent homology. Each vertex of the polytope represents a sequence that is colored according to HIV-1 subtype. Source: [100]. From Joseph Minhow Chan, Gunnar Carlsson, and Raúl Rabadán, 'Topology of viral evolution', *Proceedings of the National Academy of Sciences* 110.46 (2013): 18566–18571. Reprinted with Permission from Proceedings of the National Academy of Sciences.

continuous viral replication in the brain as a potential cause of HAD [311]. Moreover, viral recombination may occur more frequently within the populations found in the brains of individuals affected with severe HAD than in other HIV-infected individuals, further implicating unchecked viral replication as a cause of HAD.

In this section, we describe how tools of persistent homology can be used to characterize this viral recombination to study intra-host HIV evolution in patients with long-term viral progression. In particular, we are interested in understanding how recombinant viruses spread between different tissues. This can be done by comparing genomic sequences from the central nervous system (CNS) to sequences obtained from other tissues. Zigzag persistent homology [93], described in Section 2.5, provides a formalism to study and compare events across different populations.

Lamers et al. [310, 311] obtained tissue samples from the autopsies of 11 individuals who died from AIDS. They extracted genomic HIV DNA and amplified a 3.3 kb fragment stretching from env to the 3′ LTR by PCR, cloned it, and sequenced it. They published sequences of the glycoprotein gp120 ($\approx$ 1200 bp) found in the peripheral tissues of seven individuals and, for five of the individuals (Patients AZ, BW, CX, DY, GA), included sequences from the CNS. Patients AM and IV only had sequences from non-CNS tissues reported. A summary of the data is shown in Table 5.1.

Recall from Section 2.5 that zigzag persistence provides a formalism to describe "filtrations" where arrows can go in both directions; for example, when the data can be modeled by a mathematical object that first "builds up" (the "zig") and later "breaks down" (the "zag") [91, 93]. For sequences sampled from two related subpopulations, this framework provides a way to divide recombination events into four classes:

Table 5.1 *Summary of patient data: first column is the identifier of the patient, second and third columns are the number of sequences obtained from the central nervous system, fourth columns is the GenBank accession numbers*

Patient	# CNS sequences (unique sequences)	# Non-CNS sequences (unique sequences)	Accession Numbers
AZ	35 (33)	52 (48)	HM001587 – 1673
DY	107 (99)	59 (54)	HM002004 – 2169
BW	103 (99)	18 (18)	HM001674 – 1794
CX	162 (152)	47 (43)	HM001795 – 2003
GA	75 (73)	57 (55)	HM002170 – 2301
AM	—	225 (210)	HM001362 – 1586
IV	—	181 (177)	HM002302 – 2482

1. event occurring in the first population, but not the second;
2. event occurring in the second population, but not the first;
3. event detectable in either population alone (typical if the two populations are very closely related);
4. events involving both populations, detectable only in some union of their sequences and not in either population individually; this class represents the case of gene flow between genetically distinct populations.

Consider the reticulate phylogeny shown in Figure 5.28A, where the red nodes (left node in each numbered pair) are sampled from one population (e.g., geographic region or anatomical site) and the yellow nodes (right node in each pair) are sampled from a second population. Computing persistent homology identifies the recombination event as a topological loop that appears at particular scales (see Figure 5.28C). Visually, it is clear that a single recombination event has affected both populations, and can be seen from either population. Zigzag persistence allows us to recover this computationally. Starting from the first population alone, a loop is detected (Figure 5.28B). Complexes are built up (the "zig") by adding sequences from the second population (Figure 5.28C) and broken down (the "zag") by removing sequences from the first population (Figure 5.28D). The zigzag barcode captures the fact that the loop in panel B and the loop in panel D are representatives of the same homology class, indicating that the same recombination event generated them – a class 3 event. The ancestry represented in panel E contains a recombination event that brings together the red and yellow populations. The sequence of simplicial complexes starts as a single line (panel F), builds up to a square (panel G), and breaks down to a different line (panel H). As a loop

Table 5.2 *Patient status and putative recombination events indicated by persistent homology*

Patient	HAD status	Degree of neuropathology	# CNS events	# Cross-site events	# Non-CNS events
AZ	None	3	0	1	2
DY	Acute	1	2	5	1
BW	Progressive	2	3	0	0
CX	Progressive	5	8	0	1
GA	Progressive	5	5	7	8
AM	n/a	n/a	—	—	7
IV	n/a	n/a	—	—	9

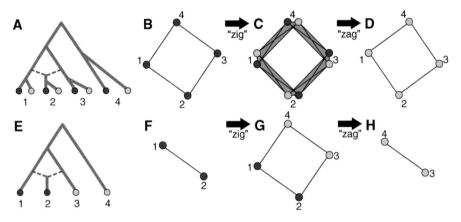

Figure 5.28 Schematic of zigzag persistence used to identify inter-population recombination (see text).

appears only when both populations are included (class 4 recombination event), this identifies exchange of genomic material between populations.

Summarizing, persistent homology was used to identify putative recombination events. Where sequences from both CNS and non-CNS sources were available, zigzag persistence was used to classify each recombination as occurring in the CNS, outside the CNS, or between CNS and non-CNS sequences (Table 5.2). The two patients exhibiting progressive HAD and the most severe neuropathology – CX and GA – also had the greatest number of recombination events localized in the CNS, suggesting that frequent viral recombination contributes to this disorder. Apart from this similarity, the viral population structure was strikingly different for the two patients: patient GA's CNS sequences were relatively more intermingled with the non-CNS sequences, with frequent recombination events occurring between the two anatomical groups. In contrast, the two groups in patient

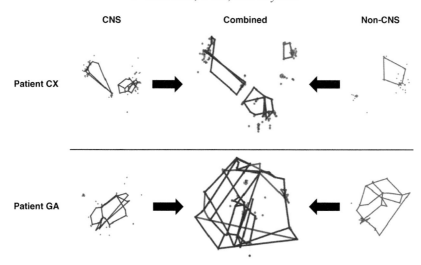

Figure 5.29 Phylogenetic networks of HIV-1 gp120 sequences obtained from patients CX and GA. Each node represents one sequence; larger nodes show sequences that were sampled multiple times. Blue nodes were sampled from the CNS; red nodes were sampled from elsewhere in the body. The position of each node is determined by the first two principal components (computed via MDS) of genetic distance (Hamming distance). The network backbone (thin gray edges) is a minimum spanning tree, and the thick red and blue edges are generators of cycles identified by persistent homology. Red cycles denote putative recombination events that involve sequences sampled fully outside the CNS; blue cycles denote events that involve some sequences from the CNS.

CX were more clearly separated. Figure 5.29 depicts phylogenetic networks of the sequences for these two patients, illustrating this difference in structure.

Since the number of sequences sampled can affect the number of cycles observed, in Table 5.3 we show $\hat{\rho}_{PH}$, an estimate of the population-scaled recombination rate, as described in Section 5.7. Again patients CX and GA stand out as having CNS populations with the highest recombination rate, suggesting that the association of HAD with severe neuropathology is not an artifact of the sampling procedure.

If the CNS and non-CNS populations are completely distinct, then the population-scaled recombination rate ρ for the combined population will equal the sum of the ρ values for each individual population. For most patients, the value of $\hat{\rho}_{PH}$ computed using all sequences is in fact less than the sum of the two $\hat{\rho}_{PH}$ values computed from the CNS and non-CNS samples. This is consistent with the two populations being partially intermingled and sharing common ancestral recombination events, such that much of the historical signal can be obtained by sampling just a single population. Patient DY was unique in that $\hat{\rho}_{PH}$ for the combined population exceeded the sum of the two individual values. Consistent with this observation,

Table 5.3 $\hat{\rho}_{PH}$ estimated from different sources

Patient	$\hat{\rho}_{PH}$ from CNS sequences	$\hat{\rho}_{PH}$ from non-CNS sequences	Sum of both estimates at left	$\hat{\rho}_{PH}$ from all sequences
AZ	0	6.7	6.7	4.4
DY	4.0	3.0	6.9	11.0
BW	5.9	0	5.9	5.3
CX	12.7	3.7	16.3	12.5
GA	12.5	27.4	39.9	35.2
AM	—	9.2	—	—
IV	—	13.1	—	—

patient DY was also the only individual in which the majority of recombination events observed occurred between representatives of the two populations ("cross-site events" in Table 5.2). These observations suggest considerable recent traffic of virus across the blood-brain barrier in this patient, perhaps borne by increased traffic of macrophages stimulated by the *Mycobacterium avium* infection that started a year prior to death. Although there is statistically significant clustering of the two populations, it is weakest in this patient compared to the others [255].

5.5 Other Viruses

Most of our knowledge of microbes relates to human pathogens, of which there are on the order of 10^3 species, representing a tiny fraction of all microbial species. It has been estimated that there are 10^{31} viruses on this planet [158, 488], constituting the largest and most diverse biological population on Earth. About 8% of our DNA is derived from remnants of viruses that once infected our ancestors. While all cells in the three domains of life store their genomes as double-stranded DNA, viruses use RNA and DNA in different forms. The taxonomy of viruses is extremely complex as there are no common structures shared by all viruses, and there is no clear evidence that all viruses share a common origin. The Baltimore classification [28], a common classification based on the type of genomic material and replication strategy, divides viruses into seven different groups.

- Group I: double-stranded DNA viruses.
- Group II: single-stranded DNA viruses. Unlike cells, these viruses use only one strand of DNA.
- Group III: double-stranded RNA viruses.
- Group IV: single-stranded RNA viruses, with genomic material encoded in the positive-sense strand.

- Group V: single-stranded RNA viruses, with genomic material encoded in the negative-sense strand.
- Group VI: single-stranded positive RNA viruses that use reverse transcription.
- Group VII: DNA viruses that use reverse transcription.

We have seen that influenza uses negative-sense RNA for its genomic material, so it is classified in the type V group. HIV is a retrovirus, using RNA and reverse transcription, and thus it is classified as type VI. An example of a type I virus is the Epstein-Barr virus, which causes mononucleosis, and which we will encounter again when talking about cancer. This classification may be neat, but it does not provide information about the origins of viruses, and two viruses belonging to the same group may have very little in common genetically, while viruses from different groups may have related genes. Such similarities could be due to a common ancestor or to different exchange modes of genomic material.

The same persistent homology approach that we used to study reassortment in influenza and recombination in HIV can be applied to study other viruses. Flaviviridae is a family of viruses comprising several different genera, including hepaciviruses and flaviviruses. Flaviviridae are positive-sense single-stranded RNA viruses (group IV), whose ability to perform homologous recombination through RNA polymerase template switching has been debated. Sporadic recombinant strains have been detected for hepaciviruses like hepatitis C [120] and flaviviruses like dengue virus [539] and West Nile virus [409]. In some of these cases, the evidence for recombination remains controversial [426]. One can use persistent homology to study the extent of recombinations in the Flaviviridae family [100]. Comparing using different measures such as the size of the longest bar (TOP) and the number of bars in the sample time (ICR), it was found that hepatitis C showed some but lower recombination than in HIV (Figure 5.30). No high-dimensional homology was found in dengue or West Nile virus, suggesting that recombination rarely occurs in these viruses.

In type V viruses, like influenza, recombination is considered to be an even less frequent event like in Newcastle or Rabies virus. Persistent homology does not identify high-dimensional classes for rabies, while the analysis of Newcastle virus confirmed a low ICR but a non-vanishing TOP.

5.6 Bacterial Evolution

Bacteria are the most common cells on Earth and even in our bodies. From marine samples, biologists estimate that there are 3×10^{28} bacterial cells on Earth. Despite being less numerous than viruses, these prokaryotes represent more than 90% of Earth's biomass [158, 488]. The bacteria in a human's gut collectively weigh about

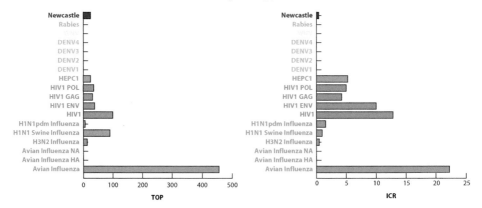

Figure 5.30 Recombination across different viruses. Left: A topological obstruction is estimated using the maximum barcode length in dimension one. Right: The rate of irreducible cycles is defined as the number of one dimensional bars in the barcode diagram divided by the time spanned by the sequence collection. Source: [100]. From Joseph Minhow Chan, Gunnar Carlsson, and Raúl Rabadán, 'Topology of viral evolution', *Proceedings of the National Academy of Sciences* 110.46 (2013): 18566–18571. Reprinted with Permission from Proceedings of the National Academy of Sciences.

a kilogram. In a gram of dental plaque there are 10^{11} bacteria. Only a small fraction of bacterial species has been characterized so far. Although large multidisciplinary efforts are under way, such as the Earth Microbiome Project (which plans to study 200,000 samples) it is unlikely that we will have a comprehensive atlas in the near future.

5.6.1 Horizontal Gene Transfer in Bacteria

Bacterial genomes vary widely in size; typically they are a few megabases long. *Mycoplasma genitalium*, an intracellular pathogenic bacterium, has one of the smallest genomes at half a megabase. *Escherichia coli*, a common bacterium living in our intestine and used in laboratories, has a genome of 4.6 megabases. Its mutation rate has been found to be 5.4×10^{-10} per base per replication, or 0.0025 per genome per replication [149, 150]. Mutation rates vary between species, but also with changes in the ambient environment. For example, it has been shown that starving bacteria have dramatically increased evolutionary rates [72, 86].

In addition to mutations, horizontal gene transfer (HGT), the exchange of genomic material in a non-vertical way, constitutes a major form of genetic innovation in bacteria. Borrowing genes through HGT allows for rapid adaptation to challenging environments [389]. As we will see, HGT has been found to be a major factor in the spread of antibiotic resistance [134]. Transfer of genetic material is well known since the work of Lederberg on bacterial conjugation in the 1940s [318] (Lederberg received the 1958 Nobel Prize for this work). Until the advent of large scale genomic studies, it was widely thought that HGT was a rare event. Now it is known that effects of HGT are found pervasively across many different bacterial species [305]. In some cases the effect of HGT is extremely dramatic, in particular when genes are imported across different domains of life. For instance, the genomes of some bacteria contain a large fraction of archaeal genes. The best known example of this borrowing is that of hyperthermophilic bacteria, which are bacteria that can tolerate temperatures near boiling, such as *Aquifex aeolicus* and *Thermotoga maritima*. In genomic analysis, HGT is usually identified through incongruent tree phylogenies, with different gene histories represented by incompatible tree topologies. The widespread effect of HGT across and within different domains of life has led some to question the existence and usefulness of representing the relationship between distant bacterial species in a Tree of Life [147].

There are three main molecular mechanisms by which HGT can occur (see Figure 5.31) [389].

- *Transformation*: the uptake of naked, free-floating DNA from the environment.
- *Transduction*: the transfer of genomic material through a virus intermediate. Viruses that infect bacteria, known as bacteriophages or phages, mediate the transduction process. The amount of DNA is limited by the size of a viral capsid, usually about 100,000 bases. Phages also can encode proteins that can help the integration of the new material into the receptor cell.
- *Conjugation*: transfer of genomic material by cell-cell contact. For this to occur, the cytoplasms of the bacteria must be connected. Bacteria often connect to each other using an appendage called a pilus. The pilus exists precisely for this role, demonstrating that HGT can be advantageous for bacteria.

HGT can be hindered by disruptions in any of the following processes: in the donor, the ability to generate genomic material in the form of free DNA or plasmids; a transportation method for the DNA, such as the existence of phages that

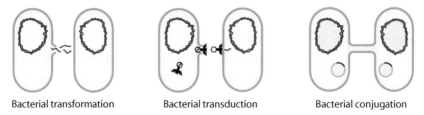

Bacterial transformation   Bacterial transduction   Bacterial conjugation

Figure 5.31 A few mechanisms of horizontal gene transfer in bacteria: transformation, transduction and conjugation.

can effectively infect both the donor and recipient; and in the recipient, the capacity to uptake and integrate the new DNA.

Experimentally, it has been shown that HGT between species decreases with increasing genetic distance [182]. In the following section, we will employ genomic data from large databases and tools from TDA to study the frequency and patterns of intra- and inter-species HGT in bacteria.

5.6.2 Pathogenic Bacteria

As previously mentioned, horizontal exchange occurs when a donor bacterium transmits foreign DNA into a genetically distinct bacterial strain; for instance, in Germany, 2011, *E. coli* acquired the Shiga toxin, typical to the *Shigella* genus, via phage-mediated gene transfer, and caused a serious outbreak of foodborne illness [435]. Control of bacterial pathogens is hampered by rampant horizontal gene transfer, which allows bacteria to acquire genes conferring resistance to commonly used antibiotics [382, 391, 497]. Genes for resistance can be transferred between strains of both the same and different species existing in the same environment. Elements of bacterial genomes demonstrating evidence of foreign origin are known as genomic islands and may be associated particularly with phenotypic effects, such as virulence or resistance to antibiotics.

Tools from topological data analysis can help to characterize the frequency and scale of horizontal gene transfer in bacteria, elucidating issues of significant public health relevance, such as the spread of antibiotic resistance in *Staphylococcus aureus* and the human microbiome's role as a reservoir for antibiotic resistance genes.

5.6.3 Multilocus Sequence Typing Analysis

Within a single bacterial species there can be many genetically distinct strains. Different strains can have important functional differences. For example, some strains may be more virulent than others and some may be more susceptible to the immune responses generated by vaccines. Multilocus sequence typing (MLST) is a method for detecting particular bacterial strains that does not require whole-genome sequencing. It relies on the fact that strains can be identified from certain representative genomic loci selected from regions within housekeeping genes. Typically the size of each locus is about 500 base pairs.

Curated MLST data from laboratories around the world is available in large online databases. Often there are thousands of strains identified within a single pathogenic species (over 10,000 in the case of *Neisseria* spp.). MLST data can be used to study horizontal exchange of genomic material in bacteria. Because different species have different loci, one can only examine horizontal exchange within species. Furthermore, because all of the selected loci exist within a few housekeeping genes, our analysis does not provide information about events involving genes other than these housekeepers.

The data used here comes from PubMLST [277]. For each of twelve bacterial species, one can construct a pseudogenome by concatenating the typed sequence at each locus. Using the Hamming distance metric, one can calculate a pairwise distance matrix between strains and compute persistent homology on the resulting metric space. In Figure 5.32, we show the persistent homology barcodes associated to the witness complex (recall Definition 2.7.3) with 250 landmark points. We plotted the H_1 barcode diagrams for *K. pneumoniae* and *S. enterica*. Based on the observed range of recombinations, one can identify two distinct species profiles: *K. pneumoniae* recombines solely at one short-lived scale, while *S. enterica* recombines both at the short-lived scale and also at another longer-lived scale. This analysis can be repeated for each species; we plotted the results as persistence diagrams in Figure 5.33. For the bulk of pathogens, there are three major scales of recombination: one short-lived scale at intermediate distances, another longer-lived scale at intermediate distances, and a third short-lived scale at longer distances. *H.*

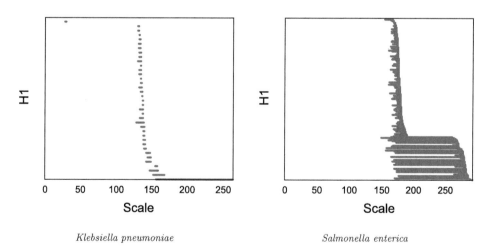

Klebsiella pneumoniae   *Salmonella enterica*

Figure 5.32 Barcode diagrams reflect different scales of genomic exchange in *K. pneumoniae* and *S. enterica*. Source: [161]. Reprinted by permission from Springer Nature: Emmett K. J., Rabadán R. (2014) Characterizing Scales of Genetic Recombination and Antibiotic Resistance in Pathogenic Bacteria Using Topological Data Analysis. In: Ślęzak D., Tan A. H., Peters J. F., Schwabe L. (eds) *Brain Informatics and Health*. BIH 2014. Lecture Notes in Computer Science, vol 8609. Springer, Cham. © Springer International Publishing Switzerland 2014.

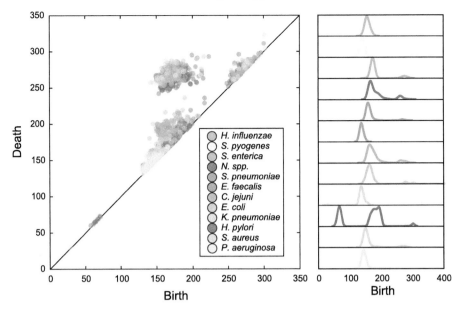

Figure 5.33 On the left, the H_1 persistence diagram for the twelve strains of pathogens selected for this study MLST profile data. Observe three scales of recombination. On the right, the birth time distribution for each strain. There is an earlier scale of recombination present in *H. pylori* not observed in the other species. Source: [161]. Reprinted by permission from Springer Nature: Emmett K. J., Rabadán R. (2014) Characterizing Scales of Genetic Recombination and Antibiotic Resistance in Pathogenic Bacteria Using Topological Data Analysis. In: Ślęzak D., Tan A. H., Peters J. F., Schwabe L. (eds) *Brain Informatics and Health*. BIH 2014. Lecture Notes in Computer Science, vol 8609. Springer, Cham. © Springer International Publishing Switzerland 2014.

pylori is a clear outlier, tending to recombine at significantly lower scales than the other pathogens.

A relative recombination rate can be defined by counting the number of H_1 loops across the filtration and then dividing by the number of samples for that species. The results of this analysis are shown in Figure 5.34, which demonstrates that there exist a wide range of recombination profiles among bacterial species. *S. enterica* and *E. coli* have the highest recombination rates, while *H. pylori* recombines at a substantially lower rate than the others. This analysis suggests that *H. pylori*'s core genome is comparatively impervious to recombination except by closely related strains.

5.6.4 Protein Family Analysis

MLST data can provide information about the exchange of genomic material in typed loci within related species. In order to study horizontal exchange between

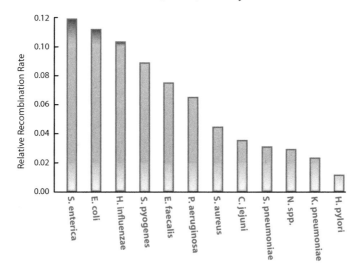

Figure 5.34 Relative recombination rates computed by persistent homology from MLST profile data. Source: [161]. Reprinted by permission from Springer Nature: Emmett K. J., Rabadán R. (2014) Characterizing Scales of Genetic Recombination and Antibiotic Resistance in Pathogenic Bacteria Using Topological Data Analysis. In: Ślęzak D., Tan A. H., Peters J. F., Schwabe L. (eds) *Brain Informatics and Health*. BIH 2014. Lecture Notes in Computer Science, vol 8609. Springer, Cham. © Springer International Publishing Switzerland 2014.

different species, one needs data that are relevant across bacterial species. One approach is to consider the presence or absence of protein families among different bacterial species. Protein families are proteins with similar sequence and function. The presence of a member of a protein family in a strain could be due to a horizontal gene transfer event between strains or species.

The presence or absence of protein families can be converted into a binary vector for each bacterial strain. One can use FigFam protein annotations in the Pathosystems Resource Institute Center (PATRIC) database, one of the most comprehensive databases for genomic annotations, including pathogenic strains [527]. When this analysis was performed FigFam contained over 100,000 protein families comprising over 950,000 unique proteins [350]. Binary vectors describing the presence or absence of protein families were used to calculate a distance matrix and compute the persistent homology in this space. Figure 5.35 shows the persistence diagram relating the scale and structure between species. Different species have a far more diverse topological structure in this space than in the MLST space, as well as a wide range of recombination scales. The large scales of exchange in *H. influenzae* suggest it is readily capable of acquiring novel genetic material from quite distantly related strains. It is known that HGT in *H. influenzae* can lead to the acquisition of virulent factors [199]. Furthermore, it has been observed that differences between

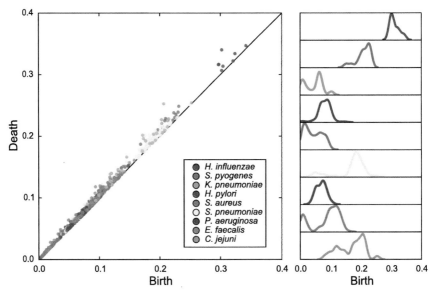

Figure 5.35 Persistence diagram for a subset of pathogenic bacteria, computed using the FigFam annotations compiled in PATRIC. Compared to the MLST persistence diagram, the Figfam diagram has a more diverse scale of topological structure. Source: [161]. Reprinted by permission from Springer Nature: Emmett K. J., Rabadán R. (2014) Characterizing Scales of Genetic Recombination and Antibiotic Resistance in Pathogenic Bacteria Using Topological Data Analysis. In: Ślęzak D., Tan A. H., Peters J. F., Schwabe L. (eds) *Brain Informatics and Health*. BIH 2014. Lecture Notes in Computer Science, vol 8609. Springer, Cham. © Springer International Publishing Switzerland 2014.

H. influenzae strains are more commonly associated to recombination than to point mutations [345].

5.6.5 Antibiotic Resistance in Staphylococcus aureus

S. aureus is a gram positive bacterium found commonly in the upper respiratory tract and nostrils. Some strains are capable of causing severe infections in high-risk populations, particularly in a hospital setting. Therefore, the emergence of antibiotic resistant *S. aureus* is a significant clinical concern. Methicillin resistant *S. aureus* (MRSA) strains are resistant to β-lactam antibiotics, which include cephalosporin and penicillin. The gene *mecA*, part of Staphyloccoccal cassette chromosome mec *(SCCmec)*, codes for a dysfunctional penicillin-binding protein 2a (PBP2a), prohibiting the β-lactam primary mechanism and causing resistance [273]. Characterizing the spread of resistance within the *S. aureus* population is clearly of critical clinical import.

To address this question, one can use FigFam annotations in PATRIC, as described in the previous section. PATRIC contains genomic annotations for 461 strains of *S. aureus*, collectively spanning 3578 protein families. One can perform a clustering analysis using Mapper [268]. By selecting as filter function the first two singular values, it can be observed that the resulting graph structure exhibits two main clusters with a thin "bridge" connecting them, as shown in Figure 5.36. These two clusters accord with previous phylogenetic studies which used multilocus sequence data to identify two major population groups [124].

142 of the 461 strains of *S. aureus* in PATRIC carry the *mecA* gene. When we color based on an enrichment for *mecA*, a stronger enrichment can be observed in the cluster on the right (Figure 5.36). This analysis would suggest that *β*-lactam

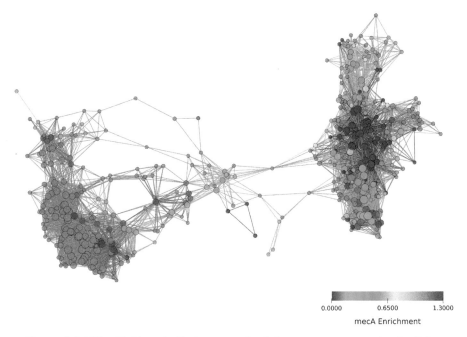

0.0000   0.6500   1.3000

mecA Enrichment

Figure 5.36 The FigFam similarity network of *S. aureus* constructed using Mapper as implemented in Ayasdi Iris. One can use a Hamming distance metric and primary and secondary metric SVD filters (res: 30, gain 4×, eq.). Node color is based on strain enrichment for *mecA*, the gene conferring *β*-lactam resistance. Two distinct clades of *S. aureus* are visible, one of which already shows significant drug resistance. The growing enrichment for *mecA* in the second clade is clinically worrisome. Source: [161]. Reprinted by permission from Springer Nature: Emmett K. J., Rabadán R. (2014) Characterizing Scales of Genetic Recombination and Antibiotic Resistance in Pathogenic Bacteria Using Topological Data Analysis. In: Ślęzak D., Tan A. H., Peters J. F., Schwabe L. (eds) *Brain Informatics and Health*. BIH 2014. Lecture Notes in Computer Science, vol 8609. Springer, Cham. © Springer International Publishing Switzerland 2014.

resistance has already become dominant in that clade, likely as a result of selective pressures. More strikingly, one observes that while *mecA* enrichment was not as strong in the second cluster, there was a distinct path of enrichment emanating along the connecting bridge between the two clusters and into the less enriched cluster. This suggests the hypothesis that antibiotic resistance has spread from the first cluster into the second cluster via strains intermediate to the two and will likely continue to appear in the second cluster.

5.7 Persistent Homology Estimators in Population Genetics

Mathematical models provide a way of generating data that can be used to tune inference procedures. In population genetics, there are simple models that can simulate the generation of mutations and recombination in populations of genomes. In Appendix B we describe some of the commonly used models of population genetics, including the Wright-Fisher, Moran, and coalescence models. In this section, we will study one of the most popular models, the coalescent model with recombination. With only two parameters, the mutation and recombination rates, one can generate large amounts of simulated data. Using this data, we will construct estimators based on persistent homology.

5.7.1 Coalescent Process

The coalescent process is a stochastic model for generating genealogies, evolutionary histories represented by lines of descent from a common ancestor, for a collection of individuals sampled from an evolving population (see Appendix B). These genealogies can then be used to simulate new, synthetic genetic sequences. Coalescence processes and the attendant coalescent theory underlie many methods commonly used in population genetics.

Starting with a sample of n individuals from a present-day population, each individual's lineage is traced backward in time by randomly choosing a member of the previous generation as the individual's parent. Two individuals may, by chance, be assigned the same parent, in which case their lineages merge. This stochastic process ends when the lineages of all sampled individuals have merged at a single most recent common ancestor.

In this process, if the total population size N is sufficiently large, then the expected time before a coalescence event, in units of $2N$ generations, is approximately exponentially distributed:

$$P(T_k = t) \approx \binom{k}{2} e^{-\binom{k}{2}t},$$

where T_k is the time that it takes for k separate lineages to collapse $k - 1$ lineages.

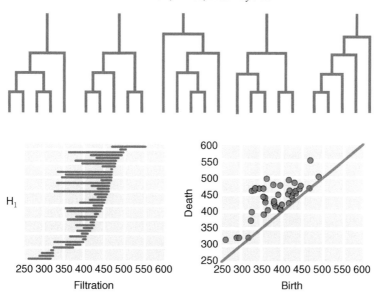

Figure 5.37 Two representations of the same topological invariants, computed using persistent homology. Left: Barcode diagram. Right: Persistence diagram. Data was generated from a coalescent simulation with $n = 100$, $\rho = 72$, and $\theta = 500$. Source: [164]. From Emmett et al., Parametric inference using persistence diagrams: A case study in population genetics, arXiv: 1406.4582 [q-bio.QM].

After generating a genealogy, the genetic sequences of the sample can be simulated by placing mutations on the individual branches of the lineage. The number of mutations on each branch is Poisson distributed with mean $\frac{\theta t}{2}$ where t is the branch length and θ is the population-scaled mutation rate. In this model, the average genetic distance between any two sampled individuals – the number of mutations separating them – is θ.

Coalescence models can be extended to include recombination events, allowing different genetic loci in a sampled individual to come from different lineages within the genealogical structure. Recombination is modeled as a splitting event in which an individual, rather than being a direct descendant of only a single parent, descends from two separate lineages – and occurs at a rate determined by a population-scaled recombination rate ρ. Thus evolutionary histories are no longer represented by a contractible tree, but, due to the combined splitting and joining actions, by an *ancestral recombination graph* which may have loops and other non-trivial, higher-dimensional topology.

5.7.2 Statistical Model

A persistence diagram generated by à coalescence simulation with recombination is shown in Figure 5.37. The information in the diagram can be used to infer the

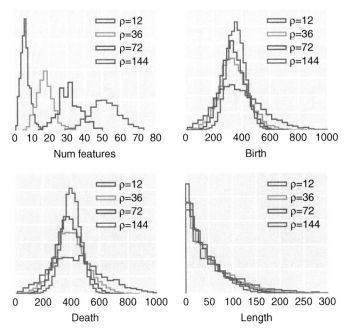

Figure 5.38 Distributions of statistics defined on the H_1 persistence diagram for different model parameters. Top left: Number of features. Top right: Birth time distribution. Bottom left: Death time distribution. Bottom right: Feature length distribution. Data generated from 1000 coalescent simulations with $n = 100$, $\theta = 500$, and variable ρ. Source: [164]. From Emmett et al., Parametric inference using persistence diagrams: A case study in population genetics, arXiv: 1406.4582 [q-bio.QM].

parameters θ and ρ (the mutation and recombination rates, respectively) that generated the data. Here, inference is based only on the detected H_1 invariants, but the idea can be readily generalized to higher dimensions. We consider the following properties of the persistence diagram: the total number of features, K; the set of birth times, $(b_1, \ldots, b_K)$; the set of death times, $(d_1, \ldots, d_K)$; and the set of persistence lengths, $(l_1, \ldots, l_K)$. In Figure 5.38 the distributions of these properties for four values of ρ are shown, keeping fixed $n = 100$ and $\theta = 500$.

It is immediately evident that the number of features K increases with ρ, consistent with the basic intuition that recombination events generate non-trivial topology in the model. The means of the birth and death time distributions depend only very weakly on ρ and are slightly smaller than θ, suggesting θ defines a natural scale in the topological space; however, higher values of ρ dramatically reduce variance of the distributions. Finally, the distribution of persistence lengths is independent of ρ.

Examining Figure 5.38, we can observe that the distribution can be approximated by $K \sim \text{Pois}(\zeta)$, $b_k \sim \text{Gamma}(\alpha, \xi)$, and $l_k \sim \exp(\eta)$. Death time is given by $d_k = b_k + l_k$, which is incomplete gamma distributed. The parameters of each distribution are assumed to be an a priori unknown function of the model parameters, θ and ρ, and the sample size, n. Keeping n fixed, and assuming each other parameter in the diagram is independent (a strong assumption), we can define the full likelihood as

$$p(D \mid \theta, \rho) = p(K \mid \theta, \rho) \prod_{k=1}^{K} p(b_k \mid \theta, \rho) p(l_k \mid \theta, \rho).$$

Simulations over a range of parameter values suggest the following functional forms for the parameters of each distribution. The number of features is Poisson distributed with an expected value

$$\zeta = a_0 \log \left(1 + \frac{\rho}{a_1 + a_2 \rho} \right).$$

Birth times are gamma distributed with shape parameter

$$\alpha = b_0 \rho + b_1$$

and scale parameter

$$\xi = \frac{1}{\alpha} (c_0 \exp(-c_1 \rho) + c_2).$$

These expressions appears to hold well in the regime $\rho < \theta$, but break down for large ρ. The length distribution is exponentially distributed with shape parameter proportional to mutation rate, $\eta = \alpha \theta$. The coefficients in each of these functions are calibrated using simulations, and could be improved with further analysis. This model has a simple structure and standard maximum likelihood approaches can be used to find optimal values of θ and ρ.

5.7.3 Coalescent Simulations

We describe results associated to the simulation of a coalescent process with sample size $n = 100$ and $l = 10{,}000$ loci. The mutation rate, θ, was varied across $\theta = \{50, 500, 5000\}$. The recombination rate, ρ, was varied across $\rho = \{4, 12, 36, 72\}$. The output of the process is a set of binary sequences of variable length (the length is dependent on θ). The Hamming metric yields a pairwise distance matrix between sequences. Computing persistent homology and using the model described in Section 5.7.2 produces estimates of θ and ρ. Results are shown in Figure 5.39, where we plot estimates and 95% confidence intervals from 500 simulations. We observe an improved ρ estimate at higher mutation rate. This is expected, as increasing θ is essentially increasing sampling on branches in the genealogy. We also observe tighter confidence intervals at higher recombination rates, consistent with the

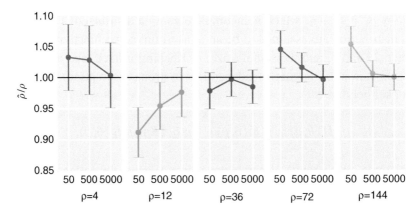

Figure 5.39 Inference of recombination rate ρ using topological information. The recombination rate ρ is estimated for five values $\{4, 12, 36, 72, 144\}$ at three different mutation rates $\{50, 500, 5000\}$. Mean estimates over 500 simulations and 95% confidence interval are shown. Source: [164]. From Emmett et al., Parametric inference using persistence diagrams: A case study in population genetics, arXiv: 1406.4582 [q-bio.QM].

behavior seen in Figure 5.38. See [259] for follow-up work on estimating recombination rates for coalescent models and further discussion of the relationship between topological invariants and population genetics.

5.8 Recombination Landscape in Humans

Sexual reproduction is a non-tree-like event, essential to ensuring genetic diversity of offspring and preserving genome integrity. Cells in sexually reproducing organisms contain two copies of most chromosomes (autosomes). Each copy differs slightly in sequence, but has the same overall structure. Humans have 22 pairs of chromosomes, as well as sex chromosomes – a pair of X chromosomes for females and an X and Y chromosome for males. Each of these 23 pairs of chromosomes is inherited from a different parent. In the process of meiosis, cells become haploid, i.e., containing only one chromosome of each pair, with different regions randomly selected from the paternal or maternal copy.

Meiosis occurs through two rounds of division. In division I of meiosis, a diploid cell containing a paternal and maternal copy of each chromosome duplicates (see Figure 5.40). Homologous chromosomes are then paired in a structure that is called a *bivalent*. This is where the process of recombination takes place. In a nutshell, meiotic recombination begins with a double-strand break in one of the parental chromosomes catalyzed by a particular protein, Spo11, and the broken strands from this chromosome are partially degraded. To repair this chromosome, the intact

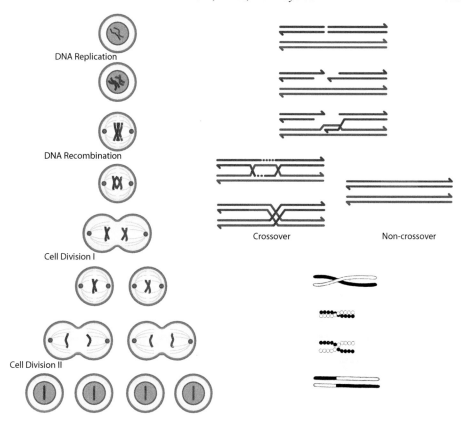

Figure 5.40 Left: Process of meiosis through two rounds of division. Top right: Cartoon of recombination. Bottom right: Illustration from the 1916 book of Morgan explaining crossovers. Source: [359].

chromosome strands are used as templates. The final recombination product could result in crossover resulting in a new chromosome generated from both parental chromosomes. Also it could lead to a non-crossover event (associated to gene conversion, or partial replacement of a DNA region by a homologous sequence), where part of the genomic material from one parent is used in the other strand. Finally, the cell divides and two of the homologues are then contained in each daughter cell. In division II of meiosis, cells divide again without further duplication of the genomic material. At the end, there are four haploid cells derived from the initial diploid cell. Each of the cells contains genomic material from the paternal, the maternal, or a recombinant of both.

Given that meiotic recombination is such a fundamental process in eukaryotic evolution, involving break and repair of genomic material, it is not surprising that it is a highly regulated process. Since the work of Morgan using the fruit fly, *Drosophila melanogaster*, as a model, we have a quantitative understanding of how often chromosomal crossovers occur in meiosis (bottom right panel in Figure 5.40, obtained from [359]). Morgan was able to establish a link between the probability of crossover and how far away in chromosomal position two different loci were. In humans, recombinations occur at an average rate of one crossover per chromosome per generation. A more quantitative measure of these rates can be obtained by estimating the probability that a crossover event will occur between two different loci in a chromosome. One defines a *centi-Morgan* (cM) distance in genomic position with a 1% chance of recombination per generation. The average rate of recombination in humans is about 1 cM per megabase.

However, genetic versus chromosomal distance approaches do not allow a high-resolution mapping below millions of bases, as it will require many generations to track many meiotic events. Pedigree and linkage disequilibrium analysis provide a much more refined view of where recombination occurs [288]. Pedigree analysis studies families of related individuals along several generations. Linkage disequilibrium (LD) is a measure of how the variability of two genomic loci is associated. If there is no recombination between loci, two mutations in the same chromosome will be always traveling together. If recombination occurs very frequently, the presence of a particular allele provides very little information about nearby mutations. The simplest measure of LD is $D_{ij} = f_{ij} - f_i f_j$, where f_{ij} is the frequency of observing two alleles i and j together, and f_i is the frequency of observing the allele i.

It has been found that recombination occurs preferentially at narrow genomic regions known as recombination hotspots [18, 37, 396]. In mammals, recombination hotspots are specified by binding sites of the meiosis-specific H3K4 trimethyltransferase PRDM9 [38, 372, 400]. However other factors play a role too. The recombination landscape in eukaryotes is actually the result of a hierarchical combination of factors that operate at different genomic scales. High-resolution mapping of meiotic double-strand breaks (DSBs) in yeast and mice [181, 294, 397, 466] reveals fine-scale variation in recombination rates within hotspots as well as frequent recombination events occurring outside hotspots [397].

Population-based recombination maps are a valuable tool in the study of recombination in humans [344, 371]. Due to the number of genomes published by such consortia as the 1,000 Genomes Project [122] and ENCODE [123], it is now possible to produce exquisitely fine-scale mapping and annotation of human recombination. Chromatin immunoprecipitation (ChIP-seq), bisulfite, or RNA sequencing methods, as well as other high-resolution data sets reveal a wide variety of distinct biological features associated with small genomic regions. These can aid in connecting locations where recombination occurs with other molecular and biological phenomena.

Establishing compelling statistical associations with such narrow and often clustered biological features, and analyzing the very large numbers of sequences in these data sets, is becoming a crucial challenge for traditional methods of recombination rate estimation (such as methods based on linkage disequilibrium). Robust and scalable methods to detect and quantify rates of recombination at different scales are particularly useful.

5.8.1 Fine-Scale Resolution of Human Recombination

The persistent homology estimators of recombination introduced in the previous section can be easily implemented on a sliding window and therefore adapted to the very long eukaryotic genomes [87, 88]. The sliding window cuts the genome into small overlapping segments. One can estimate the local recombination rate $\rho(x)$ using the persistent homology estimators of recombination rates in a sliding window centered around a genomic position x. There are different implementations of the procedure that determine the size of the window. A constant window size may not be desirable, because mutation rates can vary over different genomic regions; and thus one might get windows that contain no polymorphic sites, in which no recombination could be detected. Choosing variable window sizes to fix the number of polymorphic sites per window avoids this problem. The number of polymorphic sites per window defines the genomic scale at which recombination is observed.

Figure 5.41 captures a snapshot of the sliding window near the cytogenic band 1q24.1 of human Chromosome 1 [87]. We describe estimates of recombination rates from 647 individuals genotyped for the 1,000 Genomes Project [122], with windows containing 14 polymorphic sites. The sliding window approach assigns a finite metric space for each position x, containing 647 points, one for each individual. The one dimensional homology estimator of $\rho(x)$ based on b_1 on a sliding window allows inference of local recombination rates at x. Recombination maps reflect a landscape with peaks showing recombination hotspots and valleys showing low recombination regions.

The 1,000 Genomes Project provides genotype data of nearly 38 million single nucleotide polymorphisms (SNPs). The data is phased, meaning the sequences' locations include the specific chromatid on which they were found. The individuals genotyped by the 1,000 Genomes Project came from seven different populations: European-American, Han Chinese, Finnish, British, Japanese, Tuscan and Luhya (a Bantu ethnic group in Kenya). Each of these populations has a different recombination map that can be compared to the others. The median effective population recombination rates detected for non-African populations had $\rho \sim 0.6$ per kbp. The effective population recombination rates were substantially higher in the African population, consistent with its larger effective population size and supporting the out-of-Africa human expansion model [494]. While the recombination maps agree at a global scale, there are population-specific variations. In particular, the Luhya

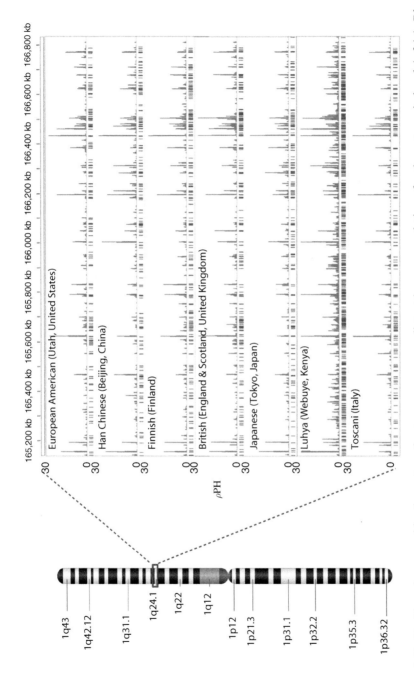

Figure 5.41 Positionwise recombination rate estimates across seven distant human populations for the cytogenic band 1q24.1. Blue bars represent recombination rates estimated with a variable-size sliding window containing 14 segregating sites. Below, red segments represent genomic regions where a 500 bp window detects recombination ($b_1 > 0$). Source: [87]. From Pablo G. Cámara et al., 'Topological data analysis generates high-resolution, genome-wide maps of human recombination', Cell Systems 3.1 (2016): 83-94. Reprinted with permission from Elsevier.

present a more unique recombination landscape. Topological methods can be used to describe gene flow across populations, where migration and admixture appear as high-dimensional loops in the evolutionary space.

The fine-resolution maps of recombination connect population maps to specific genomic locations that can inform us about specific molecular processes associated to recombination hotspots. Recently, high-throughput methods have catalogued binding sites of different proteins, epigenetic marks, and gene expression across genomes [123]. For instance, one can ask what proteins bind to the genome in locations where recombination occurs. The persistent homology recombination landscape recapitulates known proteins and epigenetic marks associated to recombination, such as the meiosis-specific histone 3 lysine 4 (H3K4) trimethyltransferase PRDM9, CpG hypomethylation and H3K4 trimethylation. Comparing persistent homology estimators of recombination to binding sites from ChIP-seq data of 118 transcription factors, these binding sites are depleted in high recombination loci on average (see Figure 5.42) [87]. In addition, this analysis led to the discovery of previously unreported transcription factors associated to recombination regions, such as members of the E2F protein family, important regulators of cell cycle progression and differentiation. These proteins bind to sites of RNA polymerase II and different regulatory subunits of the MLL/MLL1 protein complex [87].

5.9 Gene Trees and Species Trees

In the previous sections, we have described how different genomes from individual organisms are related. Sometimes, like in clonal processes, the relation between different genomes can be well represented by a phylogenetic tree. However, phylogenetic trees have been traditionally used to capture relations beyond individual organisms, describing relationships between different taxa (e.g., kingdoms, genera, or species). For instance, when Darwin in 1859 proposed his model for the origin of species he had in mind a branching process with different species as leaves (panel A in Figure 5.43). We have to remember that while genomes are ascribed to individual organisms, individuals within a species have slightly different genomes. In a strict sense, the genome of a species, such as the human genome, does not exist, only the related genomes of organisms within a species. If there is no single genome for a species, how can one construct a *species tree*, a tree where leaves are labeled by species? More formally, given a set of genomes G from organisms belonging to a set of species S, how can we find a representation between different species (or other taxa)? When does it make sense to talk about a species tree?

The construction of a species trees is not straightforward, as there are not only technical questions but also profound conceptual obstacles to this enterprise. First,

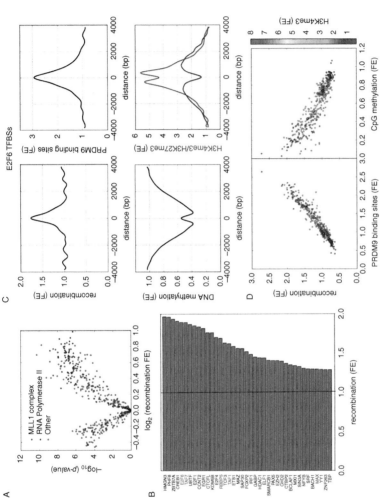

Figure 5.42 Relation of recombination to TFBSs (transcription factor binding sites). (A) Logarithm in base 2 of the recombination enrichment at TFBS, for each TF and cell line. RNA Polymerase II and TFs that form part of MLL complexes are indicated explicitly. Enrichments are taken with respect to neighboring genomic regions. (B) Recombination enrichment at TFBSs for each TF with respect to the whole-genome average. Only TFs with the highest enrichments are shown. TFs that may form part of MLL complex are indicated in red. (C) Recombination, predicted PRDM9 binding sites, CpG methylation, H3K4me3 and H3K27me3 enrichments at binding sites of E2F6. (D) Recombination enrichment against enrichment for PRDM9 binding sites (left) and sperm CpG methylation (right). Color scale represents H3K4me3 enrichment. Source: [87]. From Pablo G. Cámara et al., 'Topological data analysis generates high-resolution, genome-wide maps of human recombination', Cell Systems 3.1 (2016): 83–94. Reprinted with permission from Elsevier.

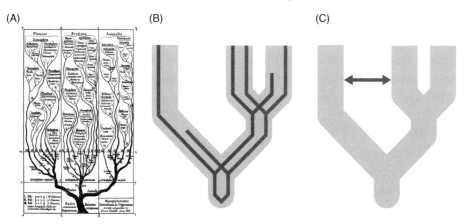

Figure 5.43 Phylogenetic trees have been traditionally used to describe the relation between species. (A) Tree of life by Haeckel, Generelle Morphologie der
Organismen (1866) with species and higher taxa assigned to branches and leaves.
In a strict sense, there is no genome of a species or any higher taxa, and different
genomes and different genomic regions could generate (slightly) different trees.
Incomplete lineage sorting is a common phenomenon that generates different tree
topologies. Source: From E. Haeckel, Generelle morphologie der organismen. All-
gemeine grundzüge der organischen formen-wissenschaft, mechanisch begründet
durch die von Charles Darwin reformirte descendenztheorie, Berlin, G. Reimer,
1866. (B) and (C) Incomplete gene sorting can occur if a locus in an ancestral
species is polymorphic (has more than two alleles). Suppose that it divides first
into two lineages and then one of those further divides into another two. The
alleles could then be fixed differently in each of the lineages. Incomplete lineage
sorting generates "gene trees" (trees from the allele) that present a different tree
topology than the species tree. These tree incompatibilities are represented in a
"fat tree" that can capture different topologies such as the ones occurring during
incomplete gene sorting (B) or represented by arrows representing horizontal gene
transfer events (C).

and the most serious obstacle, relates to the assignment of a given genome to a particular species. Even in metazoa, where phenotypic differences are significant, it is
sometimes unclear how to define species. The problem becomes acute in bacteria
and viruses. From an empirical genomic point of view, given genomic data from
two organisms, how can we determine whether they belong to the same species?
This assignment problem is linked to the definition of species. A species is commonly and informally understood as the largest group of interbreeding individuals
capable of producing fertile offspring. However, there is no consensus on a precise
species definition that can incorporate sexual and asexual organisms, that can consider more complex phenomena such as ring species (populations that can breed
with nearby populations but not those far apart), that can consider hybrids, among
others. More importantly, from a pragmatic genomic point of view, it is not unusual

to sequence the genome of an organism without knowing the mating possibilities with others. Genomic based definitions of species are based on genomic data, for instance, based on arbitrary cutoff values for species definitions, e.g., a 95% average nucleotide identity as a potential criterion to define whether two bacteria belong to the same species. In viruses, the problem is even more acute. According to the International Committee on Taxonomy of Viruses (ICTV) a virus species is a "polythetic class of viruses that constitute a replicating lineage and occupy a particular ecological niche." A "polythetic class" means a group of organisms with several properties in common but not necessarily a single defining property. Thus, in a way, the very definition of viral species is artificial.

Even if we have a good species assignment, a serious second obstacle appears. We have seen in the previous sections of this chapter that a variety of biological processes (reassortments, recombinations, horizontal gene transfers, etc.) generate genomic relations that are not well captured by trees. When these processes occur between members of different species, there will not be a tree representation. If exchange of genomic material is rare one could envision a tree with small corrections reflecting non-tree-like processes (Figure 5.43). But even if non-tree-like processes did not occur between members of different species, the history of a particular genomic region could be different from the history of another region, a phenomenon that is referred to as *incomplete lineage sorting*.

Incomplete lineage sorting could happen when lineages divide before polymorphisms fix in the population. If variant alleles are kept in the different descendant populations and fix independently in each population, the final tree of these alleles could be different from the species tree (Figure 5.43B). These tree incompatibilities are informally represented in a "fat tree" that reflects the tree topological ambiguity due to incomplete lineage sorting. The problems of defining species, of assigning species and of finding good *species summarizations* (tree or others to be defined) are linked. If organisms across different species frequently exchange genomic material it is difficult to establish clear species boundaries.

Sometimes, under the assumption that there are some smaller genomic regions where trees are good approximations, one can consider the problem of reconstructing a *species tree* as finding a good "summarization" of a set of trees. In the literature, this problem is referred to as finding the species tree from a set of smaller genomic region trees, called *gene trees*, although the latter do not necessarily refer to genes and could refer to other genomic regions. In this context, Billera, Holmes, and Vogtmann [55] applied the structure of the CAT(0) spaces described in Section 4.7. Here, the *species tree* problem is posed as the problem of finding the centroid of a set of *trees* of 12 primates, including *Homo sapiens*.

5.10 Extensions: Median Complex and Topological Minimal Graphs

Thus far, our approach has been to start with genomic data, compute a finite metric space, and relate the topological properties of the metric space to biological phenomena, such as reassortment, recombination, or horizontal gene transfer. We have constructed estimators of recombination rate based on statistical properties of persistent homology summaries. We have also deconstructed large genomes using a sliding window. Beyond these methods, there exist other constructions that can be useful. In particular, starting from the genomic data, we can increase the number of genomes by adding inferred genomes that could be associated to potential ancestors. From this extended data set, one can again define a finite metric space, whose topology could increase the sensitivity for recombination detection, at the expense of increasing the complexity and introducing spurious non-interpretable events in a biological context.

Let us consider a few simple examples for which the four gamete test (the presence of all four different alleles in two loci) indicates a non-tree-like event, and how persistent homology can or cannot detect the reticulate event.

Our first set of examples [162] consists of four genomes of length two and two bases represented by 0 and 1: $s_1 = 00$, $s_2 = 10$, $s_3 = 01$, and $s_4 = 11$. One can easily verify that the four gamete test finds an incompatibility between the first two sites, as the four possible gametes are present (Figure 5.8). This event is precluded in an infinite-sites model without recombination, where mutations in the same site are rare. Persistent homology captures this event, as it identifies a bar $[1, 2)$ in the first homology persistent group using Hamming distance. We can vary this simple example by taking four genomes of length three and two bases represented by 0 and 1: $s_1 = 000$, $s_2 = 100$, $s_3 = 010$, and $s_4 = 111$. One can again easily verify that the four gamete test finds incompatibility between the first and second sites. However, the barcode in dimension one (or higher) does not show any bar. In this simple example, it is easy to identify the reason: if s_1 is the common ancestor to other sequences, s_2 and s_3 can be considered to be the parents of s_4, a direct descendant of a reticulate event. In this case, it is easy to infer that there was an ancestral recombinant sequence, $s_r = 110$, which was not sampled in our data set (Figure 5.44A). If this missing sequence was present in our data set, we would have recovered a bar in the first homology group representing the event. This simple case shows that incomplete sampling of the process can significantly lower the sensitivity of persistent homology to detect potential recombinant events.

Another example can be found in the article by Song and Hein [476]. In this case there are five genomes with four sites: $s_1 = 0000$, $s_2 = 1100$, $s_3 = 0011$, $s_4 = 1010$, and $s_5 = 1111$. There are multiple incompatibilities between sites that can be easily identified using the four gamete test (1 and 3, 1 and 4, 2 and 3, and

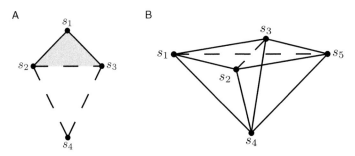

Figure 5.44 Two simple cases where the Vietoris-Rips complex applied to a distance matrix between sampled genomes fails to identify a potential recombination. (A) In this example an ancestral sequence has not been considered in the sample. If considered, that recombination is identified. (B) In more complex cases, multiple recombinations can lead to a degeneracy. Source: [162]. Reprinted with permission: © EAI European Alliance for Innovation 2016.

2 and 4). Song and Hein [476] show that at least two recombinations are needed to explain this data. When applying persistent homology using the Vietoris-Rips complex with Hamming distance as the metric, one finds that the barcode diagram does not show any event in dimension larger than zero, failing to capture potential recombinations. Close inspection shows that this is a special case, where s_4 sits at the same distance from the other four sequences. If other ancestral states could have been sampled or if s_4 had not been present in the data set, persistent homology could have detected the reticulations.

These examples show that persistent homology using Vietoris-Rips complexes applied to genetic distances is limited as a method to identify potential reticulate events. This is not surprising, as sequence data is much richer than a distance matrix; as in standard phylogenetic approaches, methods based only on distances constitute a first approximation.

Working directly with sequences provides a much more powerful data structure that can capture all potential reticulations. For instance, from our original sequence data, we can construct many distance matrices by subsampling sets of sites. If the phylogeny is truly tree-like and the infinite-sites model holds, none of these subsamples generates any non-tree-like structure, and the persistent homology barcodes for each of the subsamples should be empty for dimension bigger than zero. If the four gamete test is satisfied for a particular subset of data, the four alleles should be present and so a bar in dimension one. Subsampling sites provides a very powerful tool to increase sensitivity at heavy computational expense, as the number of potential subsets is exponential with the number of sites. There is also the problem of interpretability of the results: can we infer what could have been the recombination history, or just the minimal number of recombination events, from subsamples of data? There are several alternatives though that take advantage of the fact that homologous recombination occurs between nearby sites in the genome, or in other words, the four gamete test is more informative between nearby sites in the genome. This type of information is fundamental for interpretability and is used in most standard tests of recombination, for instance, the Hudson-Kaplan test [255].

In what follows we propose several approaches to increase the sensitivity and interpretability of persistent homology for the identification of reticulate events. The first approach, the median complex, is based on the idea of adding potential ancestral states. The second infers associated graphs, named topological minimal graphs, as explicit representations of potential histories.

5.10.1 The Median Complex Construction

In order to increase the sensitivity of persistent homology methods for identification of recombination, we apply an old insight in the field, namely that adding information about ancestral inferred states can make it easier to identify potential recombinant events. The idea is to add extra points to the original data, some of which can be mapped to potential ancestral states. The *median graph* (also called the Buneman graph) is a graph constructed with inferred median points. It was introduced as a way of capturing all maximum parsimony evolutionary trees [81] and has been the object of study for phylogenetic network inference [30, 31].

Associated to the median graph is a collection of filtered complexes referred to as the *median complex*. The persistent homology of the median complex can be computed by considering from the finite metric space consisting of the original data plus the new points imputed from the median procedure. If the original data is tree-like, there is no persistent homology information in the median complex above dimension zero; this is consistent with the persistent homology of the data [100]. But high-dimensional classes in the median complex can capture recombination events not visible in the persistent homology of the underlying complex. The major drawback to the use of median graphs is the large number of imputed points, which complicate computation and obscure the biological interpretation.

The *median sequence* $m(a, b, c)$ of three binary sequences a, b, and c is a sequence with the majority consensus at each site. For instance, take the example shown in Figure 5.46 with three sequences with three sites each: $a = 000$, $b = 110$, and $c = 011$. The median of the first site is a 0, in the second a 1 and in the third a 0, so the median sequence is $m = 010$, different from any a, b, and c. Note that the median sequence is not sensitive to sites that are specific to one of the original sequences.

Notice that in the example previously shown in panel A of Figure 5.44, the median of $s_2 = 100$, $s_3 = 010$, and $s_4 = 111$ is precisely the missing sequence

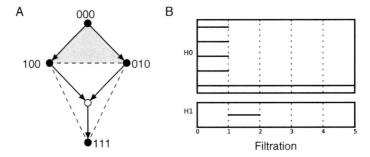

Figure 5.45 One can infer some interesting sequences by applying the median operation. The newly inferred median node (in white) can be interpreted as the missing recombinant between s_2 and s_3 and the ancestor of s_4. Adding the median sequence to the original set one can identify a new one dimensional persistent class in the interval $[1, 2)$. The median operation does not generate other new sequences. Source: [162]. Reprinted with permission: © EAI European Alliance for Innovation 2016.

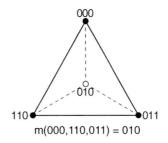

$m(000,110,011) = 010$

Figure 5.46 The median sequence is constructed for triples of sequences by taking the most common allele at each site. The process can be iterated, adding more median sequences to the original data until no new sequences can be added by this procedure (the median closure). Source: [162]. Reprinted with permission: © EAI European Alliance for Innovation 2016.

$s_r = 110$. Applying the median operation to subsets of size 3 of the augmented set of sequences does not generate any new sequences (Figure 5.45). Computing the persistent homology of the Vietoris-Rips filtration of the original set of sequences together with s_r uncovers the missing one dimensional loop in the interval $\epsilon = [1, 2)$ generated by s_1, s_2, s_3, and the newly reconstructed s_r.

For every triple of sequences one can define the median. The median of a triple may or may not be in the original set. This procedure can be repeated by adding the new median sequences to our original set of sequences S and iterating until there are no more new sequences added. The final set of original sequences and their medians, and successive medians, and so forth, constitute the *median closure* $\bar{S}$:

$$\bar{S} = \{v \mid v = m(a, b, c) \in \bar{S} \; \forall \, a, b, c \in \bar{S}\}.$$

The median closure is closed under the median operation.

Instead of constructing the Vietoris-Rips complex using the distances from the original set of sequences, one can construct the complexes using the median closure $\bar{S}$. We will refer to the resulting Vietoris-Rips filtration as the median complex.

Let us revisit our second example (Figure 5.44B). The median operation adds four new sequences, as displayed in Figure 5.47. Now persistent homology applied to the median closure identifies four one dimensional persistent intervals in the barcode diagram, all in the interval $\epsilon = [1, 2)$. In this example the minimal number of recombinations that is needed to explain the data is two, as found by Song and Hein [476]. The number of intervals found in persistent homology is now higher than the minimal number of recombinations. As a consequence, the interpretation of these homology classes as potential recombination is a central problem with the use of the median constructions.

Observe that we can compute the Vietoris-Rips complexes of two finite metric spaces: the one we previously explored with the original data (called here the leaf complex) and the median complex. If the original data is derived from a tree-like structure, the two complexes will provide no high-dimensional persistent homology. Counting bars in PH_1 for the leaf complex frequently underestimates reticulate evolution because of incomplete sampling, while counting bars in PH_1 for the median complex usually overestimates reticulate events. The median complex is in some sense an upper bound on probable recombination histories; although it does not contain within it all possible recombination graphs, as there are infinitely many

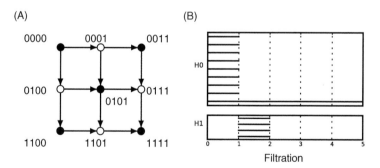

Figure 5.47 The median complex increases the power to identify potential reticulations, but complicates interpretability. When applying the median operation to the example of Song and Hein, one finds four median vertices (in white). Persistent homology identifies four one dimensional loops in this case. Song and Hein found that this data could be explained by a minimum of two recombinations. Source: [162]. Reprinted with permission: © EAI European Alliance for Innovation 2016.

complicated ancestral recombinant graphs (ARGs, see Section 5.10.2), it contains within it all maximum parsimony trees.

We now illustrate the use of the median complex via some examples using real data. In the following examples, the Vietoris-Rips complexes built from the distance matrices of the original data do not show any non-trivial homology. However, it is easy to verify using the four gamete test that there are potential recombinations. The median closure in these examples adds new median sequences that enrich the original data, increasing the power of homology to identify potential recombinations.

The first example is a classic data set in population genetics from Kreitman [309] of eleven sequences from the alcohol dehydrogenase (Adh) locus of the fruit fly *Drosophila melanogaster*. The original eleven sequences consist of 43 polymorphic sites. The median closure adds more than 30 median sequences to the original data set (see Figure 5.48). While persistent homology on the original data set of 9 sequences fails to identify any higher dimensional homology, the median complex identifies 32 bars in dimension one homology and 3 in dimension three.

Hybridization, the process of generating new species by the genetic mixing of two different species, is very common in plants. Common plants, such as wheat, are the result of hybridization and artificial hybrids are very commonly used in crops. Huber and colleagues [250] collected data from the maturase gene (*matK*) in nine species from the genus *Ranunculus*. The median closure added 23 new median vertices to the original data (Figure 5.49). Persistent homology on the original data does not identify any non-trivial class, but the median complex shows 17 one-dimensional and 3 three-dimensional classes.

5.10.2 Topological Minimal Graphs and Barcode Ensembles

In the last two decades, there have been many efforts to produce frameworks to represent non-tree-like events. Phylogenetic networks try to represent inconsistencies among trees as graphs [30, 31, 32, 33, 260, 261, 262, 263, 362]; however, the biological interpretation of these networks is often unclear. Other constructions, sometimes referred to as *explicit networks*, try to provide potential reconstruction of past events that lead to tree inconsistencies. The ancestral recombination graphs, or ARGs, provide a potential historical explanation in terms of mutation and recombination events. Mutations appear as events along the branches of the graph and recombinations appear as merges between parental branches. ARGs do not consider homoplasies due to convergent evolution [208, 209, 220].

In principle, there is an infinite number of ARGs that are consistent with the data. Population genetics models, like coalescence with recombination, can assign probabilities to them [209, 253]. Finding the minimal ARG, i.e., an ARG with the

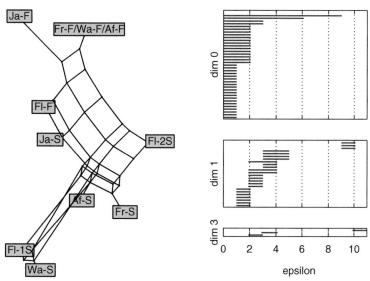

Figure 5.48 Left: The alcohol dehydrogenase (Adh) locus of the fruit fly *Drosophila melanogaster* provides a well studied set of sequences with recombination. Source: Reprinted with permission from André Karwath under the Creative Commons Attribution-Share Alike 2.5 Generic license. Right: Persistent homology on the median closure of the original data identifies one- and three-dimensional homology structures due to recombination events in the population. Source: [162]. Reprinted with permission: © EAI European Alliance for Innovation 2016.

minimal number of mutations and recombinations, is an extremely computationally intensive task. Indeed, finding a minimal ARG has been shown to be an NP-hard problem [65, 66, 522], and an infeasible approach to large data sets. There are, however, several approaches that can approximate minimal ARGs, including heuristic methods [356], branch and bound [477], galled trees [219, 221], and sequentially Markov coalescent approaches [421].

Here, we will present another approximation called a *topological ARG* or *tARG* [88], which is closely related. These capture ensembles of minimal recombination

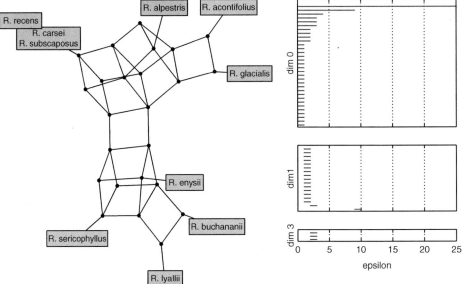

Figure 5.49 Hybridizations are commons in plants. Reprinted with permission from Walter Siegmund under the terms of the GNU Free Documentation License, Version 1.2 or any later version published by the Free Software Foundation. Left: The median closure of data collected by Huber and colleagues from the maturase gene (*matK*) in nine species from the genus *Ranunculus*. Source: Wikipedia. Right: The median complex allows us to identify 17 one-dimensional and 3 three-dimensional homology classes. Source: [162]. Reprinted with permission: © EAI European Alliance for Innovation 2016.

histories. tARGs, like minimal ARGs [220, 356], are interpretable, explicit phylogenetic representations. But unlike minimal ARGs, they can be constructed in polynomial time.

An ARG is an explicit representation of a potential history of mutations and recombinations that, starting from an ancestor, is able to generate the sampled sequences. Let us consider a sample of n sequences with m binary characters that can take values in states 0 or 1.

Definition 5.10.1. An *ARG* is a labeled directed acyclic graph N with $n+1$ external nodes, corresponding to the n sequences, and a unique root node. There are two types of internal nodes:

1. Tree nodes, of in-degree one.
2. Recombination nodes, of in-degree two.

Each node in N is labeled by an m-length binary sequence, subject to the following constraints:

1. External leaf nodes are labeled by the original sequences.
2. Tree nodes are labeled by sequences that differ from the parent node in certain positions; these represent mutations.
3. Recombination nodes have sequences attached that are formed by taking the first k sites from the sequence of one of the parent nodes and appending the last $m - k$ sites from the other parent node. These labels represent recombination of the parent sequences.

We are particularly interested in ARGs satisfying minimality conditions. A *minimal ARG* is an ARG that contains the minimal number (R_{min}) of single-crossover recombinations required to explain the binary sequence data [220].

Definition 5.10.2. An *ultra-minimal ARG* is a further restricted type of minimal ARG, that minimizes the function

$$D(N) = \sum_{r=0}^{R_{min}} d_r,$$

where d_r is the Hamming distance between the two sequences in the rth recombination. Examples of ultra-minimal ARGs are shown in Figure 5.50.

The condensed graph of an ARG is the graph resulting from collapsing edges that connect identically labeled nodes. Condensed graphs can be embedded into m-dimensional hypercubes and their diagonals.

Definition 5.10.3. A *topological ARG* (or *tARG*) associated to a set of condensed ultra-minimal ARGs $\{G_i = (V, E_i)\}$ explaining S and having the same set of vertices V is defined as the undirected graph $G = (V, E)$, with vertices V and edges $E = E_1 \cup \ldots \cup E_l$, resulting from the union of all condensed ultra-minimal ARGs.

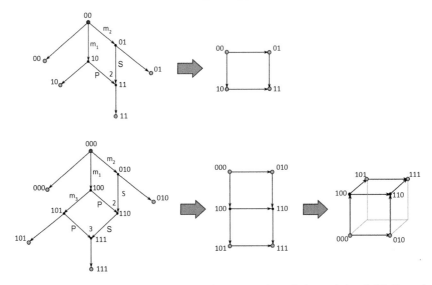

Figure 5.50 ARGs and condensed graphs. Examples of ultra-minimal ARGs and a condensed graph resulting from collapsing the unlabeled edges. The root and sampled nodes are marked in red and green. Edges in a recombination node can be annotated depending on their contribution as prefix (P) or suffix (S). Source: [88]. From Pablo G. Cámara, Arnold J. Levine, and Raúl Rabadán, 'Inference of ancestral recombination graphs through topological data analysis', PLOS Computational Biology 12.8 (2016). doi: 10.1371/journal.pcbi.1005071.

A tARG captures the possible parsimonious histories (see Figure 5.51 for examples). One advantage of tARGs over minimal ARGs is that a tARG is completely determined by its vertices, in the sense that the tARG can be computed entirely from the vertices and their labels.

Given a sample of genetic sequences, our goal is now to obtain information about the associated ultra-minimal ARGs that explain S, without explicitly constructing them (see Figure 5.52).

We now explain how to make inferences about recombination events in topological ARGs associated to sequence data by applying persistent homology. The persistent homology of the metric space determined by the original sequence data under the Hamming distance captures information about the genetic distance between recombining parental sequences. In particular, one-dimensional classes in persistent homology correspond to loops in the tARG corresponding to the data. That is, the tARG provides a framework for explaining the persistent homology barcodes.

However, the size of the barcode provides only a lower bound on the number of recombination events in the tARG, and the larger the length of the sequences, the

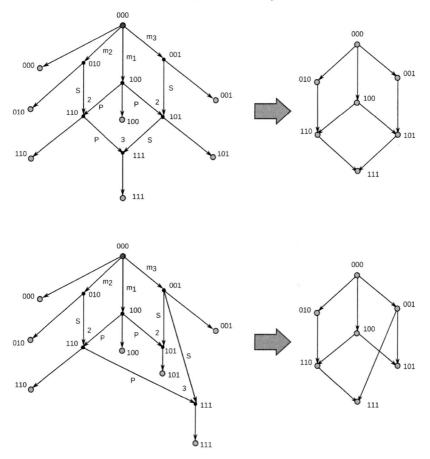

Figure 5.51 Ultra-minimal ARGs. Here we show two examples of ARGs that contain three recombinations, in this case the minimum number of recombination events. This is the minimal number required to characterize a sample of seven sequences with only three sites. Both ARGs are minimal ARGs. Source: [88]. From Pablo G. Cámara, Arnold J. Levine, and Raúl Rabadán, 'Inference of ancestral recombination graphs through topological data analysis', PLOS Computational Biology 12.8 (2016). doi: 10.1371/journal.pcbi.1005071.

worse this bound gets. A standard technique for handling this issue is to partition the sequences and reassemble local estimates. By a partition of the sequences we mean sets of subsequences specified by fixing a collection of indices $0 = i_0 < i_1 < i_2 < \ldots < i_k = m$. Now for each $0 \le j < k$, we consider the set of sequences determined by taking the characters in positions between i_j and i_{j+1} in the original sequences. Given a partition of the sequence data into distinct intervals, one can associate a barcode that captures information about recombination events with breakpoints in each interval. Taking the union of the barcodes of a partition usually captures more recombination events than the barcode associated to the union

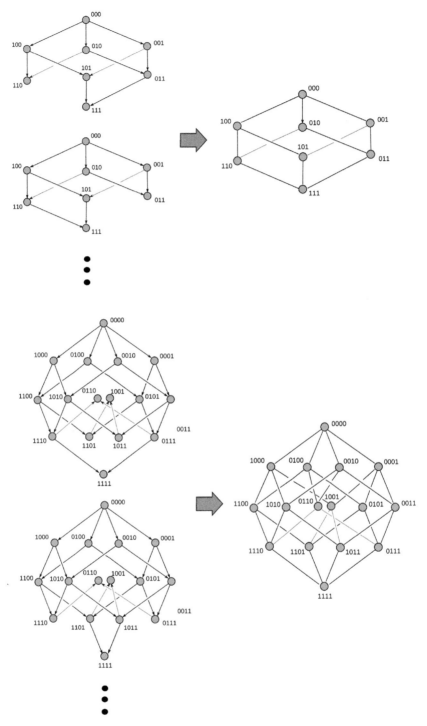

Figure 5.52 Topological ARGs. The tARG, shown on the right, can be different from the original condensed ultra-minimal ARGs, shown on the left. Source:
[88]. From Pablo G. Cámara, Arnold J. Levine, and Raúl Rabadán, 'Inference
of ancestral recombination graphs through topological data analysis', PLOS
Computational Biology 12.8 (2016). doi: 10.1371/journal.pcbi.1005071.

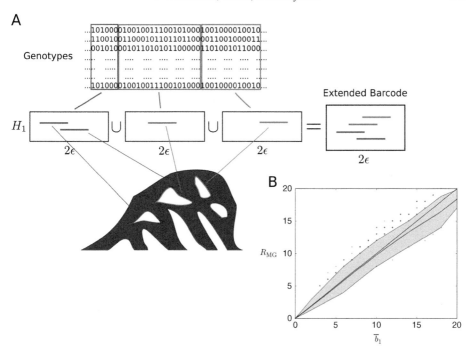

Figure 5.53 Barcode ensemble of a sample. (A) A schematic representation of the barcode ensemble of a genomic sample. Persistent homology is computed for each genomic partition of the sequences. Barcodes associated to different genomic intervals capture different recombination events with breakpoints contained within their respective partitions. The union of these barcodes builds the barcode ensemble. The total number of intervals in the barcode ensemble is denoted as $\bar{b}_1$. The genomic partitions are chosen such that $\bar{b}_1$ is maximized. (B) Comparison of the lower bounds $\bar{b}_1 \leq \overline{R}_{\min}$ and $R_{MG} \leq R_{\min}$ in coalescent simulations. Values of $\bar{b}_1$ and R_{MG} are plotted for simulated samples of 40 sequences with 12 segregating sites, sampled from a population under the coalescence model with recombination. 4000 samples were simulated in total. The colored band represents the interdecile range, whereas the central line represents the mean. The values of $\bar{b}_1$ and R_{MG} are strongly correlated (Pearson's $r = 0.98$, $p < 10^{-100}$). At high recombination rates, $\bar{b}_1$ tends to be larger than R_{MG}, as cases where $\overline{R}_{\min} > R_{\min}$ occur more frequently. Source: [88]. From Pablo G. Cámara, Arnold J. Levine, and Raúl Rabadán, 'Inference of ancestral recombination graphs through topological data analysis', PLOS Computational Biology 12.8 (2016). doi: 10.1371/journal.pcbi.1005071.

of the two genomic intervals. By systematically exploring all possible partitions of the genetic sequences in a data set, it is possible to find a partition that maximizes the total number of bars in the barcodes, referred to as the *barcode ensemble* and denoted by $\bar{b}_1$ (see Figure 5.53). A detailed explanation of the algorithm to compute barcode ensembles can be found in [88]. In simulated data, $\bar{b}_1$ is a good

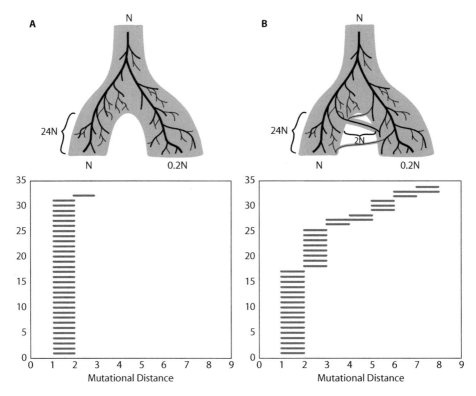

Figure 5.54 Barcode ensemble of two divergent sexually reproducing populations. The case in (A) assumes the two populations are completely isolated. All
recombination events present in the barcode ensemble involve genetically close
parental gametes. The case in (B) allows migration between the populations at
a low rate. Some of the recombination events present in the barcode ensemble
involve distant parental strains, leading to larger death times ϵ_d. The total number
of detected recombination events is similar in both cases and uniform across the
entire genome. Intervals with the location of the recombination breakpoints are
indicated for each recombination event, where positions refer to segregating sites.
Source: [88]. From Pablo G. Cámara, Arnold J. Levine, and Raúl Rabadán, 'Inference of ancestral recombination graphs through topological data analysis', PLOS
Computational Biology 12.8 (2016). doi: 10.1371/journal.pcbi.1005071.

approximation of the minimum number of recombination events, as Myers and
Griffiths described [370]. Barcode ensembles not only provide $\bar{b}_1$, but also richer
information that bounds the genetic distances between recombining sequences.

Let us consider two cases of sampling two sexually reproducing populations
with effective population sizes N and $N/5$ that diverged $24N$ generations ago.
In the first case, the two populations were completely isolated from each other
(Figure 5.54A). In the second case, a migration occurs between the two populations at a low rate (Figure 5.54B). We can detect the presence of gene

flow by studying the barcode ensembles. Whereas the number of total detected recombination events was very similar, reflecting the fact that both examples had the same recombination rate, migration was reflected in the existence of large scale loops. Specifically, these correspond to migration events followed by recombination.

The phenotypic variation and geographical distribution of finches on the Galapagos Islands inspired Darwin's theory of the origin of species. It is believed that these finch species originated from a common ancestor 1.5 million years ago [406]. Recently, genetic information was collected from 15 different species of finches from the Galapagos archipelago and the Cocos Islands [312]. With information about homozygous single-nucleotide variants in a nine megabase genomic region of 112 finch samples, we computed a barcode ensemble. The one dimensional barcode ensemble (Figure 5.55A) indicates 13 potential recombination events. Interestingly, the majority of these events involve individuals from multiple species and include *Certhidea* samples (Figure 5.55B), the *Certhidea* being the most ancestral lineage in the data set. Barcode ensembles provide support for genetic introgression, meaning the acquisition of genetic material from one species by another through hybridization [312].

5.11 Summary

- Vertical evolution is the direct transmission of genetic material from parent to offspring.
- Horizontal evolution refers to other modes of acquisition of genomic material that are not vertical. Phylogenetic trees cannot represent these events, which are better summarized by a network with loops (reticulate evolution). At all levels of life (virus, bacteria, eukaryotes) there are reticulate events. Viruses have recombination and reassortment. Bacteria have transformation, transduction and conjugation. Eukaryotes have recombination.
- Finite metric spaces can be constructed by comparing genomic sequences. Persistence homology captures properties of these spaces. In particular, if genomic data is derived from trees, the only non-trivial homology is in dimension zero. Or equivalently, there is a topological obstruction to constructing trees when homology is found in dimensions higher than zero.
- The number of loops tells us how frequently reticulate events occur. We have seen, for instance, that HIV has a high rate of recombination relative to other viruses, and that *H. pylori* has a low rate of horizontal gene transfer among bacteria.
- The scale of bar tells us how different the species are, and persistent homology generators identify individuals whose ancestors were involved in reticulate

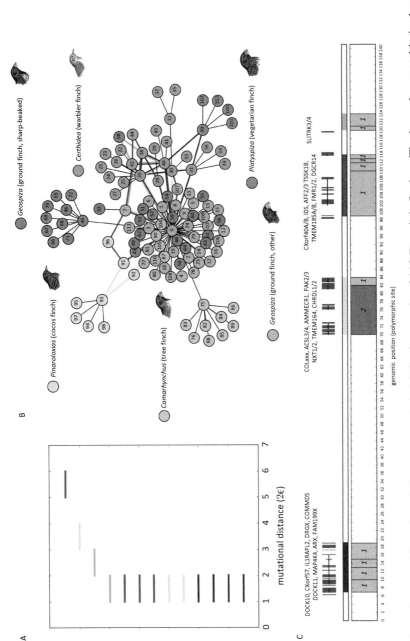

Figure 5.55 Barcode ensemble and partially reconstructed tARG of a sample of 112 Darwin's finches. The barcode ensemble is shown in (A), based on 140 homozygous SNPs present in a 9 megabase scaffold. In total, 13 recombinations or gene flow events were captured in the barcode ensemble at different genetic scales. Bars are colored according to the position of the corresponding recombination breakpoint in the genome, as depicted in (C). We also indicate the number of recombination events detected at each genomic interval, as well as some of the orthologous genes – distinct alleles thought to be passed down from a common ancestor – present in regions where recombination events were detected. The reconstructed tARG is presented in (B). Loops in the reconstructed tARG are outlined using the same color code as (A). We also include leaf nodes that do not participate in recombination, clustering them with a nearest-neighbor algorithm based on genetic distance. Edge lengths are arbitrary. Source: [88]. From Pablo G. Cámara, Arnold J. Levine, and Raúl Rabadán, 'Inference of ancestral recombination graphs through topological data analysis', PLOS Computational Biology 12.8 (2016). doi: 10.1371/journal.pcbi.1005071.

events. Barcodes in dimension one and above provide valuable information about size and frequency of genomic exchange events.

- In segmented viruses, like influenza, reassortment can lead to the formation of novel viruses by combining segments from different parental strains. Other viruses, like HIV, recombine generating high diversity.
- Persistent homology can be used to estimate actual recombination rates by fitting models involving topological features to values generated by simulations (e.g., created by the coalescent model of evolution). These models can be used to estimate recombination rates across the human genome.
- A persistent homology sliding window approach in large genomes provides fine detail on recombination rates in specific locations in the genome.
- Other constructions, like median complexes and barcode ensembles, increase the sensitivity for identification of non-tree-like events.

5.12 Suggestions for Further Reading, Databases, and Software

Here is a recommendation of a few books that explore topics related to the ones described in this chapter.

- *Gene Genealogies, Variation and Evolution*, by Jotun Hein, Mikkel Schierup and Carsten Wiuf [237], is an excellent primer in coalescence theory, with dedicated chapters on population genetics models, coalescence, ancestral recombinant graphs, and linkage disequilibrium. There is also a chapter on applications to human evolution, population structure, and migrations.
- *Phylogenetic Networks*, by Daniel Huson, Regula Rupp and Celine Scornavacca [262], provides a very clear survey of methods for inference of phylogenetic networks.
- *Viruses*, by Arnold Levine [328], is a lucid introduction for the neophyte to the world of viruses, including a description of molecular biology and historical accounts of HIV, influenza, and some other common human pathogens.
- *The Evolution and Emergence of RNA Viruses*, by Eddie Holmes [246], is a beautiful account of different evolutionary aspects of RNA viruses, insightful and very complete, with very interesting speculations about the deep origins of RNA viruses.
- *Principles of Virology: Molecular Biology*, by Jane Flynt, Lynn Enquist, Vincent Racaniello, and Anna Skalka [179] is a very clear two-volume description of the main principles of viral entry, replication, propagation, etc. Highly recommended for the reader who wants to delve deeper into how viruses operate.

- Recent review of applications of topological data analysis to genomic data have been written by P. Cámara [87] and the authors of this book [502].

Data used in the examples in this chapter and related data can be found in diverse databases.

- Influenza genomes annotated by subtype, day and geographic location can be found in the Influenza virus resource, in NCBI (National Center for Biotechnology Information, `www.ncbi.nlm.nih.gov/genomes/FLU/`), the Global Initiative on Sharing All Influenza Data (GISAID, `http://platform.gisaid.org/`), and the Influenza Research Database (IRD, `www.fludb.org/`).
- HIV genomes and immunological data can be obtained in Los Alamos HIV database (`www.hiv.lanl.gov/content/sequence/HIV/mainpage.html`). HIV genomes used in the study of HIV associated dementia can be found with consecutive GenBank accession numbers HM001362 to HM002482.
- For other viruses, Los Alamos HIV database also has compiled annotated genetic data from Hepatitis C Virus (HCV) and Hemorrhagic Fever Viruses (HFV) Databases (mostly Ebola). More general information can be found in the National Center for Biotechnology Information viral genome database (`www.ncbi.nlm.nih.gov/genome/viruses/`).
- The 1,000 Genome Project data can be found in The International Genome Sample Resource (IGSR, `www.1000genomes.org`).
- MLST bacterial data can be found in PubMLST [277].
- Data sets and software used along this chapter can be found in `https://github.com/RabadanLab`.

Here is a list of some software that can be used to detect recombinations and breakpoints.

- RDP in `http://web.cbio.uct.ac.za/~darren/rdp.html`
- GARD in `www.hyphy.org/`.
- BARCE in `www.topali.org/`.
- Simplot in `https://sray.med.som.jhmi.edu/scroftware/simplot/`
- PhylPro in `https://cran.r-project.org/web/packages/stepwise/index.html`
- Recco in `http://recco.bioinf.mpi-inf.mpg.de/`
- MaxChi in `http://web.cbio.uct.ac.za/~darren/rdp.html`.
- Chimaera in `http://web.cbio.uct.ac.za/~darren/rdp.html`.
- GeneConv in `http://web.cbio.uct.ac.za/~darren/rdp.html;https://www.math.wustl.edu/~sawyer/geneconv/`.

- 3seq Substitution in `http://web.cbio.uct.ac.za/~darren/rdp.html`, `http://mol.ax/software/3seq/`.
- PhiPack in `www.splitstree.org/`.
- SiScan in `http://web.cbio.uct.ac.za/ darren/rdp.html`; `http://mateo.fourment.free.fr/software.html`.
- TREE in `https://github.com/MelissaMcguirl/TREE`.

6

Cancer Genomics

This chapter aims to introduce the reader to an important problem in biology and in our own lives: cancer. Massive efforts have been dedicated to providing large-scale cancer-related molecular data to the scientific community. In this chapter, we provide the reader with some of the background material and concepts necessary for starting work in this area. First, we will walk the reader through a brief history of cancer genetics, from its origins, through the molecular biology revolution and, finally, the modern age of genomics. This history will illustrate the path leading to our current understanding of cancer as a molecular disease caused by mutations. We will also explain the most common types of genomic alterations found in cancer. Then, we will go through some recent examples on how topological data analysis techniques have been used in cancer research: identifying molecular markers associated with patients in breast cancer, distinguishing benign moles from melanomas, and studying the response of various cancers to different drugs.

6.1 A Brief History of Cancer

In March 1953, Carl O. Nordling, a Finnish architect with an inclination to statistical problems, published an article in the *British Journal of Cancer* [386]. He noted a common observation that most of us have unfortunately experienced through our relatives and friends: the incidence of cancer increases with age. Apart from some specific pediatric tumors, cancer is uncommon in children and adolescents (with a typical incidence of $\sim 18/100,000$ per year), and becomes more common in adults (increasing to $\sim 500/100,000$ per year) [472].

Nordling plotted the logarithm of cancer-associated death rates versus the logarithm of age using data from males from four different countries: the United States, United Kingdom, France and Norway. Interestingly, the mortality data fitted similar

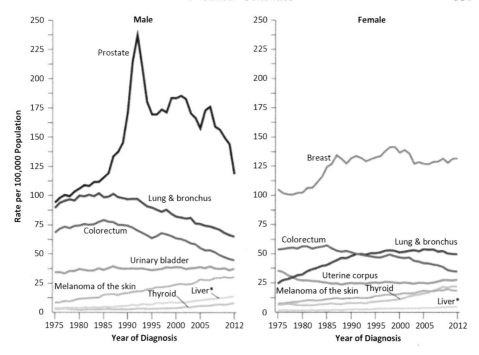

Figure 6.1 Age-adjusted incidence rates of different tumors for males (left) and females (right). Source: Surveillance, Epidemiology, and EndResults (SEER) Program, National Cancer Institute. The large fluctuations in the prostate incidence are due to over-diagnosis associated to changes in diagnosis procedures, including the PSA (Prostate Specific Antigen) test. General trends can be associated with change of habits (such as in lung cancers). The incidence of other tumors, including melanomas, seems to be steadily increasing. Source: [457]. From Rebecca L. Siegel, Kimberly D. Miller, Ahmedin Jemal, *Cancer Statistics, CA: A Cancer Journal for Clinicians*, Volume 66, Issue 1, pp 7-30, Jan 2016 © 2016. Reprinted with Permission from John Wiley and Sons.

straight lines in all four data sets (Figure 6.2), suggesting that the mortality of cancer followed a rule $\sim t^{\alpha}$, where t is the age and $\alpha \sim 6$. Nordling's work was carried out in a time where the main causes of tumors were still unclear, and conflicting evidence suggested both exogenous and endogenous elements contributing to carcinogenesis.

Several lines of work in the beginning of the twentieth century suggested that particular chemical compounds could induce cancers in laboratory animals. Katsusaburō Yamagiwa and Kōichi Ichikawa induced skin carcinomas in rabbits by painting their ears with coal tar [542], showing its carcinogenic properties. In the middle of the twentieth century, chemists were systematically cataloging compounds by their ability to cause tumors in mice. In addition to chemical compounds, it was clear that radiation could also increase the incidence of tumors.

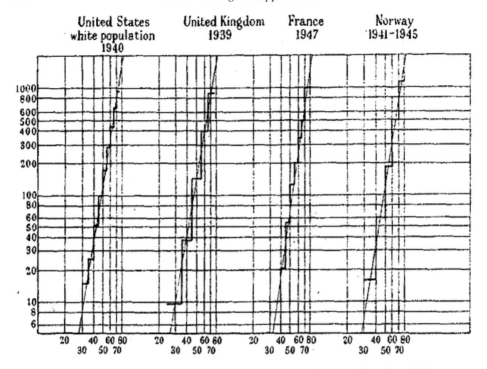

Figure 6.2 Nordling analysis of cancer mortality rates in males from four different countries. Source: [386]. Reprinted by permission from Springer Nature: *Nature, British Journal of Cancer*, A new theory on the cancer-inducing mechanism, C. O. Nordling, 7.1, 1953. © 1953.

The beginning of the atomic era was marked with scientists, health workers, and survivors of nuclear attacks exposed to high doses of radiation, leading to a remarkable increase in the incidence of unusual tumors. How did these diverse classes of agents all lead to cancer?

Theodor Boveri was a German biologist who studied the organization of cellular genomic material into chromosomes. In 1902 he hypothesized that cancers originate from alterations in the genomic material of a single normal cell [68] (see Figure 6.3). In 1927, Hermann J. Muller received a Nobel Prize for showing that X-rays could induce mutations in flies [364] (see Figure 6.4). Before publishing his observations in 1953, Nordling was also aware that "*the original cancerous cell is nothing but an ordinary cell affected by genetic mutation of some kind.*" Nordling reasoned that, if a single mutation was sufficient to cause cancer, and there was a constant rate of mutations, then tumor incidence should be independent of age. However, if multiple mutations were needed, the incidence should increase with age. Thus, he proposed a simple model that explained the "universal" $\alpha \sim 6$

Figure 6.3 Cancer genomes can be unstable. Instead of 23 pairs of similar chromosomes, the chromosomes in cancer cells come in different copy numbers, with amplified and deleted regions. Source: [270]. From Aniek Janssen et al., Chromosome segregation errors as a cause of DNA damage and structural chromosome aberrations, Science 333.6051 (2011): 1895-1898. © 2011. Reprinted with permission from AAAS.

Figure 6.4 A few pioneers in the history of cancer genetics. On the left, Theodor Boveri hypothesized that cancer could originate from alterations of the genomic material of normal cell. (The History Collection / Alamy Stock Photo). In the center, Katsusaburo Yamagiwa showed that some chemical compounds could cause tumors in laboratory animals. (Pictorial Press Ltd / Alamy Stock Photo.) On the right, Hermann Muller showed in 1927 that X-rays could cause mutations in flies. (INTERFOTO / Alamy Stock Photo.)

by assuming that tumors were caused by around seven independent mutations. Nordling discussed variations on the universal factor, including hormonal-related tumors in women and childhood tumors. These observations were replicated in greater detail and different mathematical models were proposed in subsequent years [16].

6.2 Cancer in the Era of Molecular Biology

Some of the first breakthroughs in the molecular biology of cancer came through an unexpected route: viruses. As we showed in the previous chapter, viruses are small particles that replicate inside cells. Some viruses replicate and kill the host cell, while others lead to uncontrolled cell proliferation. At the turn of the twentieth century, sporadic observations reported transmissible tumors in animals. It was the work of Peyton Rous in 1909 at Rockefeller University that opened the field of tumor virology. Studying a sarcoma in a hen of "light color and pure blood," Rous showed that it was able to transmit the tumor to other chickens [438]. In 1911, Rous filtered the tumor cells and showed that inoculation of the cell-free filtrate in other chickens was sufficient to recreate the tumor [437]. The Rous sarcoma virus (RSV) was the first definite oncovirus (tumor causing virus) identified, playing a crucial role in the successive developments in the molecular understanding of cancer. Peyton Rous was awarded the Nobel Prize in 1966 for his work, more than 50 years after the original discovery.

This is where the modern history of molecular biology and cancer research starts. The work on RSV was revisited at the end of the 1950s by Renato Dulbecco, Harry Rubin, and Howard Temin. They observed that normal chicken cells could be "transformed" into tumor cells in a Petri dish. The transformed cells resembled tumor cells but could be studied in a simpler environment than a living animal. Although the genomic material of RSV is RNA, Temin showed that the viral genetic material persisted in the transformed cells in the form of DNA [493]. Temin proposed that after RSV infected a cell, it generates DNA that is able to replicate as the cell's normal DNA. The enzyme responsible for the RNA to DNA conversion (reverse transcriptase) was simultaneously identified by Temin and David Baltimore [29, 492] in 1970. Viruses like RSV that carry reverse transcriptase are called *retroviruses*. Dulbecco, Baltimore and Temin shared the Nobel Prize in 1975. (See Figure 6.5.)

The possibility of studying tumors in vitro and the development of new technologies led to a young generation of molecular biologists that expanded on the relation between oncoviruses and cancer. RSV is a small virus that contains only four genes, one gene more than other related, non-transforming retroviruses. This led to the speculation that the extra gene encoded a protein src (for sarcoma), that was somehow responsible for the transformation. The puzzle was finally solved in 1976 by Michael Bishop and Harold Varmus, when they found that homologous copies of the transforming gene were present in different bird genomes [484]. (See Figure 6.6.) The discovery that the genes that were responsible for transforming normal cells were present in untransformed cells was surprising. However, it was not a new idea. Several researchers, including Robert Huebner and George Todaro, had previously suggested that cancer could be the result of the activation of silent

Figure 6.5 The beginning of the understanding of the molecular biology of cancer through oncoviruses. From left to right: Peyton Rous (Keystone Pictures USA / Alamy Stock Photo), Renato Dulbecco (Science History Images / Alamy Stock Photo), and Howard Temin. (Courtesy of the University of Wisconsin-Madison Archives (ID 16555))

genes present in normal cells. These activated genes, or oncogenes, were derived from our own normal genes. The original gene, called the proto-oncogene, is activated through diverse genetic and exogenous mechanisms leading to malignant transformation. The work of Bishop and Varmus established that a normal cellular gene, in this case *c-src* (c- for cell), was the ancestor of the transformed gene found in the RSV oncovirus, *v-src* (v- for virus). This work demonstrated that cancers could be due to alterations in our own genes, and that some viruses, like RSV, could hijack these genes leading to transformation. This discovery was awarded a Nobel Prize in 1989. This was a fascinating discovery that led to the quest for oncogenes in our own cells.

The findings of Bishop and Varmus led to speculation that perhaps other mechanisms could activate oncogenes as well. In 1982 three independent investigators, Robert Weinberg, Michael Wigler and Mariano Barbacid, cloned the first oncogene, *RAS* [203, 398, 444, 456].

A related line of research was carried out on a different type of virus, SV40 (Simian virus 40). SV40 is a double-stranded DNA virus (unlike RSV). The genome of this virus is small, containing only five different genes. Two of these genes, called the T-antigens, are expressed early after infection. SV40 was first identified in primary cells isolated from kidneys of monkeys that were used for the production of the Salk poliovirus vaccine. [1] Although monkey cells die when infected with the virus, occasionally cells from other species (mice and hamsters)

---

[1] The story of the discovery of this virus is fascinating but outside the scope of this book. We recommend reading chapter 5 of the book of A. Levine on viruses [328].

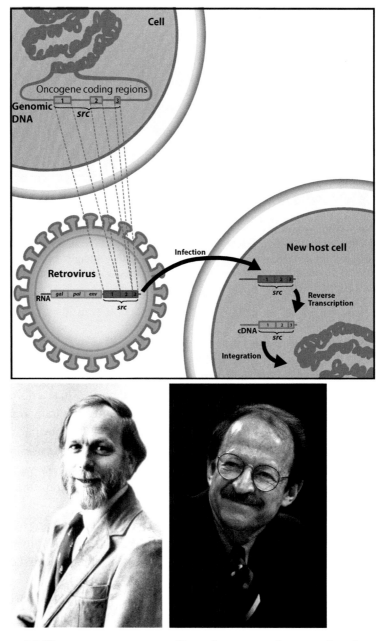

Figure 6.6 Viruses can cause cancer. Retroviruses can take genes from host cells and integrate them into other cells. Some of these host genes can lead to the transformation of infected cells. For the discovery of the cellular origin of retroviral oncogenes, Michael Bishop (left, Science History Images / Alamy Stock Photo) and Harold Varmus (right, Richard Ellis / Alamy Stock Photo) were awarded the Nobel Prize in 1989. Source: Nobel Prize webpage.

are transformed. In the rare case where SV40 genomic material integrates into the host cell genome, it transcribes the T-antigens, generating a cancerous cell. In 1979, Arnold Levine, Lionel Crawford, David P. Lane, and Lloyd Old identified a human 53 kilodalton protein (named p53) bound to the large T protein [139, 313, 331]. In the late 1980s, it became clear that p53 inhibited cell proliferation. It was found that p53 was inactivated in a large fraction of human cancers (near 50%), and that mice with inactivated copies of the *p53* gene grew a variety of tumors. These proteins that become inactivated in tumors are called tumor suppressors. These results suggested that tumors can be associated to oncogenes, like *RAS*, becoming activated in tumors, and to tumor suppressor genes, like *p53*, becoming inactivated.

6.3 The Standard Model of Tumor Evolution

We now briefly summarize the standard model of tumor evolution as put forth by Peter Nowell in 1976 [388]. We have many cells in our body, around 4×10^{13}. As time passes and cells divide, mutations accumulate randomly. If any of these cells acquires the right combination of mutations, it could lead to uncontrolled cell proliferation. In this model, all cancer cells are derived from a single cell, the original clone (see Figure 6.7). Chemical carcinogens or radiation could accelerate the processes of tumorigenesis and progression, leading to earlier onset. Some of these mutations could lead to activation of oncogenes or inactivation of tumor suppressors. These mutations confer selective advantage, and subsequent clones replace ancestral ones, accelerating the progression.

Mutations contribute to cancer formation and progression by activating and inactivating key fundamental cellular processes [227]. These altered processes include indefinite replicative potential (cells can replicate without apparently diminishing their replication potential), evasion of programmed cell death (*apoptosis*), stimulation of self-sufficient growth signals (cells can replicate without receiving external inputs), inactivation of antigrowth signals, abnormal metabolism, activation of signals for blood vessel formation (*angiogenesis*), invasion, evasion of immune response, and spread of tumor cells to other organs (*metastasis*).

Mutations accumulated in our bodies throughout our lives are termed *somatic*. These are not all the mutations that could contribute to cancer. Mutations that are inherited (*germline mutations*) could increase the risk of developing certain cancers. For instance, Frederick Li and Joseph Fraumeni characterized families with mutations in *p53*, whose members developed multiple tumors at an early age. Other familial diseases have also helped in the identification of different genes implicated in cancer.

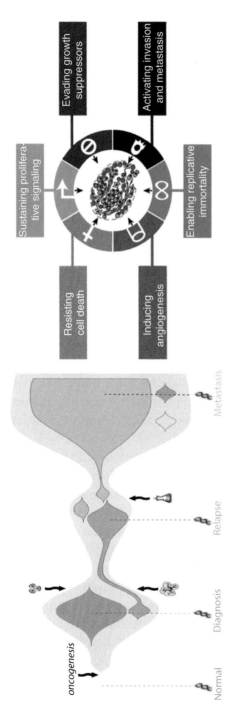

Figure 6.7 The origins and evolution of cancers. Left: Cancer understood as a clonal process. A single clone starts acquiring mutations that lead to uncontrolled growth. The cartoon represents subsequent nested clonal expansions that characterize different phases of tumor evolution (time runs from left to right). We have more than 10 trillion cells in our bodies that accumulate mutations over time. These mutations lead to clonal expansions. Tumors are diagnosed when the number of cancerous cells is sufficient to be detected with current technologies, or when associated clinical symptoms become apparent. Tumors are treated, but other clones could emerge, leading to resistance to therapy and/or metastasis. Source: [546]. Reprinted by permission from Springer Nature: Springer: Zairis S., Khiabanian H., Blumberg A. J., Rabadán R. (2014) Moduli Spaces of Phylogenetic Trees Describing Tumor Evolutionary Patterns. In: Ślęzak D., Tan A. H., Peters J. F., Schwabe L. (eds) *Brain Informatics and Health*. BIH 2014. Lecture Notes in Computer Science, vol 8609. Springer, Cham. Right: Some of these mutations activate and deactivate important key steps necessary for oncogenesis and tumor progression. The figure on the right represents the summary by Hanahan and Weinberg of the main mechanisms that are altered in tumor proliferation. Source: [227]. From Douglas Hanahan and Robert A. Weinberg, Hallmarks of Cancer: The Next Generation, *Cell*, Volume 144, Issue 5, pp 646-674. Reprinted with permission from Elsevier.

Although primitive, the model of clonal evolution through somatic mutations is a first approximation to how tumors evolve. However, it is far from
comprehensive, as many other factors are known to play a crucial role, including surrounding cells (microenvironment), epigenetic mechanisms, and immune
response.

6.4 Cancer in the Era of Genomic Data

The process of identifying oncogenes and tumor suppressors, so laborious and
painstaking in the 1980s, has been changed dramatically by the development of
high-throughput sequencing techniques in the first decade of the twenty-first century. Genomic material from samples obtained from tumors could be sequenced
and compared to the normal tissue. Studies of cohorts of patients with different
tumors started to show the distribution of somatic alterations associated with particular tumors. The process of finding somatic alterations in cancer starts with the
collection of tumor samples and matched normal tissue from patients. Figure 6.8
shows a standard procedure for how these mutations are read. The DNA from each
sample is extracted, fragmented, and enriched for certain genomic regions of interest. For instance, one could be interested in sequencing some genomic region with
an oncogene or a tumor suppressor. Some popular procedures involve the selection
of all genomic regions that contain coding genes, also known as the exome. This
is called whole-exome sequencing or WES. The capture of the genomic regions

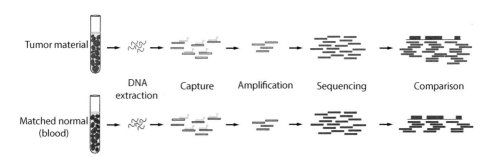

Figure 6.8 High-throughput sequencing allows us to read the somatic alterations
in tumor cells. Samples are obtained from tumor and matched normal tissue from
the same individual. The DNA is extracted and fragmented. In the case of whole-
exome sequencing, different regions are captured using complementary probes.
The captured material is then amplified and sequenced. The final results are files
containing reads (small sentences of nucleotides) that can be aligned to a reference genome and annotated for diverse variants. The results from the tumor and
normal tissue are compared for the identification of diverse alterations present in
the tumor tissue but not in the matched control.

of interest is achieved through various techniques usually relying on complementary oligonucleotides. Other protocols aim for whole-genome sequencing (WGS) without enrichment for particular regions. Finally the captured DNA is sequenced, generating a large collection of sequences consisting of the four characters (A, C, G, and T), each roughly 100 nucleotides in length. These sequences are referred to as reads.

The next procedure is the alignment of these reads to a reference (haploid) genome. Sequence alignment is the procedure of finding the best match between two different genomic sequences. Similarity between the two sequences should account for the possibility of one sequence not matching some of the nucleotides in another sequence. In our particular case of interest, reads from the tumor and matched normal DNA are aligned into the human genome. The comparison between the tumor and normal alignments allows the identification of diverse genetic alterations (see Figure 6.9).

For instance, the reads from the tumor sample could report a particular base change that are not present in the matched normal tissue. This is an example of a *somatic point mutation*. The difference could be a small number of bases in the tumor that are not present in the normal (small insertion) or present in the normal but not in the tumor (small deletion). Sometimes, large portions of the genome (from thousands of bases to whole chromosomes) are lost or amplified in the tumor (Figure 6.9). These events can be identified by either an unusually low or high number of reads in the tumor sample, respectively. Other types of alterations such as translocations involve the generation of new genes by the joining of two distant genomic regions. There could also be genomic material that maps into viruses or bacteria that could be present in the tumor sample.

In the sections that follow we will briefly walk the reader through some of the most common somatic alterations in cancer.

6.4.1 Point Mutations

Somatic point mutations are on average the most common somatic alterations across many different tumor types, although there is a large range of variation in the absolute number (see Figure 6.10). Pediatric tumors have in general a lower number of mutations than adult tumors, in concordance with the observations by Nordling in the 1950s [386]. Some tumors like melanoma or lung tumors present large numbers of somatic point mutations associated with the exposure to different carcinogenic environmental factors, like UV radiation or tobacco smoke, respectively. There is a fraction of tumors with mutations inactivating the DNA damage repair pathways that accumulate large numbers of mutations, sometimes referred to as hypermutant tumors.

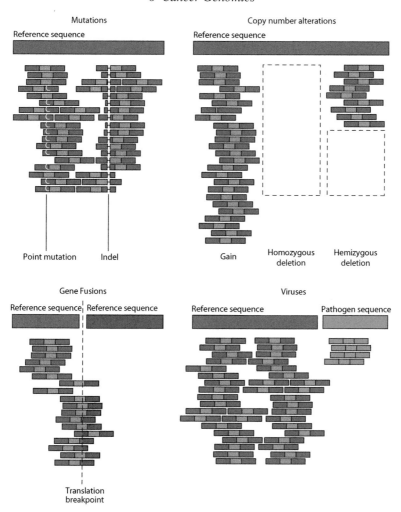

Figure 6.9 Diverse alterations that can be read using genomic technologies. Point mutations are identified when reads aligned to a particular genomic location report a different base than that in the reference genome and the matched normal. Indels, or small insertions and deletions, are reported when the best alignment of a set of reads mapped to a genomic locus skips or inserts a few bases that are not present in the reference and matched controls. Copy number variations can be identified by two complementary methods: statistically significant difference in number of reads in the tumor versus normal, and loss of heterozygosity. A gain (a few extra copies) or amplification (more than 10 extra copies) can be inferred when there are more reads mapped to a particular genomic locus in tumor than normal. Heterozygous (1 copy) or homozygous (both copies) losses occur when there are fewer reads. Loss of heterozygosity (LOH) is the loss of some variants that are heterozygous (50% one allele and 50% the other) in the normal but change the allele frequency in the tumor. Translocations can be identified through pairs of reads mapping to distant locations in reference locations. Finally, there can be reads that do not map to the human genome. These can be the result of the presence of other organisms in the tumor.

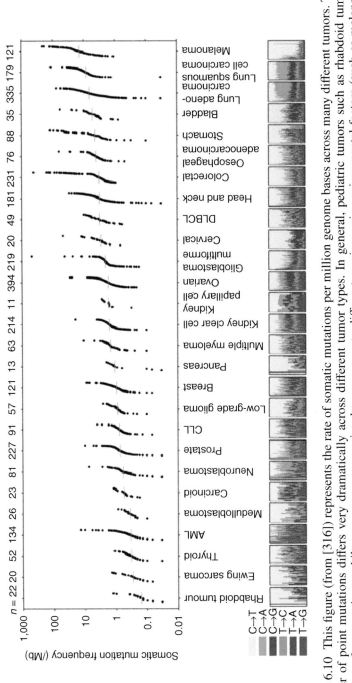

Figure 6.10 This figure (from [316]) represents the rate of somatic mutations per million genome bases across many different tumors. The number of point mutations differs very dramatically across different tumor types. In general, pediatric tumors such as rhabdoid tumors show very few mutations, while some tumors associated to exposure to different carcinogenic environmental factors (such as melanoma and UV radiation) present large numbers of somatic point mutations. Source: [316]. Reprinted by permission from Springer Nature: *Nature*, Mutational heterogeneity in cancer and the search for new cancer-associated genes, Lawrence, Michael S., et al., 499.7457 (2013): 214-218. © 2013.

Case	Genomic Position	Ref/Var	gene	AA	Tumor Frequency	Depth	tf_lower	tf_upper	Normal Frequency	Depth
TCGA-14-3477	chr17:7518257-7518257	G/A	TP53	P250L	98	241	92	99	0	271
TCGA-06-0747	chr7:55221823-55221823	C/T	EGFR	A289A,A289	97	3987	95	97	1	123
TCGA-14-1453	chrX:76760970-76760970	C/T	ATRX	R1803H,R176	96	250	89	99	0	133
TCGA-06-2563	chr7:55189204-55189204	C/T	EGFR	R252C,R252C	95	2784	93	96	1	163
TCGA-06-2565	chr7:55178574-55178574	G/A	EGFR	R108K,R108K	94	4239	93	95	0	142
TCGA-06-0129	chr17:7518263-7518263	C/T	TP53	R248Q	94	84	78	99	0	84
TCGA-06-6389	chr17:7518915-7518915	T/C	TP53	Y220C	93	30	54	100	0	52
TCGA-12-0618	chr17:7577568-7577568	C/T	TP53	C238Y,C238Y	90	103	74	99	0	81
TCGA-02-2483	chr17:7517845-7517845	C/T	TP53	R273H	90	59	64	99	0	60
TCGA-14-3477	chr4:54831633-54831633	G/A	PDGFRA	D400N	88	1813	85	91	0	160
TCGA-14-0865	chr4:54834499-54834499	A/G	PDGFRA	N468S	88	541	81	93	0	77
TCGA-06-2570	chr17:7519249-7519249	G/C	TP53	Q136E	88	75	67	99	0	70
TCGA-14-3477	chr4:54831657-54831657	G/C	PDGFRA	E408Q	87	1180	82	90	0	110
TCGA-06-0237	chr7:55210075-55210075	T/G	EGFR	L62R,L62R,L6	85	1282	80	89	0	200
TCGA-06-0221	chrX:76944376-76944376	G/A	ATRX	Q177*,Q139	85	88	63	96	0	119
TCGA-06-0221	chr17:7578512-7578513	TC/-	TP53	TK(140-139)-	83	68	55	98	0	78
TCGA-06-1804	chr10:89643813-89643813	G/A	PTEN	G44D	81	16	28	99	0	118
TCGA-14-1456	chrX:76758774-76758774	C/G	ATRX	Q1843H,Q18	76	41	38	96	0	77
TCGA-06-0125	chr7:55200537-55200537	G/T	EGFR	G598V,G598	73	4429	71	76	0	148
TCGA-26-5132	chr13:47851731-47851731	C/T	RB1	R445*	73	44	37	93	0	67
TCGA-02-2483	chrX:76735929-76735929	C/-	ATRX	W2001-,W19	71	42	35	93	0	36
TCGA-12-0670	chr7:55189237-55189237	A/C	EGFR	T263P,T263R	70	3733	67	73	0	589
TCGA-26-5133	chr17:26691654-26691654	G/-	NF1	S2309-,S228	70	82	43	89	0	156
TCGA-12-0619	chr17:7578440-7578440	T/C	TP53	K164E,K164E	70	64	41	89	0	120
TCGA-12-0656	chr1:115058052-115058052	T/A	NRAS	Q61L	69	1151	63	74	0	1361
TCGA-19-2619	chr1:155100785-155100785	G/A	NTRK1	Q46Q,Q76Q	68	84	42	86	0	66
TCGA-06-2558	chr17:7517822-7517822	C/G	TP53	D281H	65	55	35	87	0	68
TCGA-06-2559	chr10:89682973-89682973	G/T	PTEN	R159S	64	103	41	81	0	137
TCGA-06-2558	chr10:89682884-89682884	C/T	PTEN	R130*	63	129	42	79	0	269
TCGA-06-2561	chr12:25289551-25289551	C/T	KRAS	G12D,G12D	60	105	38	77	0	50
TCGA-12-0707	chr17:26700287-26700287	A/G	NF1	I2405V,I2384	57	21	15	89	0	119
TCGA-12-0656	chr10:89710832-89710832	C/T	PTEN	R335*	56	715	48	64	0	912
TCGA-14-1455	chr4:54825917-54825917	T/C	PDGFRA	C235R	54	3684	50	57	0	385
TCGA-06-5417	chr17:7518988-7518988	G/A	TP53	R196*	54	87	32	72	1	98
TCGA-06-5417	chrX:76665474-76665474	T/C	ATRX	H2254R,H22	51	185	36	65	0	230
TCGA-19-5950	chr11:107677665-107677665	A/G	ATM	Y1753C,Y405	51	123	33	67	0	134
TCGA-06-5417	chr3:180418785-180418785	G/A	PIK3CA	E545K	48	99	29	67	0	111
TCGA-12-0692	chr13:28597589-28597589	T/A	FLT3	K772N	48	108	29	66	0	39
TCGA-26-1442	chr2:208821357-208821357	C/T	IDH1	R132H	47	55	23	71	0	88
TCGA-06-2570	chr2:208821357-208821357	C/T	IDH1	R132H	46	155	30	60	0	132
TCGA-19-5954	chr3:180399317-180399317	C/T	PIK3CA	R4*	46	50	21	70	0	63
TCGA-06-0128	chr2:208821357-208821357	C/T	IDH1	R132H	42	145	26	58	0	146
TCGA-06-0140	chr13:47849145-47849145	C/T	RB1	Q436*	42	140	26	59	0	141
TCGA-12-0656	chr17:26583330-26583330	T/G	NF1	L1104R,L110	41	292	29	53	0	270
TCGA-76-4925	chr1:155116543-155116543	G/A	NTRK1	K689K,K719H	41	247	28	54	0	216
TCGA-12-0616	chr4:55592080-55592080	G/A	KIT	P468P,P468R	40	131	24	58	0	106

Figure 6.11 An example of somatic point mutations in glioblastomas from The Cancer Genome Atlas. This is the most common brain tumor type in adults. Each row reports a particular somatic mutation. Different columns annotate the mutations: the patient identifier, the genomic position, the type of mutation, gene, amino acid change, frequency of the mutation in the sample, how many reads were mapped to the location, lower posterior estimates of frequency in tumor, upper posterior estimates of frequency in tumor, frequency of the alteration in the matched normal sample, and number of reads in the normal mapped to the position.

The typical genomic information associated with point mutations can be displayed in a table similar to Figure 6.11. Each row represents a single point mutation. The first column contains the (de-identified) code of a patient. The second column captures the chromosomal location where the somatic mutation is found.

For instance, the first row tells us there is a mutation in position 7,518,257 of chromosome 17. This mutation changes the amino acid 250 of the *TP53* gene from a proline (P) to a leucine (L). We all have two copies of chromosome 17 in our cells; however this particular mutation, despite not occurring in normal cells (0% frequency), was at a frequency of 98% in tumor cells. The columns referred to as depth represent the number of sequence reads that cover that genomic position, a proxy of the relative amount of DNA in a particular genomic locus. A mutation that occurs in one of the copies of *TP53* and was found in all tumor cells should be present in 50% of reads as there are two copies in each cell. The fact that this number is close to 100% indicates that the non-mutated allele is probably lost (see the next paragraph). This is one of the most common mechanisms of inactivating tumor supressors where both alleles are inactivated through point mutations or deletions.

The second row is also illustrative. It reports a mutation in position 55,221,823 of chromosome 7. This mutation does not change the amino acid where it is located (alanine) but despite being absent in the normal tissue is reported in 97% of the sequence reads covering this particular region. Closer inspection of the depth shows an extraordinary increase from 123 reads to 3987 in tumor, suggesting a 60 fold amplification of the mutated allele. It is known that this genomic region contains an oncogene (*EGFR*) that is frequently amplified in glioblastomas. The synonymous mutation (a mutation that did not change the amino acid) was carried over in the amplification process.

A close inspection of this list reveals several important features. There are several genes recurrently mutated in glioblastomas (*TP53*, *ATRX*, *EGFR*, *PTEN*, *PDGFRA*, among others). Some of these mutations appear amplified (*EGFR* or *PDGFRA*) while others are accompanied with loss of the non-mutated allele (*TP53*). Other genes present inactivating mutations, suggesting that they could function as tumor suppressors (*ATRX*, *RB1*, *PTEN*).

6.4.2 Copy Number Alterations

A second type of somatic alteration, common across many different tumors, is *copy number alterations*. As cells replicate, genomic regions can be lost (deletion) in one allele (hemizygous) or both alleles (homozygous); on the other hand, extra copies of a genomic region can be incorporated (amplification). These alterations can range from a few bases to the whole chromosome. Chromosomal regions that are amplified or deleted that contain oncogenes or tumor suppressors could be selected in the process of tumor development.

Two types of information are usually considered here for the characterization of copy number alterations. The first is the loss or increase of genomic material in tumor sample versus the normal counterpart, indicating a deletion or amplification, respectively. The second is the changes of allele frequencies of heterozygous positions. Let us explain what we mean. Across the genome of our normal cells there are positions that differ between the two chromosomes. For instance, it could be that in one position in our chromosome we observe a C while in the same position in the other chromosome we observe a T. These positions are called heterozygous. If one of the two regions in the chromosomes is lost or amplified, that will generate an imbalance between the two alleles (C or T), one becoming more frequent than the other. This phenomenon is called loss of heterozygosity (LOH). Copy number losses or amplifications can be identified by changes in the amount of genomic material covering a particular genomic region and a corresponding LOH.

Figure 6.12 shows two examples of these phenomena in chronic lymphocytic leukemia patients. On the left, there is a loss of one of the regions of chromosome 17 (in the p arm) reflected by loss of tumor material in that region and LOH in polymorphic position in this area. On the right, there is an example of a different phenomenon where no significant difference of genomic material is present in chromosome 20, but a significant LOH is observed across a whole chromosomal arm. This is an example of a phenomenon called copy neutral loss of heterozygosity, where one of the genomic regions is lost and the corresponding region of the other chromosome is duplicated.

6.4.3 Gene Fusions and Translocations

Gene fusions are the result of translocations, where two distant genomic regions, within or between chromosomes, are joined together. Translocations can take two different genes and generate a new one containing features from both parental genes. The most famous example of gene translocation is the Philadelphia chromosome, the result of a reciprocal translocation between chromosomes 9 (q34) and 22 (q11); see Figure 6.13. First described in 1960 by Peter Nowell [387], these two chromosomal regions contain two genes, the Abelson murine leukemia viral oncogene homolog 1 (or *ABL1*) (in chromosome 9) and the breakpoint cluster region gene (or *BCR*). *ABL1* is a potent tyrosine kinase whose activity is highly regulated. When fused to the new gene, *BCR-ABL* becomes a potent oncogene with uncontrolled activity. This is a very common translocation in chronic myelogenous leukemia. *BCR-ABL* fusions are the target of imatinib, one of the major successes of targeted cancer therapy [152].

A second type of translocation involves the juxtaposition of a very activating promoter, a region in the genome that controls the transcription of a given gene, next to a potent oncogene without affecting its coding domains. One example of this phenomenon is the *MYC* oncogene, a potent oncogenic transcription factor suspected to be active in more than 50% of all tumors. This activation occurs in a variety of ways, including translocations activating the expression of *MYC* in different B-cell lymphomas, leading to dysregulated expression of a normal protein [131].

Genomic approaches such as the ones described above are mapping the landscape of translocations across many different tumor types. When reads or pairs of reads are partially aligned to two different genomic regions, one can infer that these two regions have been joined in the tumor. Sequencing the RNA instead of the DNA allows one to read which novel expressed transcripts have appeared in

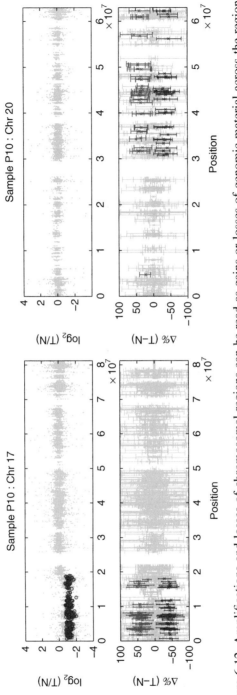

Figure 6.12 Amplifications and losses of chromosomal regions can be read as gains or losses of genomic material across the region and loss of heterozygosity. Top figures represent the logarithm in base 2 of the ratio between the amount of genomic material covering different genomic sections in the tumor versus the normal control. The lower part is an estimate of the allele frequency of the heterozygous position in the tumor samples. Red and blue bars represent posterior estimates of frequencies that differ from the expected 50%. On the left, we have an example of a loss of the p arm of chromosome 17 in a chronic lymphocytic leukemia patient. On the right, we have an example of a copy neutral loss of heterozygosity, where no significant loss or gain of material can be found but there is a systematic LOH in a whole chromosomal arm.

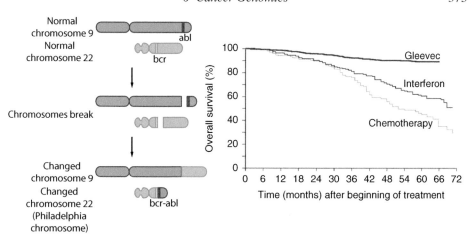

Figure 6.13 Left: The Philadelphia chromosome is the result of a reciprocal translocation between chromosomes 9 and 22 bringing together two genes *BCL* and *ABL1*. The new fusion protein is a potent oncogene and a common alteration in B-cell acute lymphoblastic leukemias, and virtually all chronic myelogenous leukemia. Right: *BCR-ABL* fusions are the target of imatinib (Gleevec or Glivec), one of the most effective targeted therapies in the recent history of cancer. Source: Reprinted by permission from Springer Nature: Springer Nature, *Nature Medicine*, Perspectives on the development of imatinib and the future of cancer research, Brian J. Druker, 15(10): 1149-52 © 2009.

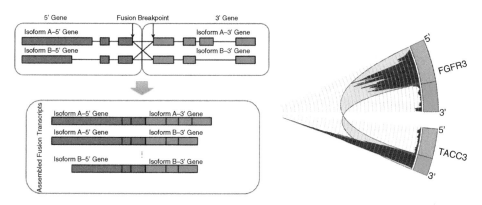

Figure 6.14 Left: Identifying gene fusions by reconstructing novel transcripts. Right: Unbiased characterization of gene fusion events are allowing us to map and catalog many potent oncogenic fusions in different cancers. For instance, *FGFR-TACC* fusions appear in 3% to 5% of glioblastomas.

the tumor (left panel of Figure 6.14). Using this information, one can characterize the most common fusion events across many different tumors. For instance, *FGFR-TACC* fusions [461] are found in nearly 5% of all glioblastomas (right panel of Figure 6.14), but also in many bladder and lung tumors.

6.4.4 Viruses

Somatic mutations accumulate during the entire lifespan of each individual. However, they are not the only important contributors to cancers. The relationship between viruses and cancer has been the subject of a long and tortuous investigation. As we explained in the introduction to this chapter, some of the major discoveries in the understanding of the molecular mechanisms of cancers came through the study of transforming viruses. However, it was not until the 1960s that the relation between some viruses and specific human cancers became evident. The World Health Organization recently estimated that nearly 20% of human cancers are caused by or associated with infections [399]. At present seven viruses have been shown to be strongly associated to human cancers

1. **Epstein-Barr virus** (Human Herpes Virus 4, HHV-4 or EBV). In 1958, Denis P. Burkitt, a surgeon working in Uganda, described a fast growing tumor type affecting the jaws of children with a median age of five years [84]. In 1963 a specimen of this tumor (Burkitt's lymphoma) was sent to London where Michael A. Epstein and Yvonne Barr identified a new virus, now known as the Epstein-Barr virus [165]. EBV is a common virus that for some reason is associated with Burkitt's lymphomas in children from Equatorial Africa but rarely in the rest of the world. Figure 6.15 represents the somatic mutations and virus

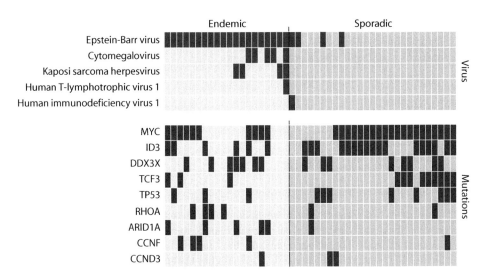

Figure 6.15 Burkitt's lymphomas occur in children in Equatorial Africa and rarely in other parts of the world. Each column is a separate patient; the colored rectangles indicate the presence of particular viruses or mutation. The endemic tumors are always associated with the EBV virus while the association is rare in sporadic cases. In addition to viruses, somatic point mutations occur in key oncogenes and tumor suppressors. Source: [2].

associated with endemic Burkitt's lymphomas from Uganda (left) in comparison to sporadic cases from the United States (right) [2]. Endemic Burkitt's lymphomas are associated with somatic mutations in distinct genes and the EBV virus. Since then, EBV has been found in many other tumors, including nasopharyngeal carcinomas, gastric cancers and specific types of peripheral T-cell lymphomas.

2. **Human T-cell lymphotropic virus type 1** (HTLV1) is a retrovirus that has been associated with a rare type of T-cell lymphoma (adult T-cell lymphoma or ATL), discovered in Japan at the end of the 1970s [243]. The virus is rare in most populations in the world, with higher prevalence in Japan, the Caribbean, and some populations in South America. Transmission is through contaminated blood products or direct mother to child transmission.

3. **Human papillomavirus** (HPV) is a type of non-enveloped double-stranded DNA virus (a Group I virus in Baltimore's classification) with a genome size of around 8000 nucleotides. HPV has been associated to a significant fraction of oral, cervical, vaginal, vulvar, penile and anal cancers. Harald zur Hausen received the Nobel Prize in 2008 for the discovery that human papilloma viruses cause cervical cancer. It has been estimated that more than half a million people get HPV-related cancers every year. Fortunately, the development of the HPV vaccine could prevent the development of these cancers.

4. **Hepatitis B virus** (HBV) is a member of the Hepadnaviridae family of viruses. HBV is one of the smallest enveloped viruses that infect mammals (viral particle of diameter of 40 nm). The genome, of only 3200 nucleotides, is made of partly double-stranded and partly single-stranded DNA. Chronic hepatitis, caused by HCV or HBV infection, can lead to hepatocellular carcinomas. Fortunately, a HBV vaccination has been available since 1981.

5. **Hepatitis C virus** (HCV) is a positive-sense single-stranded RNA virus that we briefly encountered in the previous chapter. Like HBV, chronic infection is associated to high risk of developing liver cancer.

6. **Kaposi's sarcoma-associated herpesvirus** (HHV-8) is a double-stranded DNA virus belonging to the herpeviridae family, with a large genome of 170,000 bases. It is associated to rare lymphoproliferative disorders, such as primary effusion lymphoma, multicentric Castleman's disease, and Kaposi's sarcomas, a rare tumor in immunosuppressed patients. Kaposi's sarcomas were first reported in 1872 by a Hungarian physician working in Vienna, Moritz Kaposi, as a rare condition in some Mediterranean populations. During the AIDS epidemics in the 1980s, near 50% of AIDS patients reported Kaposi's sarcomas. It was identified in 1994 by Yuan Chang and Patrick S. Moore [101].

7. **Merkel cell carcinoma virus** (MCCV) is a small (5400 bases), non-enveloped, double-stranded DNA virus. It was identified in 2008 by the same team that

identified HHV-8 [175]. The integration of the genomic sequence of this virus has been found in 80% of a rare, highly aggressive type of skin cancer, Merkel cell carcinomas.

6.5 Differential Gene Expression Analysis in Cancer

Each cell generates a different number of RNA copies for each gene, and many of these RNAs (the coding RNAs) will be translated into proteins. The RNA expression of a cell, the number of RNA copies of each gene, is correlated then with the amounts of proteins that are being produced. This relation is not linear, as different RNAs can be translated at different rates and proteins can have very different lifetimes. In addition, many RNAs are not translated into proteins; these are the so-called non-coding RNAs. Non-coding RNAs play many functions in the cell, such as regulation of other transcripts or as scaffolds for proteins, but many of these functions remain uncharacterized.

The activation of different pathways in tumors is reflected in changes of expression of different genes. For instance, there are important oncogenes that are transcription factors, such as *MYC*, which activates transcription of a large number of genes involved in the cell cycle. Looking at the RNA expression of transcriptional targets of *MYC* informs us about the activity of this protein. The expression of all the genes in a cell provides extremely useful information about transcriptional programs that are active in the cell.

Gene expression in cancer has been used for many purposes, for instance, to examine which transcriptional programs are activated (like cell proliferation) or inactivated (like apoptosis) in tumors. From the transcriptionally active genes one can infer the activity of different transcription factors. Another standard technique is to study the mechanisms of action of a particular drug or the effect of a mutation by comparing the expression profiles of cells before and after treatment. A different use that we will return to is the classification of patients based on the transcriptional programs of the tumor cells. These studies proceed by collecting RNA from large collections of tumors and classifying patients according to their transcriptional profile. These clusters then can be associated to other phenotypes, such as drug response or survival.

Most of the RNA expression data in tumors considers a population of cells (bulk), including tumor cells and surrounding cells (microenvironment) that are associated with the tumor. The measured total transcription abundance is then the sum of the transcription abundance of the cells of the sample, which may be a complex mixture. Many tumors present cells that are different at the genomic, transcriptomic or cellular level; this is referred to as tumor heterogeneity. In addition, stromal cells, the cells surrounding the tumor, may be quite diverse. Cells also vary their transcriptional state over time, as they differentiate, replicate, or engage with the environment. In sum, bulk transcriptional data reflects a mixture of cells and transcriptional programs. Single cell sequencing techniques are now used to disentangle these effects. In the next chapter, we will introduce the reader to several of these techniques and some examples of their use in the context of cancer.

6.6 The Space of Glioblastomas

As tumors evolve, their genomes accumulate mutations. Sequencing tumors provides a way of understanding the molecular mechanisms driving this process, as well as the mechanisms of resistance to therapies and potential therapeutic alternatives tailored to the genetic background of specific tumors. Glioblastoma (GBM) is one of the most common and most aggressive types of brain tumors. Median survival after initial diagnosis is little bit more than a year. The standard of care consists of surgery followed by radiotherapy and an alkylating agent, temozolomide (TMZ). Tumors invariably recur, leading to a fatal outcome. How these tumors evolve, the effect of the therapies, and the mechanism of relapse in these tumors is unclear.

To study how GBM evolves, Wang et al. sequenced longitudinal tumor samples in 114 GBM patients, both at relapse and at diagnosis [520]. They also sequenced tumor-matched normal samples. Comparison of the mutational profile in the three samples provides mutations that are in common (founder mutations), those specific to diagnosis and those specific to recurrence. Mutations that are specific to diagnosis could be associated to sensitivity to therapy, and mutations associated to recurrence could inform us about the mechanisms of resistance. From each of the 114 samples, we have three numbers corresponding to the three different types of mutations. From each triplet one can draw a simple phylogenetic tree of three branches: one representing the mutations in common, another branch representing the mutations specific to diagnosis, and the final branch representing the mutations acquired in the recurrent tumor. The mutational story of each patient is then represented by a tree, and the genomic information of the 114 patients is a forest (Figure 6.16). Now the question of how tumors in different patients evolve can be understood in terms of the metric spaces of phylogenetic trees described in Section 4.7.

We can now show the forest representing our data project as points in $\mathbb{P}\Sigma_3$ (Figure 6.16), as described in Section 4.7.2. Here, the upper corner represents the

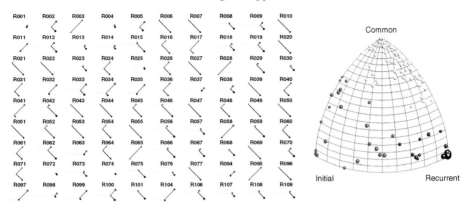

Figure 6.16 Genomic information of tumors from a patient generates a tree representing the different phases of tumor evolution. Cohorts of patients can be represented by a forest (left). In yellow are mutations that are in common in diagnosis and relapse, in red the ones that are specific to diagnosis and in black those that are specific to relapse. The forest can be mapped to points in evolutionary moduli spaces (right). Machine learning and statistical techniques can be applied to classify patient histories and to associate different mutational profiles or clinical outcomes. Source: [520]. Reprinted with permission from Springer Nature: Wang, Jiguang, et al. "Clonal evolution of glioblastoma under therapy." Nature Genetics 48.7 (2016): 768-776.

fraction of mutations that are common to both samples, the left corner represents the fraction exclusive to the untreated sample, and the right corner represents the fraction exclusive to recurrence.

If we want to see if there are different patterns of how tumors evolve in different patients, we can perform clustering. We describe the application of three clustering algorithms to this metric space: *k*-means clustering, spectral clustering, and density-based spatial clustering (DBSCAN). In order to ensure stability of the results, they were cross-validated using Monte Carlo simulations. Unsupervised clustering of the different phylogenies identifies three clusters. The yellow group represents the limiting case where few mutations are lost from diagnosis. This is similar to the classical model of linear tumor evolution, where mutations accumulate in clones that drive recurrence. The abundance of points far from the right edge of the diagram suggests that in most patients, the dominant clones prior to treatment appear to be replaced by new clones that do not share many of the same mutations. If many mutations in the initial sample are lost at recurrence, this suggests that the clone dominant at recurrence originated (i.e., diverged from the clone dominant at diagnosis) a relatively long time before the initial sample was taken. This is an interesting finding as it suggests that a different clone to the one that caused the initial tumor is responsible for the recurrent tumor.

Interestingly, patient histories identified in the black cluster correspond to particular trees with very long branches associated to relapse tumors. These long branches, associated with a phenomenon called hypermutation, were present only in patients treated with TMZ, and patients in this cluster are associated with longer survival (more than two years). All these patients harbor mutations in the mismatch repair pathway, mostly inactivating mutations in *MSH6*. The MSH6 protein plays an essential role in repairing damaged DNA, by fixing potential mistakes in the replication of DNA. These tumors cannot effectively repair the damage caused by the therapy (TMZ), accumulating many more mutations in branches of the tree associated to the relapse. These mutations are also different from the other mutations. Hypermutated recurrent tumors are highly enriched with C to T (and G to A) transitions, occur in a CC/GG motif, and are associated with the expression of the hypermutated genes.

6.7 Cross-Sectional Data in Cancer and Patient Stratification Using Expression Data

One important question both from the basic biology and clinical points of view is how molecular data, mutations, fusions, and expression can classify patients into different subtypes. From the basic biology point of view, this is important because common molecular features could tell us about the molecular patterns that are associated to particular patients. These molecular patterns can in turn reveal the specific pathways that are activated or deactivated in tumors, and the specific alterations that are related to these pathways. From the clinical point of view, the problem of patient stratification, or classifying patients into different sets, is an extremely relevant one, as molecular data can tell us whether patients could be sensitive or resistant to a particular therapy, and what molecular features are associated to progression or metastasis. Clinical questions can be translated into a problem of understanding the shape or structure of molecular data associated to a large cohort of patients. Information on many patients is usually referred to as cross-sectional data. The "dual or transpose problem," looking at different genes in the space of patients, is probably the more interesting biologically, as genes differentially regulated in a group of patients reveal common deregulated pathways (Figure 6.17).

All of the molecular data discussed in the previous sections could be used to classify patients or genes. The data can be discrete and sparse, such as somatic point mutations, with a binary value that indicates if the mutations are present or not. Expression data, in contrast, is a very rich source of information as it associates each transcript with a real value that corresponds to the copies of mRNA present

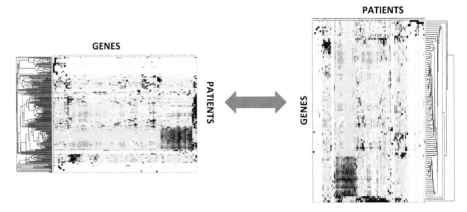

Figure 6.17 Hierarchical clustering of expression point cloud data corresponding to peripheral T-cell lymphoma patients. The data can be seen as two "dual" point clouds. In the space of genes each point is a patient and clustering of points corresponds to clustering of patients. On the dual space, the space of patients, each point is a gene and clustering of points corresponds to clustering of genes.

in the tumor. Most of the data corresponds to the ensemble of cells present in the sample and as such represents a complex mixture of expression levels from different tumor cells, and even different types of non-tumor cells that are also present in the sample. The typical structure of expression data is a point in a very high-dimensional real vector space, as there are typically on the order of 22,000 potential transcripts. Each patient represents a point in this space, and the cross-sectional data corresponds to a point cloud. The question of stratification is usually posed as a clustering problem: how many groups of patients are there presenting similar expression profiles?

Expression-based classification of tumors has been a dominant theme for research since the first microarray experiments and there is an extensive literature on the topic [236, 512] that we do not have the space to discuss. All these earlier approaches are in some way or another based on the idea of clustering patients and genes (see Figure 6.18). It could be, however, that the point cloud data does not have a nice cluster structure. Indeed, that is generally the case due to many biological and technical factors. Not every tumor activates or suppresses different pathways with the same strength, resulting in a more continuous structure from suppression to activation. There is also a common phenomenon of non-tumor cells infiltrating the tumor sample. These and other factors contribute to generating large continuous structures that sometimes are not correctly represented by clusters.

In [383], the authors studied the point cloud data associated with 295 breast cancers, given microarray gene expression data and normal breast tissue. Expression data was normalized and represented by Mapper. As we saw in Section 2.8, Mapper represents the cluster structure of the inverse images of a function on the data. In

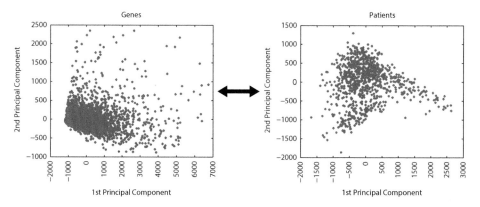

Figure 6.18 Expression point cloud data corresponding to 881 breast cancer patients from The Cancer Genome Atlas Consortium.

this study the function used was provided by a measure of the deviation of the expression data of the tumor samples compared to the expression in normal controls. Clusters of points of overlapping intervals in the image of this function were represented as nodes, and edges corresponded to shared points between different clusters (see top left, Figure 6.19). Blue colors correspond to samples with close similarity to normal tissues (left part of the figure). On the right hand side the samples diverge into two branches. The lower branch, named in the study as c-MYB+ tumors, constitutes 7.5% of the cohort (22 tumors). These tumors are most distinct from the normal tissues and are characterized by the high expression of particular genes, including *c-MYB, ER, DNALI1* and *C9ORF116*. Hierarchical clustering fails to identify this particular subset of tumors (see bottom left Figure 6.19), and segregates these tumors into separate clusters with low confidence. Interestingly, these tumors do not correspond to a previously reported breast cancer expression subtype. This new class of tumors, c-MYB+ tumors, is characterized by very good survival and no metastasis.

To validate these observations in an independent cohort, we looked at samples with high expression of *DNALI1* and *C9ORF116* (more than a 2-fold overexpression) in 960 breast invasive carcinomas from The Cancer Genome Atlas. Of these 960 tumors, 32 have expression in these genes and show excellent survival (right Figure 6.19), confirming the observation of [383]. The tumors do not contain *TP53* mutations and deletions, and are associated with *GATA3* mutations, suggesting a distinct mutational and expression subtype.

A different approach was taken by L. Seemann, J. Schulman and G. Gunaratne [450]. They looked at the expression data of 202 glioblastoma patients from The Cancer Genome Atlas. Expression-based clustering has identified four groups of glioblastoma expression profiles: classical, mesenchymal, proneural and neural [71, 515]. This partition into four groups has been questioned on several grounds.

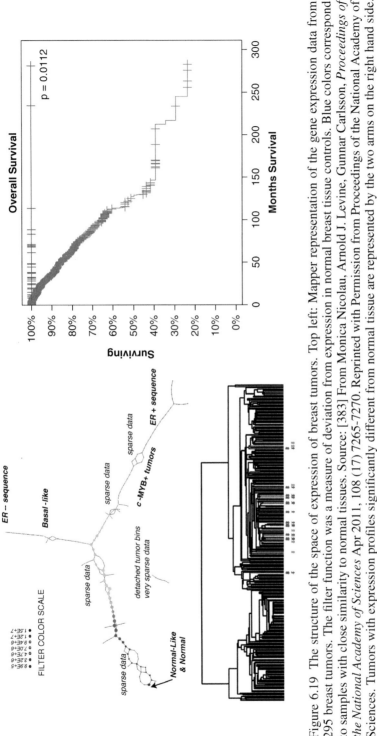

Figure 6.19 The structure of the space of expression of breast tumors. Top left: Mapper representation of the gene expression data from 295 breast tumors. The filter function was a measure of deviation from expression in normal breast tissue controls. Blue colors correspond to samples with close similarity to normal tissues. Source: [383] From Monica Nicolau, Arnold J. Levine, Gunnar Carlsson, *Proceedings of the National Academy of Sciences* Apr 2011, 108 (17) 7265-7270. Reprinted with Permission from Proceedings of the National Academy of Sciences. Tumors with expression profiles significantly different from normal tissue are represented by the two arms on the right hand side. The upper arm is characterized by low expression of the estrogen receptor (ER−). The lower branch contains samples with high expression of c-MYB+. These c-MYB+ tumors cannot be identified using standard clustering (in lower left figure hierarchical clustering split c-MYB+ tumors, represented in red). Independent validation using 960 breast invasive carcinomas from The Cancer Genome Atlas of two of the highest expressed genes in c-MYB+ tumors, DNALI1 and C9ORF116, shows very good prognosis for these tumors.

First, several groups have shown that sampling from different genomic regions does generate different profiles [197, 478]. This heterogeneity has also been seen using single cell expression data [401]. The expression profile also changes over time, before and after therapy [520]. Thus, there are significant concerns about the reliability of classifications of tumors based on expression profiles; such classification might not even make sense in highly heterogeneous tumors. This motivates the search for different approaches that can recover more complex structures.

An approach using persistent homology was explored in [450] to stratify patients is based on a hierarchical partition of samples. A first step was dimensionality reduction. A common problem in all expression-based studies is that the number of genes whose expression is considered is usually much larger than the number of samples, or patients in this case. However, many genes are not expressed, and many others have a similar pattern of variation. This suggests that the dimensionality of the gene space is effectively much lower. Seemann and colleagues propose to reduce first the number of genes. In particular, zeroth dimensional persistent homology was used to cluster genes with similar expression profile. Similarity between different patient profiles was computed using Pearson correlation, or more specifically, $d_{ij} = 1 - \text{corr}(e_i, e_j)$, where e_i is the expression profile of patient i. Using the persistent homology of the associated Vietoris-Rips filtration, 30 genes were identified as characterizing the two long-lived clusters. Patients were then represented by projection onto these genes. The resulting clustering analysis produced novel predictions for genes implicated in GBM. This work represents just the first step towards the use of topological methods for more nuanced classification of tumors.

6.8 Cross-Sectional Data in Cancer and Identifying Driver Genes in Cancer

We have described how different genomic alterations could contribute to cancer formation and progression. We have also shown how we can identify these alterations using genomic technologies. However, not every alteration in a tumor plays a role in its clonal history. For instance, many tumors contain tens of thousands of point mutations, but only a handful of those have a role in oncogenesis and progression (termed the *driver alterations*). How can we identify a few driver alterations within the large background of other irrelevant alterations (the so-called passengers)? Most of the ideas for identifying the most relevant players are based on the concept that, if a gene is found mutated more than would be expected from random variation in a sufficiently large cohort of patients, these alterations have probably been selected along the history of the tumor. Figure 6.20 shows a common representation, known as circos plot, of somatic alterations from 150 glioblastoma exomes. The external annotation refers to different chromosomes. In the interior,

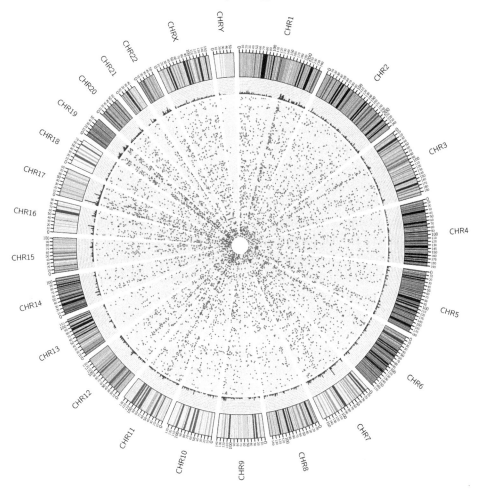

Figure 6.20 A representation of the mutations that occur in 150 glioblastoma patients. Each patient is represented by a concentric circle and the angle represents the chromosomal position where the mutation occurs. Genes frequently mutated across many patients are captured by the histogram in the external part of the representation.

in light green there is information on the 150 tumors represented on concentric circles [183]. Protein changing mutations are represented as red dots. Finally, between the external depiction of chromosomes and the mutations there is a histogram, representing the number of times that a particular gene has been mutated in the cohort. Recurrent mutations occur in chromosome 7, containing an oncogene (*EGFR*), and in chromosome 17, containing a tumor suppressor (*TP53*).

Recurrence-based methods are the standard approach for identifying driver genes in cancer. However, we know that there are alterations that are not very

frequent but could have a strong impact on the tumor development. An alternative approach for identifying genes with a clear phenotypic effect is to look for
alterations that share a common phenotypic profile, such as expression. Obviously
not every driver gene alteration should have a strong impact in expression, and not
every alteration with strong impact in expression is a driver alteration. But the association of the genetic and phenotypic effects is strong indication that the alteration
has an effect. Using the expression profiles of many patients and the correlation
as a similarity measure, one can apply Mapper, using the distances to the first two
nearest neighbors as a filter function. Now, one can see if patients with a particular
alteration present a similar expression profile. Figure 6.21 represents expression
data from 512 low grade glioma samples.

The Mapper representation finds three distinct groups with strong statistical
association with mutated genes.

- A group enriched in *CIC*, *NOTCH1* and *IDH1* mutations.
- A group enriched in *IDH1* and *TP53* mutations.
- A group enriched in *EGFR* and *PTEN* mutations.

These subtypes capture the recent glioma classification [381] into *IDH* wild-type
cases, *IDH* mutant with a co-deletion of chromosomal arms 1p and 19q, called *IDH*
mutant-codel, and finally patients with mutations in *IDH1* without the co-deletion
1p/19q (non-codel). Codel tumors have a good prognosis and are associated with
mutations in *IDH1*, *CIC*, *FUBP1*, and *NOTCH1*. The non-codel tumors harbor
mutations in *TP53*. The lower grade gliomas without an *IDH1* mutation clinically
resemble high grade gliomas (glioblastomas), and are associated to *EGFR* and
PTEN deletions.

In summary, by studying the localization of different mutations in the expression
space, one is able to identify genes that are associated with specific expression
profiles. These expression profiles are associated with different tumor subtypes that
have well-defined clinical characteristics.

6.9 The Tissue of Origin of Melanomas

Melanoma is the most aggressive form of skin cancer, with a five-year survival
rate of 98% for early-stage tumors compared to 63% and 16% for regional lymph
node and distant metastasis, respectively [472]. A strong risk factor for melanoma
is UV exposure, which typically causes an abundance of somatic mutations on
the order of 10 mutations per megabase [12]. This mutational burden far exceeds
that of many other solid tumors and complicates the process of separating the
passenger from the driver mutations. Interestingly, it has recently been shown

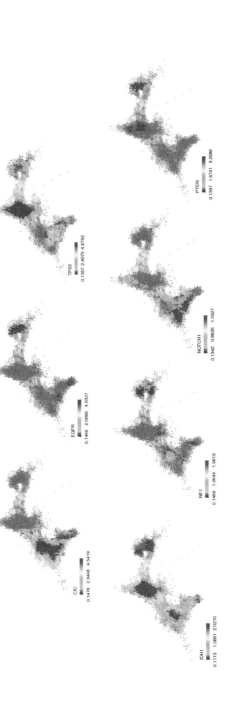

Figure 6.21 Identifying cancer driver genes using Mapper. If certain genes induce a common expression profile across patients, this should be reflected as similarity in the Mapper representation.

that higher mutational loads in melanoma actually lead to greater immunogenicity, better response to immune checkpoint inhibitors like CTLA-4 antibodies, and improved survival [471].

A hallmark feature of melanoma tumorigenesis is the phenomenon of oncogene-induced senescence, in which a typically strong oncogenic mutation – usually *BRAF* or *NRAS* in melanoma – triggers a "fail-safe mechanism" through the p19-p53 or p16-pRb pathways leading to senescence (stop proliferation) rather than cell proliferation. The result is a benign growth called a nevus, known in lay terms as a mole. The most prevalent types of acquired nevi are the common acquired nevus and the dysplastic nevus, both of which have an increased risk of progression to cancer, generally through the loss of tumor suppressor genes like *PTEN*.

A number of genomic studies have explored the mutational and regulatory landscape of metastatic melanoma [245, 514], and The Cancer Genome Atlas (TCGA) has made publicly available an abundance of genomic, transcriptomic, epigenetic, and clinical data for primary and metastatic melanoma. Despite this progress, the key transcriptional events that govern the development from normal skin to nevus to primary and finally metastatic melanoma remain uncharacterized.

The continuous nature of the transformation and the fact that tumor samples are "contaminated" with normal cells of different kinds leads to continuous point cloud data with structures that are difficult to discern with standard clustering techniques. We now describe the results of applying Mapper to transcriptional data from patient-derived samples from melanomas to identify substructures that were predictive of a number of features, including survival, tumor attributes, and gene modules. Comparison of these results to other standard methods of unsupervised exploratory data analysis, including hierarchical clustering and principal component analysis (PCA), is illuminating [334].

The full spectrum of transcriptional changes throughout the progression of melanoma has yet to be fully elucidated. Expression data was collected from 122 punch biopsies of four different subtypes of tissue: 51 primary melanoma (PM), 27 common acquired nevus (CAN), 15 dysplastic nevus (DN), and 29 normal skin (NS). 13 DN were also matched to 13 NS. Initial processing of data included mapping and aligning, calculating counts per transcript using subread, and normalizing using trimmed mean of M-values [434]. An initial approach was to identify differentially expressed genes (DEG) distinguishing PM, CAN, DN, and NS using a standard linear model [469]. This analysis identified the molecular signature of 4862 genes. We then calculated Z-scores derived from the normalized log counts per million for the set of DEG. We next calculated the pairwise distance matrix samples using a distance associated to the correlation ($d = 2(1 - r)$, where r is the Pearson correlation). We then provided this pairwise distance matrix as input for unsupervised analysis with

(1) principal component analysis (PCA),
(2) hierarchical clustering, and
(3) Mapper, using the Euclidean distance as the metric and distances to the first
     two nereast neighbors as the filter function.

PCA of these genes confirmed general separation of these tissue types
(Figure 6.22).

Hierarchical clustering was also performed with the Euclidean distance metric
and average linkage (Figure 6.22). This method also demonstrated overall effective
clustering between all subtypes except for DN and NS. While 13 DN and 13 NS
derived from the same patient, this fact alone did not explain their clustering. A
separate analysis showed that DN clustered with NS likely due to a similarly low
melanocytic content in both tissue types. Interestingly, two subclusters of PM separated far from the rest of the PM cohort, one low-thickness subcluster that clustered
with DN and NS and one high-thickness subcluster.

The Mapper representation provided a richer analysis of the data that shared
some features of the PCA and hierarchical clustering stories, but also contrasted
in other unexpected ways. We were able to capture a rich topological network that
not only demonstrated the separation of tumor subtypes, but also suggested CAN
as a further outgroup from the rest of the other subtypes (Figure 6.23).

Interestingly, there were two general subclusters of DN and NS, one to the right
and to the bottom of the PM subnetwork. On closer inspection of the members
comprising these two subclusters, the right subcluster contained 7 DN and 4 NS,
while the bottom subcluster contained 3 DN and 24 NS, suggesting that Mapper
was better able to distinguish between DN and NS. Of the 13 matched DN, 4 in the
right and 3 in the bottom subcluster were found next to their matched NS. However,
6 of the matched DN were placed in the right subcluster and were separated from
their matched NS in the bottom subcluster. These findings suggest that Mapper, to
some extent, was able to distinguish the tumor subtype.

Although the Mapper representation was built based on differential expression
of subtypes, the resulting structure reflects biological and phenotypic attributes
beyond just subtype. Coloring by tumor thickness of PM, we can see that there
is a coherent progression of tumor thickness away from the bottom and right
DN and NS subclusters where the outer flares approach a higher tumor stage
(Figure 6.23).

Recall that hierarchical clustering identified a subcluster of PM that was closer
to the DN and NS clade but apart from the rest of the PM. Mapper resolved this
inconsistency by readily demonstrating that these lower thickness PM were closer
to the rest of the PM subnetwork but were close to the DN and NS subclusters, as
well.

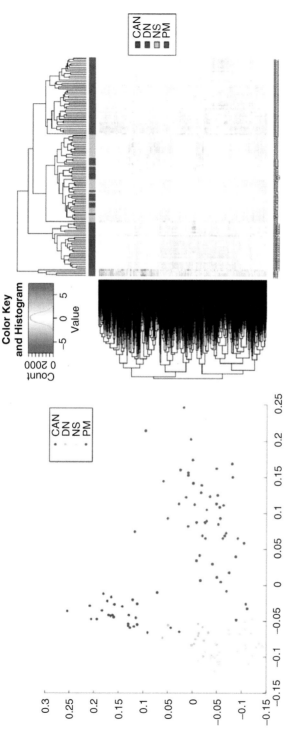

Figure 6.22 Expression analysis of skin, nevi, and melanomas. Left: Principal component analysis of expression of 51 primary melanoma (PM), 27 common acquired nevus (CAN), 15 dysplastic nevus (DN), and 29 normal skin (NS). The analysis shows two flares corresponding to primary melanomas and nevi emanating from normal type core. Right: Hierarchical clustering (Euclidean distance metric with average linkage) artificially separates some primary melanomas and fails to separate normal tissue from dysplastic nevi.

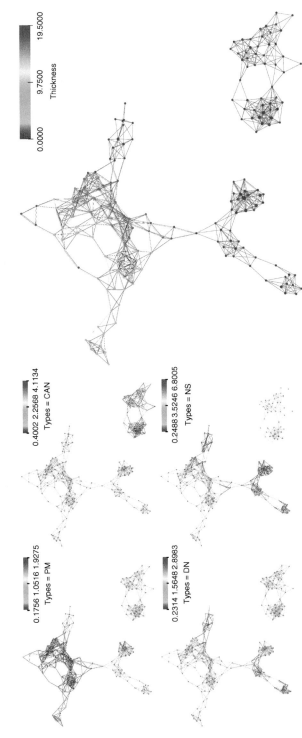

Figure 6.23 Mapper representation of skin, nevi, and melanomas. Distance metric was Euclidean and the filter function was the distances to the first two nearest neighbors. Left: Mapper graphs of tumor subtypes primary melanoma (PM, upper left), common acquired nevus (CAN, upper right), normal skin (NS, lower left), and dysplastic nevus (DN, lower right). Right: Mapper graphs of subtypes colored by tumor thickness of primary melanomas.

The Cancer Genome Atlas (TCGA) also provided a ready source of public RNA sequencing data of both 93 primary (PM) and 352 metastatic tumors (MM). The metastatic tumors included 72 regional cutaneous/in-transit/satellite metastasis (RCM), 215 regional lymph node metastasis (RLNM), and 65 distant metastasis (DM). Similar to the previous analysis, we first applied a standard linear model [469] to identify 695 genes that were differentially expressed between primary and metastatic melanoma. PCA of the resulting DEG data showed some separation of PM away from all subtypes of MM along both the first and second components. Unsurprisingly, subtypes of MM are not clustered separately as DEG analysis was performed with the label MM versus PM as the primary covariate (Figure 6.24).

Hierarchical clustering provided a similar picture as PCA. Three major clades were identified, including a predominantly PM cluster and a predominantly RLNM cluster. The third cluster was heterogeneous (Figure 6.24). Again, Mapper was able to provide a portrait of the shape of the data in richer detail. Not only was Mapper able to separate PM from MM, but it was also able to identify distinct clusters of RCM and RLNM. Interestingly, DM and PM occupied distinct domains of the same cluster (Figure 6.25).

Beyond subtype differentiation, the inherent structure of the topological network reflected underlying biological structure as well. Coloring by time to death after diagnosis as well as by living status at end of study, we identified two subclusters with better survival, one among the PM cohort with greater number of survivors at the end of study and a subgroup within the RLNM cohort with longer time to death among individuals that died at end of study (figure 6.25).

6.10 Association between Drug Sensitivity and Genomic Alterations

In the previous chapters we have described different kinds of somatic alterations and how these alterations stratify patients and define prognostic markers. We have seen that cancers in different patients are the result of different evolutionary histories and environmental exposures. The phenotypic evolution of tumors, their growth, how they metastasize, and how they respond to therapies will depend on these factors. The overall goal of precision cancer approaches is to find ways of linking the genomic and environmental data of a tumor to specific therapies.

We would like to end this chapter with a description of some applications of topological data analysis to understanding how genetic information could be used in connection to drug sensitivity. Methods that model and predict therapeutic sensitivity of cancer can be extremely useful in the development of more effective treatments. Somatic genetic alterations in cancer have been linked with the aberrant

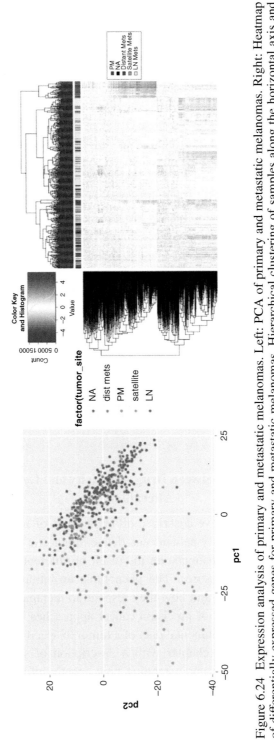

Figure 6.24 Expression analysis of primary and metastatic melanomas. Left: PCA of primary and metastatic melanomas. Right: Heatmap of differentially expressed genes for primary and metastatic melanomas. Hierarchical clustering of samples along the horizontal axis and hierarchical clustering of genes along the vertical axis. Euclidean distance metric with average linkage was used.

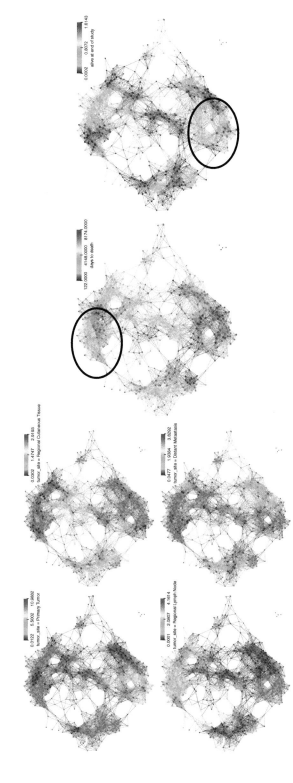

Figure 6.25 Mapper characterization of melanoma expression data from The Cancer Genome Atlas. The distance to the first two nearest neighbors was used as a filter function. Left: The four figures that the physical site location is related to expression in the Mapper graph. Upper left: primary melanoma (PM). Lower left: regional lymph node metastases (RLNM). Upper right: regional cutaneous tissue metastases (RCM). Lower right: distant metastases (DM). Right: Mapper graph of TCGA showing time until death after diagnosis (left) and whether the patient was alive at the end of the study (right). Circled are two subclusters of longer survival.

behavior of signaling pathways, which has led to the development of therapies targeted at these pathways.

While recent advances in sequencing technology make it possible to obtain a wealth of data on the genomic and transcriptomic profiles of specific tumors, our ability to translate this information into improvements in clinical outcomes is limited by our lack of understanding in two areas. First, we do not understand the function of most mutations. Second, we do not know which drugs are best suited for targeting the pathways those mutations affect or participate in. We use Mapper in a computational approach for genome-based drug sensitivity prediction to determine the therapeutic impact of recurrent gene alterations and the role of tumor heterogeneity in drug resistance across a range of cancers.

We have performed initial analysis of the Cancer Cell Line Encyclopedia (CCLE), which contains genomic characterizations of a large panel of cancer cell lines. In the left part of Figure 6.26, we have analyzed expression across cell lines using Mapper. Each node is a set of cancer cell lines that share similar expression profiles, and gene expression is used to construct a similarity metric. The filter function was the map to $\mathbb{R}^2$ specified by the distances to the first two nearest neighbors. Using this approach we first identify the overall network structure of cancer cell lines. Then, we identify specific genes that are characteristic of certain cell lines. For example, we identify that *PDGFRA* is expressed with high specificity in glioblastoma and neuroblastoma.

In the right part of Figure 6.26, we perform an analysis on the same data set, transposed. Here, each node is a gene, and the expression across different cell lines is used to construct a distance metric for the graph. We color genes to show the average expression across central nervous system cell lines. We observe a large cohort of genes centrally expressed across most cell lines. Flares emanating from the main set of genes correspond to genes with unique patterns of expression in particular cell lines. For example, central nervous system specific genes are localized in the flare on the right side of the network. Furthermore, we can localize specific genes within the network to identify neighbors with correlated expression, as we do with *EGFR* in the figure inset.

These networks can be used in a useful manner not only for representing these data sets, but also for predicting drug sensitivity based on genomic alterations. For example, we can perform clustering and feature selection based on these representations of the sample space. Using drug sensitivity information across each cell line from CCLE, specific neighborhoods of sensitivity can be identified. Then, these neighborhoods can be modeled for enrichment of specific alterations. In this way, we incorporate genetic background into the prediction of drug sensitivity. In Figure 6.27, we show a representation in the cell line space, highlighting a common mutation across many tumors in *BRAF*. We showed in the previous section that the

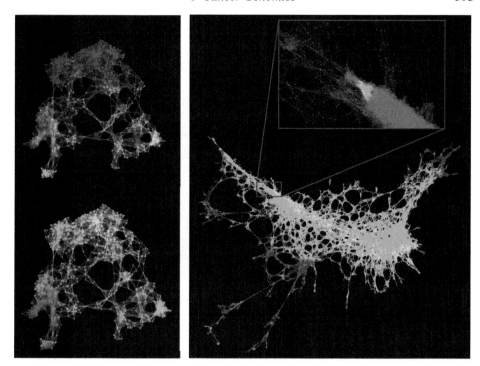

Figure 6.26 Mapper graphs of expression data from the Cancer Cell Line Encyclopedia (CCLE). Left: These images show two distinct patterns of gene expression of targets across a large variety of human cancers. Each node is a set of cell lines sharing a similar expression profile. On the top left, *PDGFRA* is expressed with high specificity in glioblastoma and neuroblastoma (warm colors in the uppermost and lower right nodes). The bottom left image shows the expression of *EGFR* across several tumors (warm colors in nodes in the left portion of the image). Right: Dual representation, where nodes are composed of genes that share expression across CCLE cell lines. Normalized correlation was used to generate the metric, and the filter functions are the first two principal components. Coloring shows average expression in central nervous system cell lines. Inset localizes *EGFR* within the larger network.

BRAF mutation V600E is frequently found in a large fraction of melanomas, but specific point mutations in *BRAF* are also found across many other tumors including 100% of hairy cell leukemias [498], 57% of Langerhans cell histiocytosis [26], and 36% of thyroid papillary cancers [297]. In the representation, we see that *BRAF* mutant cell lines co-localize in a specific region of the Mapper representation. Recently, different drugs have been developed to target specific alterations in *BRAF*. We represent on the right panel of Figure 6.27 the cell lines that are sensitive to PLX4720, a specific V600 mutant *BRAF* inhibitor. PLX4720 is the precursor of PLX4032 (Vemurafenib), a specific *BRAF* inhibitor approved by FDA for treatment of late-stage melanoma.

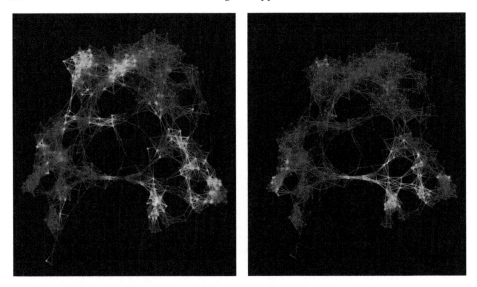

Figure 6.27 Left: Mapper graphs of expression data from the Cancer Cell Line Encyclopedia (CCLE). Nodes are composed of genes, features are the expression across CCLE cell lines. Norm correlation metric with principal component lenses is used. Each node is a set of cell lines with similar expression profile. For representation purposes, the network is then colored based on the expression of specific genes of interest. In this example, coloring shows cell lines with *BRAF* mutations. *BRAF* is a gene mutated in different tumors, including melanomas. Right: Coloring shows cell lines sensitive to PLX4720, a compound with specific action on mutant *BRAF*.

Cellular heterogeneity reflects both clonal heterogeneity and genetic instability; thus, it can be impacted by anticancer therapy on several levels. First, new selective pressures are expected to favor relatively treatment-resistant clonal subpopulations over sensitive ones, therefore limiting clonal diversity. Second, genotoxic treatments may elevate genomic instability, thereby potentially increasing cellular genetic diversity. Despite its clinical importance, the potential impact of cancer therapy on cellular genetic heterogeneity is largely unknown. Topological data analysis (TDA) methods to model phenotypic and genetic determinants of drug resistance *in silico* can generate testable hypotheses for addressing drug resistance. Very recently, the role of clonal heterogeneity in tumors and the impact on therapy has been studied using topological data analysis in [321, 322].

6.11 Summary

- Cancer is the result of mutations in our cells. Nowell first formalized the notion of cancer as an evolutionary process in which genetic instability creates variation that is sieved by natural selection [388]. This instability leads to a menagerie of somatic mutations, including substitutions, indels, copy number variants, methylation aberrations, translocations, and gene fusions.

- Cancer evolution is largely a clonal process, where an initial tumor cell population proliferates, often through abnormal mitosis, and different cell lineages are created by sequential mutations that confer greater or lesser fitness.
- Fitness is determined by the "hallmarks of cancer": the tumor cell's abilities of evasion of apoptosis, self-sufficiency in growth signals, insensitivity to anti-growth signals, sustained angiogenesis, limitless replicative potential, and tissue invasion and metastasis [227].
- The mutations at each step of a clonal expansion endow the cancer cell with variable fitness. Indeed, these mutations can be divided into driver lesions that further tumor progression and passenger lesions that are byproducts of the mutagenic environment of the cancer [222]. Driver genes provide potential candidates for oncogene addiction, in which the survival of a tumor cell becomes increasingly dependent on the continuing function of a lesion [536].
- Recent technological developments are generating large scale molecular data from many different tumor types in many patients. This data includes systematic characterization of mutations, expression and methylation profiles, and many other kinds of information. Problems in studying the molecular mechanisms of cancer initiation and progression, prognosis, or sensitivity to therapies can be translated into understanding the structure of these data sets.
- Mapper has been used to study molecular data in cross-sectional studies to understand the transcriptional similarity between the tumors in different patients. These studies classify patients based on transcriptional similarity and identify subsets of patients with differential survival.
- Molecular data can be collected along the progression of a tumor or at different locations in a metastatic process. Mapper has been used to understand the relation between molecular data and physical or temporal information, and progression of the disease.
- Molecular characteristics and responses to drugs vary enormously across patients (patient heterogeneity) and also within a tumor of a single patient (tumor heterogeneity). Integrating molecular and drug response data is one of the main challenges in developing precision therapeutics for cancer patients. Topological data analysis can uncover the structure of these data sets, linking molecular features to drug response characteristics.
- As we will see in the next chapter, single cell technologies are generating molecular data across many cells in different biological systems, including tumors. How these cells are different and how these differences relate to tumor evolution and drug response constitute a fascinating problem that is now beginning

to be explored. The structure of these single cell tumor data sets remains poorly characterized; unsupervised techniques, including topological data analysis, can potentially identify fundamental molecular mechanisms driving tumor initiation progression. As these technologies improve and larger single cell cancer data sets become available, we expect a greater need for these methods.

6.12 Suggestions for Further Reading and Databases

There are a few books that we recommend to the neophyte.

- *The Emperor of all Maladies* [363] is a Pulitzer prize winning book that narrates the history of our understanding and treatment of cancer.
- Robert Weinberg's *The Biology of Cancer* [531], provides a comprehensive overview of cancer research, explaining the main molecular mechanisms that have been identified.
- Arnold Levine's book on viruses [328], is a nice introduction to viruses, including historical accounts on the discovery and mechanisms of oncoviruses.

There are extensive databases that provide a large variety of different data sets associated with multiple cancers.

- The Cancer Genome Atlas constitutes a large US-based effort between the National Cancer Institute and National Human Genome Research Institute to characterize genomic/transcriptomic/epigenetic changes together with clinical annotation in 33 types of cancer. (http://cancergenome.nih.gov)
- The International Cancer Genome Consortium constitutes a worldwide effort to generate a comprehensive genomic, transcriptomic and epigenomic description in 50 different major tumor types. https://dcc.icgc.org.
- The Cancer Cell Line Encyclopedia (CCLE) project provides a detailed genetic characterization of a large panel of human cancer cell lines together with responses to different drugs. www.broadinstitute.org/software/cprg/?q=node/11
- The Genomics of Drug Sensitivity in Cancer (GDSC) is a resource for therapeutic and genomic characterizations of a large panel of cell lines. www.cancerRxgene.org

7

Single Cell Expression Data

In multicellular organisms cells can have different genomes, and distinct cell types have different expression profiles. Humans, for instance, are composed of more than 40 billion cells [53] forming distinct organs, tissues, and cell types. This genetic and transcriptomic variability has important phenotypic consequences. An example of genetic variability is evident in some of our immune cells, T-cells and B-cells, which rearrange and mutate sections of their genome. These mutations and rearrangements lead to a large repertoire of B- and T-cell receptors providing the means to fight the gamut of potential pathogens. Our gametes contain half of the genomic material of somatic cells after carrying out meiotic recombination. Even for two cells that share the same genome, the expression profile can vary dramatically. The changes in expression from a stem cell to terminally differentiated cells are the result of a carefully orchestrated program of cellular differentiation.

As cells transit through different states of differentiation and the cell cycle, different transcription programs are activated and deactivated. A population of cells contains, in general, a representation of a diverse set of transcriptional programs, and expression profiles from these cells represent an average that may not correctly represent the underlying diversity. Single cell RNA sequencing provides the opportunity to accurately map these transcriptional states. In single cell RNA-seq experiments, each cell can be represented by a point in a very high-dimensional space, whose dimension is typically the number of expressed genes (several thousands). Due to the high-throughput nature of the data (measures involving tens of thousands of genes and thousands of cells), single cell analysis requires methods that are able to deal with large amounts of very high-dimensional data. In addition, these methods should preserve the continuous character of the data, as cellular differentiation can be thought of as a biological continuous process: there is usually a continuous set of states interpolating between a stem-like state and any of the fully differentiated states descending from it.

We will argue in this section that the condensed representations produced by Mapper, applied in the previous chapter to the analysis of cancer cross-sectional data, satisfy these two requirements precisely.

7.1 Introduction to Single Cell Technologies

Recently, single cell sequencing has emerged as a new high-throughput method to access the genome, the epigenome, and the transcriptome of hundreds or thousands of individual cells. These technical developments have been the confluence of several techniques, including the following.

- **Single cell isolation methods**. The first step in sequencing RNA or DNA from single cells is to generate a suspension of single cells, which can be challenging for particular tissues and cell types. Once in suspension, individual cells are isolated by serial dilution, micropipetting, optical tweezers, etc. Although these techniques are effective in isolating single cells, they are not scalable for isolating thousands of single cells. Scalable techniques for single cell isolation remains an active area of research where the most popular techniques include fluorescent activated cell sorting (FACS) and microfluidic devices.

- **Methods for amplification of DNA and RNA from single cells**. A variety of methods have been described to amplify genomic material from single cells, including polymerases from different organisms. For RNA, one of the common techniques for single cell RNA-seq is Smart-Seq, which amplifies full transcripts using a retroviral reverse transcriptase, a switching mechanism at the 5′ end of the RNA transcript, and then amplifies the CDNA [419]. CEL-Seq uses in vitro transcription as an amplification protocol, avoiding some of the exponential amplification artifacts from PCR [232]. Drop-seq and inDROP are two related but independently developed methods based on micro droplets [302, 338]. Each micro droplet contains a cell barcode and primers together with a captured single cell. The approach allows study of the transcriptome of thousands of cells. Several techniques have been described to amplify DNA material from single cells. PCR based methods (degenerative oligonucleotide PCR, or DOP-PCR) use random or degenerate primers, providing low coverage of whole genomic regions. More popular methods are based on multiple displacement amplification (MDA) using DNA polymerases from a phage (Φ29) or, more rarely, from a thermophilic bacterium, *Bacillus stearothermophilus* (Bst polymerase).

There is a frenzied competition in the development of methods for isolating single cells, amplifying RNA and DNA, and sequencing. In the next few years, this will result in dramatic changes in the type of data available, the quality of the data, and the throughput. The availability of single cell data has led to the development of computational methods to study diverse biological processes. Among other things, single cell transcriptomics has enabled more detailed studies of cellular differentiation processes in developmental biology [47, 408, 431, 505] and cancer biology [401, 499]. Single cell analysis has the power to identify different types of minority cells that are eclipsed within larger populations, to identify transitions between different states to draw transcriptional trajectories, and to find specific markers and transcription factors for the different cell types and states.

Ideally, one would like methods for studying single cell transcriptomic data that do not rely on previously known information and thereby allow the discovery of potentially novel biology. The number of cells in these experiments is on the order of thousands which is frequently comparable to the number of genes studied. Recall from the discussion in Chapter 3, to get a good sample of a truly high-dimensional object (here the dimension is the number of genes), one needs a number of points (cells) exponential in the dimension. This is one of the reasons that most approaches to the study of single cell data are based on dramatically reducing the dimensionality of the space through the selection of a few known markers, applying standard dimensionality reduction techniques (e.g., using PCA or *t*-SNE [13]), or looking for specific low-dimensional features (such as reconstructing trajectories or bifurcating points [223, 452]).

An alternative strategy based on ideas from topological data analysis tries to derive a low-dimensional space that can capture some of the biologically interesting properties (such as number of cell types, or trajectories); we use a Reeb graph to describe the data. Recall from Section 1.12 that a Reeb graph is a one-dimensional object that can capture some of the low-dimensional features of the data. As discussed in Section 2.8, Reeb graphs can be approximately inferred from the data using the Mapper algorithm. A generic pipeline for analyzing single cell expression data begins by filtering out low quality cells, based on standard criteria such as the ratio of mapped to unmapped reads, and normalizing the data to account for differences in the length of the transcripts and the total amount of RNA sequenced, as determined by spikes in reads or other methods [483]. The remaining high quality cells are represented as points in a high-dimensional space, of dimension given by the number of different transcripts present in the samples. This

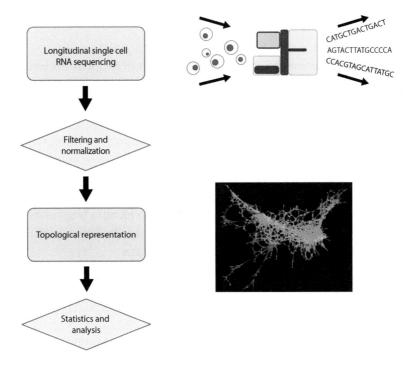

Figure 7.1 Processing of single cell transcriptomic data.

space is endowed with a metric in a standard way, for instance using Pearson's correlation as a measure of similarity. Applying Mapper with various choices of filter function then produces a graph representation; this yields a low-dimensional condensed representation that tries to preserve salient local relations between cells in the high-dimensional space (Figure 7.1).

7.2 Identifying Distinct Cell Subpopulations in Cancer

Our first example of the use of single cell genomic data is in cancer (see Figure 7.2). As we previously discussed, cancers are (among other factors) the result of the accumulation of somatic mutations and epigenetic changes that lead to uncontrolled cell growth. Not all cells in a tumor share the same genetic, transcriptional, epigenetic, morphological, and phenotypic profile, a fact that is usually described as tumor heterogeneity. Two populations of cells that share a dominant clone could have very different phenotypes, as minor populations can be incentivized to grow or become resistant to specific therapies, leading to the long time evolution of these tumors.

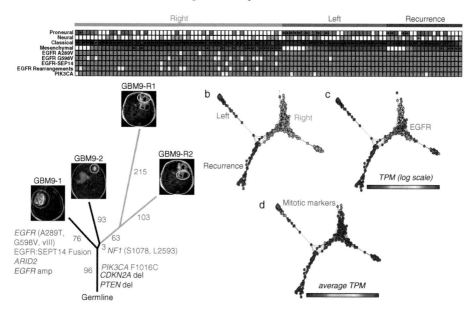

Figure 7.2 Single cell RNA-seq allows the spatial and temporal study of the structure of tumors. This is a particular case of a patient with two focal glioblastomas, on the left and right hemispheres. After surgery and standard treatment, the tumor reappeared on the left side. Genomic analysis (on the left) shows that the initial tumors were seeded by two independent, but related clones. The recurrent tumor was genetically similar to the one on the left. The expression profiles from single cells from the two foci at diagnosis and the relapse recapitulate the clonal history. Transcriptionally and genetically, the recurrence resembles the left parental tumor. A small subset of the cells in the initial left tumor show a similar transcription profile as the recurrent tumor, suggesting that the resistant population originated from a subclonal population in the original tumor. Source: [320]. From Jin-Ku Lee et al., Spatiotemporal genomic architecture informs precision oncology in glioblastoma, *Nature Genetics* 49.4 (2017): 594-599. © 2017. Reprinted with permission from Springer Nature.

Single cell techniques provide the means to study heterogeneous cell populations. The following example studies the mutational and transcriptional profile of a multicentric glioblastoma. Multicentric glioblastomas represent tumors that occur in multiple discrete areas in the brain. In this particular case, at diagnosis, the tumor presented two focal points, on the left and on the right brain frontal lobes. After surgery, chemoradiotherapy, and *EGFR* targeted therapy, the tumor recurred on the left side. Different samples were taken from the initial left and right loci and two samples at recurrence. The history of this tumor was then reconstructed using genomic sequencing from each of the biopsies. The genetic characterization shows that the right tumor shares most but not all genetic alterations with the left tumor, indicating a common origin for the two clones that seeded the left and right tumors.

The two loci at diagnosis show distinct clonal and subclonal alterations, indicating that there were two independent founding clones for each location. The recurrence samples were genetically similar to the original tumor in the same side.

Although the recurrent tumor shared many alterations with the parental tumor in the left section, the recurrent tumor had also acquired other alterations in the course of the progression.

To study this case in further detail, single cell RNA-seq was performed on cells from the two primary tumors and the recurrent tumor. The current standard for classification of glioblastomas based on expression identifies four subtypes, neural, proneural, classical and mesenchymal [515]. When single cells are classified into these four types, the heterogeneity becomes evident, with the right initial tumor being composed of a majority of classical cells. Both the left initial and recurrence tumors showed more heterogeneous cell populations involving three different subtypes (classical, proneural and mesenchymal). This classification does not provide any information on how related the cells responsible for the relapse are to any of the original tumors. All three cell populations show a minority of cells in active cellular division, as indicated by the upregulation of mitotic genes.

Using Mapper, one can appreciate a more continuous structure that recapitulates the clonal and genetic history. The tumor on the right appears to be transcriptionally distinct from the left tumor and the recurrence tumor. Expression profiles from cells in the recurrence tumor resembled the originating initial tumor. This is an important finding, as it shows a continued progression at the expression level, with a few cells at diagnosis having a similar pattern as cells at relapse. It also shows that *EGFR* mutation is a subclonal event, occurring only in the tumor at diagnosis that is not responsible for the relapse. This observation illustrates the problem of clonal heterogeneity for targeted therapies: tumors with heterogeneous populations of cells containing different alterations are less sensitive to specific therapies which target a subpopulation.

7.2.1 Clonal Heterogeneity from Single Cell Tumor Genomics

The recent development of single cell transcriptomics and genomics is providing an opportunity to study the role of clonal heterogeneity in tumors [159, 378, 401] and to identify small, previously uncharacterized cell populations [214]. The single cell approach to studying complex populations brings with it new challenges associated with the large number of sampled genomes. Another rapidly maturing technology in modeling tumor population dynamics is that of patient-derived xenografts. Patient-derived xenografts, or PDX, are generated by transplanting tumor tissue into immunodeficient mice. With different rates of success depending on tumor type and specific samples, these tumors are able to proliferate in the mice, and they can be passed from one animal to another. While not completely recapitulating tumors in humans with an intact immune system, they capture many in vivo properties of tumors, allowing tumor evolution studies with and without therapy.

Single cell genomics provide the opportunity to understand clonal dynamics in PDX models, connecting different cell populations that are established at different times. Subclones are selected to set up different passages. Eirew et al. studied single-nucleus deep-sequencing from different passages of breast cancer PDX [159]. This study collected single cell data from 55 informative sites from a primary breast cancer tumor and three subsequent mouse passages. These sites were selected using the union of bulk DNA sequencing data across different samples, which excludes, of course, specific alterations in single cells.

Since single cell data from the primary tumor was not available, we generated eight cluster-representative sequences using 27, 36, and 27 single nuclei from the first, second, and fourth passages. From these, we subsampled 3000 trees from all possibilities and projected the data into $\mathbb{P}\Sigma_4$. First, we included trees relating the germline sequence, a randomly selected cluster-representative sequence from the primary tumor, and randomly selected single-nucleus sequences from the initial two xenograft passages. Then, in the second analysis, we included trees relating a randomly selected cluster-representative sequence from the primary tumor and randomly selected single-nucleus sequences from three consecutive xenograft passages.

The results (Figure 7.3) showed consistent linear evolution from primary tumor through the first two xenograft passages. However, significant heterogeneity of tumor clones is observed upon the fourth mouse passage. The first time window (purple) is completely contained within the topology corresponding to linear evolution, unlike the second (gold) which is centered on the origin and extends into all three possible topologies. The point cloud for the second time window displays a higher standard deviation than the first (10.49 versus 8.69), and its centroid is essentially a star tree. The high degree of genotypic heterogeneity giving rise to the second time window distribution is suggestive of a clonal replacement event between the time points of Xenograft 2 (X2) and Xenograft 4 (X4). Many of the prevalent alterations before X4 disappear during the final passage, and many new mutations rise to dominance. This raises interesting questions about the long-term fidelity of PDX vehicles to the genetics of their ancestral primary tumors, which theoretically they serve to mimic.

7.3 Asynchronous Differentiation Processes

One of the most interesting applications of topological data analysis is related to single cell expression profiles along a particular differentiation process. For instance, during the process of differentiation, one can observe how stem cells

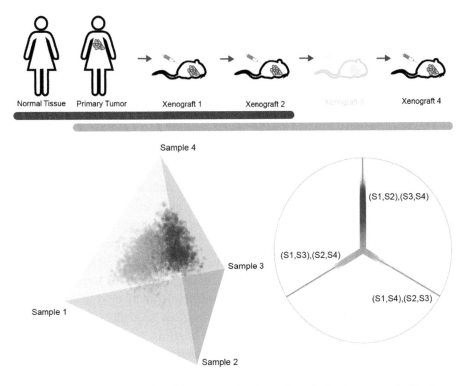

Figure 7.3 Emerging clonal heterogeneity in patient-derived xenograft. Single cell analysis of tumor evolution in a breast cancer derived xenograft model. Single-nucleus deep-sequence data obtained from passages 1, 2, and 4. Single cell data from the primary tumor was not available, however, Eirew et al. identified eight distinct clusters. These data were used to generate two $\mathbb{P}\Sigma_4$ spaces. Source: [545]. From Zairis et al., Genomic data analysis in tree spaces, arXiv: 1607.07503 [9-bio.GN].

evolve into multiple differentiated cells [431]. Single cell RNA-seq samples the transcriptional programs of cells moving along differentiation trajectories. But of course, not all cells move at the same time, and while some retain the characteristics of the original state, others quickly differentiate into final states. In experiments where time information is available, one can organize the process and assign a pseudo-time, so that the transcription data correlates with time. This pseudo-time information is extremely useful as it can organize different transcriptional programs along the differentiation process.

Ideally, one would like to reconstruct evolutionary trajectories in the high-dimensional expression space, and provide a representation that preserves the high-dimensional similarity. One of these representations can be obtained using Mapper. Once the Mapper representation has been established, one can associate a time to different states along the graph (Figure 7.4). We can define a root node as

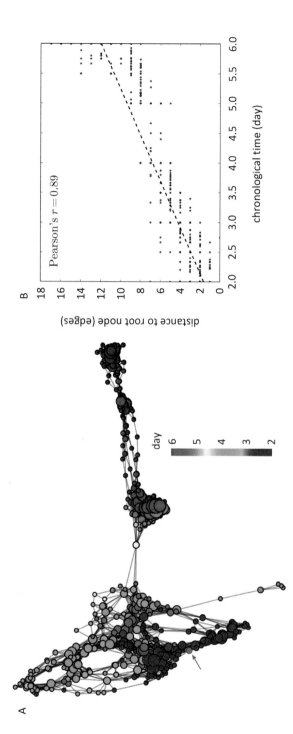

Figure 7.4 When single cells are obtained at different time points one can assign a pseudo-time to each transcriptional state. The pseudo-time orders transcriptional states for the root node to the most differentiated states. Source: [431]. From Abbas H. Rizvi et al., Nature Biotechnology 35, 551–560 (270). © 2017 Nature. Reprinted with Permission from Springer Nature.

the node that maximizes the correlation between the distance in the Mapper graph and time. A pseudo-time can then be inferred by calculating the distance in the graph. In Figure 7.4, the representation is marked with a red arrow. In differentiation, this node corresponds to the most undifferentiated transcriptional state. As expected, the distance along the graph from the root node is associated with the differentiation state.

Different genes are expressed at different stages while others are not expressed or not particularly associated with the progression. One can define the centroid of the expression of a particular gene in the representation to quantify the measure of dispersion of its expression. A relatively simple way to do this while matching to the experimental time is to fit a linear relation between the distance in the representation from the root node to a particular node β, d_β, and the average time of sampling cells associated with that node, $\langle t_\beta \rangle$, (right of Figure 7.4):

$$d_\beta \sim a_0 + a_1 \langle t_\beta \rangle.$$

From there one can define the centroid μ_i for the expression e_i of a particular gene i, measured in time units as:

$$\mu_i = \frac{1}{a_1}\left(\frac{\Sigma_\beta d_\beta e_{i,\beta}}{\Sigma_\beta e_{i,\beta}} - a_0\right).$$

In a similar way, one can define the dispersion, σ_i, as

$$\sigma_i = \frac{1}{a_1}\left(\sqrt{\frac{\Sigma_\beta (d_\beta - a_1\mu_i - a_0)^2 e_{i,\beta}}{\Sigma_\beta e_{i,\beta}}}\right).$$

Centroids and dispersions are a way to assign different transcriptional programs to different differentiation states. In the following, we show how topological data analysis could be used for studying single cell transcriptomic data in differentiation processes.

7.4 Differentiation in Human Preimplantation Embryos

One of the most fascinating biological processes is the development of a metazoon from a single cell: an exquisitely orchestrated organization of transcriptional programs that gives rise to different tissues and cell types, in a particular spatiotemporal fashion. Fertilization occurs with the fusion of parental gametes (egg with a sperm) which creates a zygote. Before the implantation of the embryo in the mother's uterus (six days after fertilization in humans), the original zygote cell undergoes successive replications (Figure 7.5). In the first stages, the zygote divides exponentially (2, then 4, then 8 cells, etc.) to generate the morula. During this process the preimplantation embryo is surrounded by a protein shell called the zona pellucida, that precludes premature attachment to the oviduct walls, where the

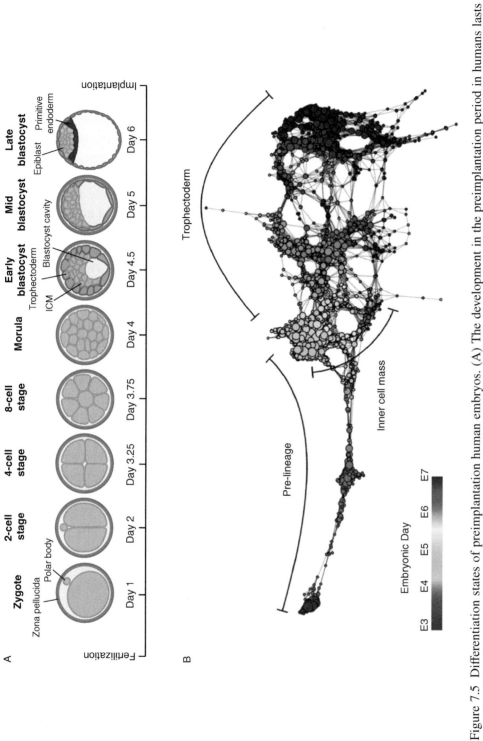

Figure 7.5 Differentiation states of preimplantation human embryos. (A) The development in the preimplantation period in humans lasts a week, where the zygote divides exponentially to generate the morula, which further differentiates into a blastocyst. (B) Mapper graph from single cell data. Source: [431]. From Abbas H. Rizvi et al., *Nature Biotechnology* 35, 551–560 (270). © 2017 Nature. Reprinted with Permission from Springer Nature.

whole process takes place. When there are about 32 cells, the blastomeres generate a cavity by accumulating fluid in the intercellular space. The generation of this cavity generates the blastocyst. Subsequently, the cells on the outside of the blastocyst differentiate into the trophectoderm, induced by the expression of a combination of transcription factors. The cells in the interior of the morula form the inner cell mass that further differentiates into the epiblast and the primitive endoderm. The late blastocyst is composed of three different cell types: the trophectoderm, primitive endoderm, and epiblast. The cells in the trophectoderm lead to the development of and interaction with the placenta, the primitive endoderm forms the amniotic sac where the embryo resides during pregnancy, and the epiblast further differentiates into the three germ layers (endoderm, mesoderm, and ectoderm). Finally, the blastocyst growth disrupts the zona pellucida, leading to the implantation of the zygote into the uterine wall.

To study this process, our final example is a single cell RNA sequencing data set of 1529 cells collected from 88 preimplantation human embryos [408]. The data set captures the process of differentiating embryonic cells at different times and characterizes the segregation between trophectoderm and the inner cell mass lineages. In these examples, multidimensional scaling (MDS) was used as the auxiliary filter function for the condensed representation and Pearson's correlation distance was used as the metric. Analysis of the Mapper graph shows how the cells progress from a highly homogeneous expression pattern corresponding to the morula formation to an intermediate state; this is followed by the establishment of specific transcriptional programs of expression of lineage-specific genes, coinciding with the blastocyst formation. The inner mass cells present a more homogeneous transcriptional program with high expression of embryonic-specific growth factors and receptors (such as TDGF1 and PDGFRA), while the trophectoderm is associated with GATA transcription factor genes expression [431].

This is a nice example of how topological data analysis applied to single cell expression data can recapitulate the history of the first stages of human differentiation. By studying the specific cell populations, one can hope to recover the successive combinatorial transcriptional programs that define this process.

7.5 Summary

The application of topological based approaches to single cell data is at a nascent stage. The technology, methods and many of the ideas reviewed here will be rapidly evolving in the next few years.

- Genomic technologies have recently been applied to single cells to study a diversity of biological problems, including heterogeneity in cancer, mapping transcriptional programs along development, and identification of rare species.

- Evolution in cancer occurs by changes in the genome and the transcriptional states. The spatial and temporal diversity maps to transcriptional states.
- In the differentiation processes, expression profiles of single cells can capture transition states, bridging the differences between the undifferentiated and differentiated populations.
- Topological data analysis methods, such as Mapper, can identify transcriptional profiles and infer the continuous relationship between related states. This is of particular relevance to the study of transition states.
- Topological data analysis can be complemented with temporal information of the biological processes, allowing the identification of different transcriptional states.

7.6 Suggestions for Further Reading, Databases, and Software

- Single cell genomic and transcriptomic studies are relatively recent, and dramatic developments appear almost every other month in both the technology and analysis sectors. Recent reviews worth noting include one by Yong Wang and Nicholas E. Navin [524] and one by Stephen Quake and colleagues [192].
- On the computational side, different approaches for dimensionality reduction have been applied, including multidimensional scaling (MDS), independent component analysis (ICA), and t-distributed stochastic neighbor embedding (t-SNE). A nice review of these techniques can be found in [483]. As is often the case, the computational techniques are developed within particular applications, including resolving spatial/expression structures [445], studying B-cell development [47], transcriptome dynamics of skeletal myoblasts during differentiation [504], and early development of mouse embryos [340], among many others.
- Interesting applications of single cell genomic and transcriptomic technologies beyond the few examples described in this chapter can be found in lineage decision making [451], understanding tumor heterogeneity in cancer [377], the discovery of new species in the tree of life [430], etc.

The software (and documentation) for analyzing time evolution using single cell data can be found at `http://github.com/RabadanLab/SCTDA`. An online database and exploration tool for some results in neuronal development can be found at `http://rabadan.c2b2.columbia.edu/motor_neurons_tda`.

8

Three-Dimensional Structure of DNA

Chemically, DNA is a long polymer. This polymer is packed inside the nucleus of the cell by binding to sets of specific proteins called nucleosomes. Chromatin refers to the combined structure of DNA and these DNA binding proteins. One can visualize the three-dimensional structure of DNA as a long polymer winding around at different scales. The three-dimensional structure of chromatin plays a crucial role in a large variety of fundamental biological processes, including replication and expression. For instance, to be transcribed (i.e., to generate RNA from this DNA), the local structure of DNA has to be accessible to different proteins that bind to specific locations. Distant genomic locations can be brought together and coregulated by the same transcriptionary machinery. In this way, the structure of chromatin regulates expression of RNA and influences the expression of proteins by different cells. Thus the three-dimensional structure of DNA determines why two cells containing the same genome can function in very different ways. For instance, a motor neuron and a B-cell share the same genome but their behavior, function, and the proteins they express are vastly different.

In an eukaryotic cell there are well-studied structures that organize chromatin (Figure 8.1). The two DNA strands are 2.5 nanometers wide and coil in the form of a helix of 10.4 base pairs per turn. 146 bases of DNA can wrap around nucleosomes, a structure of four proteins (histones). Histones pack into 30 nanometer filaments in a highly compact way (heterochromatin). At even larger scales, on the order of a million bases, these structures are assembled into different territories, as topological associated domains or TADs. In [330] Lieberman-Aiden et al. suggested specific large conformations, called *fractal globules*. At still larger scales, it has been postulated that different chromosomes are located at specific three-dimensional chromosomal territories. The location of genomic regions within these territories is associated to transcriptionally active domains [375]. Due to the nature of the relevant biological processes, including DNA replication, repair, and transcription, DNA presents a highly dynamical nature.

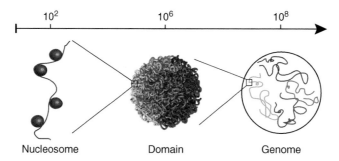

$$10^2 \qquad\qquad 10^6 \qquad\qquad 10^8$$

Nucleosome   Domain   Genome

Figure 8.1 Summary of some structures found in chromatin organization. Nucle-osomes are sets of proteins that bind DNA at 150 base pair scale. At much larger scales, on the order of megabases, chromatins organize into different territories bound by specific proteins. At even larger scales different structures have been proposed, including the so-called *fractal globule* structure. At still larger scales chromosomes can be found in separate chromosomal territories. Source: [330]. From Erez Lieberman-Aiden, et al., Comprehensive mapping of long-range inter-actions reveals folding principles of the human genome, Science 326.5950 (2009): 289–293. © 2009 Reprinted with permission from AAAS.

In the previous chapters, we have explored the use of high-throughput sequenc-ing technologies to read the genome of organisms. There are preliminary indica-tions that these techniques can also be used to infer three-dimensional properties of DNA across different genomic scales. We will describe here genome wide chro-matin conformation techniques. Data generated in this way provides information about genomic locations that are in close proximity in three dimensions. We will follow reference [163] to describe how topological techniques can be used to infer and quantify three-dimensional structural properties of DNA. In particular, persis-tent homology provides a natural framework for summarizing these properties. We will first demonstrate the efficacy of persistent homology techniques in data from simulated polymers, and then in the circular bacterial genome of *C. crescentus* and a human lymphoblastoid cell line. We believe that this will be a fertile area for the application of topological data analysis in the future.

8.1 Background

In the last few years there have been extraordinary developments in providing genome wide information on the three-dimensional structure of DNA [25, 138, 330]. Chromosome conformation capture technologies give a variety of methods to study the three-dimensional organization of chromatin inside the nucleus of a cell. These methods identify genomic locations that are in very close physical proximity (in space).

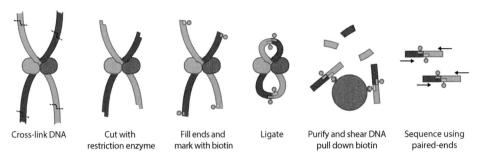

| Cross-link DNA | Cut with restriction enzyme | Fill ends and mark with biotin | Ligate | Purify and shear DNA pull down biotin | Sequence using paired-ends |

Figure 8.2 Hi-C protocol to map the three-dimensional chromatin structure. First, nearby regions in DNA are cross-linked in formaldehyde. Then DNA is fragmented using a restriction enzyme. A biotinylated residue is incorporated and ends are ligated. The DNA is sheared and the junctions are pulled down with streptavidin beads that bind strongly to biotin. The purified fragments are then sequenced, leading to information on close proximity DNA fragments.

Chromosome conformation capture technologies vary depending on the extent of the regions interrogated. Hi-C protocols use high-throughput sequencing techniques to provide genome wide maps of DNA interaction. A common Hi-C protocol has been summarized in Figure 8.2: DNA fragments in close proximity are cross-linked in formaldehyde, DNA is fragmented, nearby fragments are ligated to form close loops. These loops are then sequenced. Mapping sequence reads into a reference genome, one can identify specific genomic locations in close proximity. The final result is summarized in a contact data matrix that quantifies the genomic locations that are in close three-dimensional proximity [138]. Specifically, if there are n locations, the contact matrix C is an $n \times n$ matrix such that the entry $C_{ij} = C_{ji}$ encodes the proximity between locations i and j. Further processing involves denoising and normalization of the raw data matrix [25].

There are many caveats with the use of this procedure. For instance, we are assuming that different cells present the same three-dimensional structure. Contact matrices represent the average over many different cell configurations, and might not represent any specific configuration. Recently chromosome conformation capture techniques have been applied to single cells [374, 375, 486]. These studies have found that despite a large degree of variability, megabase scale domains are well maintained across individual cells.

8.2 TDA and Chromatin Structure

We are interested in learning global properties of the three-dimensional structure of DNA: how DNA folds at different scales. Hi-C data provides the means to capture information about the three-dimensional information of DNA as off diagonal elements in the contact matrix. Specifically, once correctly normalized, one assesses the proximity between different genomic regions using the contact matrix. Small loops in DNA can be detected as close to the diagonal non-zero elements in the contact matrix.

Topological data analysis provides a natural language to identify and quantify loops in the three-dimensional structure. One can apply persistent homology to the similarity matrices derived from Hi-C data to try to detect the shape of the DNA. In the following discussion, we focus on PH_1, the one-dimensional persistent homology barcodes, which represent one-dimensional physical loops in DNA.

Recall that each element of the barcode is an interval $[b_i, d_i)$, where b_i is the smallest scale and d_i is the largest scale where the class is present. Following the work of MacPherson and Schweinhart [339], one can define the size of a particular persistent homology class as the mean of the birth and death:

$$x_i = \frac{b_i + d_i}{2}.$$

The values of x_i represent a sample of the distribution of different folding scales inferred from the Hi-C data. Of particular interest in these analyses are large scale interactions (larger than 100 kilobases) that can represent different biologically meaningful structures. For instance, it has been observed that transcription happens in transcription factories, specific three-dimensional locations in the nucleus that accrue large protein complexes involved in transcription [265]. Typically transcription factories encompass tens of RNA polymerases together with a variety of proteins involved in RNA processing, such as helicases, transcription factors, splicing factors, etc. Transcription factories can be physically observed by electron microscopy. Other types of structure can be observed where promoters (the region in a gene associated with transcriptional initiation of a gene, located near the transcription start site of the gene) interact with enhancers (a region in the genome that enhances the transcription of a particular gene) at long distances, sometimes at megabase scales. Long range interactions in chromatin are known to occur in DNA repair and replication among many other processes.

Different biologically distinct structures represent different looping structures in DNA. For instance, a one-jump loop can be observed on promoter-enhancer interactions, while multi-jump loops (multiple loops) from diverse regions can be associated to a transcription factory (Figure 8.3). So the scale and structure of loops can provide biologically relevant information.

While persistent homology provides interesting information on the size and number of cycles, one may be interested in specific details concerning a particular

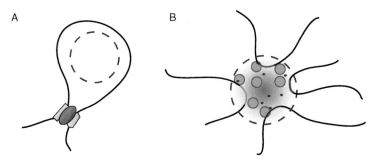

Figure 8.3 The scale and structure of loops can provide biologically relevant information. Here are two examples of known structures associated to transcriptional regulation. (A) Enhancers are genomic loci that regulate the expression of genes that could be located in different genomic locations. Loops in DNA can put enhancers and promoters in close proximity. (B) Other examples can be found in transcription factories containing RNA polymerases, splicing proteins, and other proteins involved in transcription and RNA processing. These are regions that can be linked to distant genomic locations. Source: [163]. Reprinted with permission: © EAI European Alliance for Innovation 2016.

loop in the class. This is important to identify genomic regions of interest and to link structure to function, for instance, linking specific structures to regulation of specific genes. If we identify a class, how can we select a particular member? One obvious criterion is to identify the cycle in the homology class that has minimal size. In the context of Hi-C data and contact maps, Emmett and colleagues proposed [163] to identify minimal cycles as corresponding to the shortest length along the genome being homologically independent to other classes born at smallest scales [449]. However, it has been shown that finding the minimal cycle is an NP-complete problem [113], and so approximation techniques are necessary.

8.3 Simulations

The molecule of DNA can be treated as a polymer with specific biophysical properties. As such it can be modeled as a long homogeneous flexible fiber with interactions within specific sites (see, for instance, [148]). Simulations allow evaluation of the inference procedures used from Hi-C and other types of chromosomal conformation data.

In [163] a 50 megabase chromatin polymer was simulated by considering a polymer formed by 1000 smaller monomers of a few nucleosomes each (Figure 8.4). Specific interactions representing protein-mediated interactions were incorporated by hand at ten random positions in the polymer. A contact map was constructed

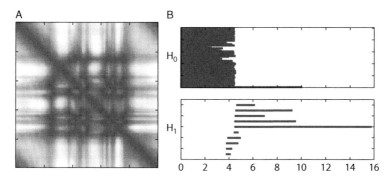

Figure 8.4 Simulations of DNA as a polymer. DNA can be simulated as a long polymer consisting of a large number of monomeric units interacting at specific places. Here, we show the data of a 50 Mb polymer with 10 fixed loops at random positions in the genome consisting of 1000 monomeric units. (A) The average of 5000 simulations allows construction of a contact map. (B) Using persistent homology in a similarity matrix derived from the contact map one can clearly identify the ten loops as ten long bars in dimension one persistent classes. Source: [163]. Reprinted with permission: © EAI European Alliance for Innovation 2016.

using 5000 conformations (Figure 8.4). The contact map was transformed into a similarity matrix $d = 1 - \rho$, where ρ is the Pearson correlation between two genomic positions. The one-dimensional homology groups clearly identify ten long bars, corresponding to the interacting position in the polymer. Polymer simulations provide a nice way to optimize the identification of interactions from contact maps.

8.4 The Topology of Bacterial DNA

We now explore the persistent barcode diagrams derived from real data in two very different systems: a bacterium and an eukaryotic cell.

The typical size of a bacterial genome is a few megabases, which if linearly stretched would reach more than 1 mm. However, bacteria are only a few micrometers long, meaning that the genome has to be compacted 1000 fold. Although bacteria do not have a proper nuclear membrane, there is an irregular shaped region, called the nucleoid, that aggregates most of the genomic material. There are several mechanisms of packing the bacterial DNA genome into the cell. The first mechanism is negative DNA supercoiling. The DNA is a double helix that twists every 10.5 bases. If a segment of DNA of length L (in bases) is circularized by pasting the two ends, there will be a number of turns expected that create a relaxed DNA $\sim L/10.5$. DNA supercoiling occurs when there is an over (positive) or under (negative) winding of DNA (see top panel of Figure 8.5). In general most DNA presents a negative supercoiling. In bacteria, negative supercoiling generates plectonemic

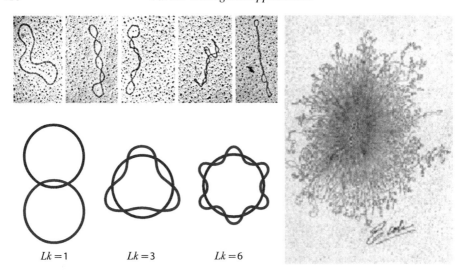

$Lk = 1$ $Lk = 3$ $Lk = 6$

Figure 8.5 Left: Diverse levels of twisting generate supercoiling in circular DNA. Source: [306]. From Kornberg A. 1980 DNA replication, p. 29. San Francisco, CA: W. H. Freeman. Supercoiling can be quantified by the linking number between the two strands of DNA if ends are joined. In a relaxed DNA configuration one should expect a turn every 10.5 bases. So in a length L DNA fragment, the expected linking number is $L/10.5$; deviations from this linking number lead to different levels of supercoiling. Right: Plectoneme emanating from an *E. coli* nucleoid core. Source: [211]. © Designergenes Posters Ltd; in memory of Dr Ruth Kavenoff 1944–1999.

loops. A simple way of characterizing the level of supercoiling is by the linking number of the two strands of DNA (see bottom panel of Figure 8.5). Nice work relating the topology of links and knots to the structure of DNA has been carried out by different groups [20, 80, 126, 178, 308, 415].

The second mechanism of compacting bacterial DNA is by topological domains, supercoiled domains insulated from each other. Topological domains vary in size but are of order 10 kb, indicating that a typical bacterium genome contains hundreds of topological domains. The structure of each domain is kept independent by protein and RNA complexes that work as boundary elements. Of special significance is the structural maintenance of chromosome (SMC) condensin complexes and topoisomerases. SMC proteins form part of a highly conserved complex from bacteria to human that bridge different chromosomal loci working as a high level scaffold. Topoisomerases work as enzymes that can cut the DNA strains, modifying the DNA topology, changing the winding and unlinking DNA loops.

Macrodomains are much larger structures (of almost a megabase) that encompass many topological domains and suggest highly structured regions with reduced

DNA mobility. Understanding the structure of bacterial DNA, the boundary elements, the macrodomain structure, the functional characterization, transcription, and replication remains a very active area of research. For a nice review on bacterial chromosomal organization see [523].

Caulobacter crescentus is a gram negative bacterium ubiquitously found in water. The genome has a length of 4 megabases in a circular chromosome coding for near four thousand genes. Hi-C interaction data from *Caulobacter crescentus* was examined as published in [317]. This paper found several structures, including chromatin interaction domains at scales of 100 kilobases and smaller plectonemes. A simple model of nesting these chromatin structures was proposed where plectonemes were arranged in a brush-like fashion and topological domains encompassed several plectonemes (see panel A of Figure 8.6). The contact matrix binned at a ten kilobase resolution from a wildtype *Caulobacter* cell is represented in panel B of Figure 8.6. Panel C shows the barcode diagram in dimensions 0 and 1 computed from the associated similarity matrix. The one-dimensional barcode shows a very interesting bimodal structure corresponding to a large number of smaller loops and a few larger ones, as shown in the bimodal distribution of panel D of Figure 8.6.

To specify genomic locations associated to particular one-dimensional persistent homology classes, one can identify small cycles within each class. These representative cycles are depicted in Figure 8.7. The bimodal distribution shown in panel D of Figure 8.6 shows two types of loops. The smaller loops are located close to the diagonal (as expected) and could be related to structural maintenance complexes [523]. More interestingly, the larger ones (of approximately 100 kb) connect extremely distant genomic locations in particular locations, suggesting specific genomic locations associated to large range interactions in *Caulobacter crescentus*.

8.5 The Topology of Human DNA

If stretched end to end, the roughly 6 billion bases of the human genome would stretch nearly two meters, yet they are able to occupy a volume of only a few cubic micrometers within the cell nucleus. Even more surprising (or perhaps not surprising at all), this million-fold compression is highly non-random, and exhibits a complex hierarchical structure which impacts genome function intimately by regulating gene expression. This multiscale pattern exhibits increasing levels of complexity: from nucleosomes every 150 bases, to interactions between promoters at the megabase scale, to topologically associated domains at the 10 megabase scale, and finally, to the organization of the chromosomes [138]. Chromatin conformation is dynamic, and changes continuously throughout the cell

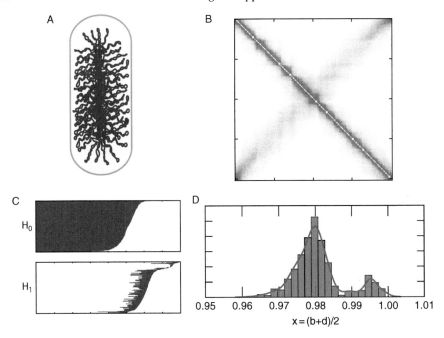

Figure 8.6 Persistent homology study of bacterial DNA structure. (A) Cartoon model of the structure of DNA in *Caulobacter crescentus*. The genome is contained in a large circular chromosome. At smaller scales there are compact domains and plectonemes, supercoiled DNA loops emanating from a central circular fiber. (B) The contact map reveals an off-diagonal structure reflecting the circular nature of the bacterial chromosome. Persistent homology maps (C) indicate a more refined structure shown as distribution of H_1 bar sizes showing a bimodal distribution of DNA folding patterns (D). Source: [163]. Reprinted with permission: © EAI European Alliance for Innovation 2016.

cycle under the influence of a diverse range of chromatin remodeling proteins. Chromatin architecture can also be impacted by post-translational modifications of histones, including but not limited to methylation and acetylation of specific residues on histone tails. Many data sets of human cell Hi-C data have been published [274, 330, 420]. In [163], Emmett *et al.* applied the persistent homology pipeline to study the three-dimensional chromatin structure of a healthy human lymphoblastoid cell line published in [330]. Figure 8.8 shows the contact map of chromosome 1 binned at 1 megabase resolution. On the right of the same figure, the barcode diagrams for dimensions zero, one and two are given. The one-dimensional persistence diagram reveals an interesting pattern of short and long range interactions, as shown in the bimodality of sizes in Figure 8.9. These results support previous observations [330] regarding topological associated domains of size 10 megabases.

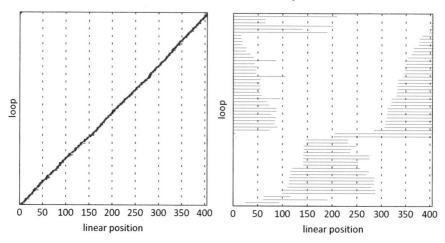

Figure 8.7 Genomic position of minimal cycles associated to persistent homology classes in dimension one. As previously shown, loops can be divided into two types. On the left, smaller loops distribute uniformly across the genome represented as the diagonal. The right panel shows the genomic positions associated to larger loops which clearly show two large interacting domains. Source: [163]. Reprinted with permission: © EAI European Alliance for Innovation 2016.

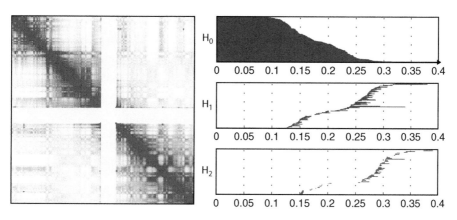

Figure 8.8 Hi-C data for chromosome 1 from a human lymphoblastoid cell line. On the left we show the contact map representation. The white band in the middle represents the centrosome where information is not available. On the right, persistent homology barcode diagrams in dimensions zero, one, and two reveal long-range interaction patterns. Source: [163]. Reprinted with permission: © EAI European Alliance for Innovation 2016.

8.6 Summary

The application of topological approaches to studying the three-dimensional structure of DNA is still in the early stages. We expect that the technology, methods, and many of the ideas reviewed here will be evolving very rapidly in the next few years.

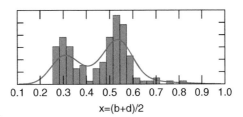

x=(b+d)/2

Figure 8.9 The one-dimensional persistent homology barcode of a human lymphoblastoid cell line shows a clear bimodal distribution related to topological associated domains of an approximate size of 10 megabases.

- Cells with the same genome can have vastly different form and function. The three-dimensional architecture plays an important role regulating important biological processes.
- Chromatin structure in the nucleus of cells presents structure at different scales, from hundreds of bases (nucleosomes) to topological associated domains at megabase scale, to chromosomal territories.
- These structures are associated to biologically functional processes, such as RNA transcription.
- Recently chromosomal conformation capture techniques are generating genome wide contact maps reporting large scale interations.
- Topological data analysis techniques, in particular persistent homology, are a natural language to study contact maps and infer the size and number of loop structures in the genome.

8.7 Suggestions for Databases and Software

- The Mirny group has produced software for polymer models `http://bitbucket.org/mirnylab/openmm-polymer` that allow one to perform simulations, and recreate contact maps.
- The data from the work of Lieberman-Aiden et al. [330] can be found at `http://hic.umassmed.edu/welcome/welcome.php`. The 3D Genome Browser at Penn State allows one to browse existing Hi-C data sets and to visualize user-generated Hi-C data sets `http://promoter.bx.psu.edu/hi-c/`.
- Large collections of Hi-C data sets can be found at GEP DataSets `www.ncbi.nlm.nih.gov/gds/?term=hi-c`.

9
Topological Data Analysis beyond Genomics

In this last chapter we will briefly introduce several recent interesting applications of TDA to diverse biological problems beyond the genetics and genomics work we have focused on in this book. In the first part of this chapter we will explain how TDA can be used to study ordered data, referred to here as series data. Series data is frequently found in many biological applications; for instance, when studying the evolution of a biological organism or population, where data is ordered in time, or when looking at genomic data along a chromosome, where data is ordered by chromosomal location. Examples of time series data with periodic patterns can be found in the cell cycle, or the phenotypic changes in immune genes following infection and recovery [501].

Next, we will discuss TDA techniques for studying graphs, or networks. Networks are standard representations of complex biological systems with different components interacting. For example:

1. The set of interactions between different proteins within an organism is traditionally represented by a graph where nodes represent proteins and edges physical interaction.
2. Transcriptional networks are captured by graphs, where nodes represent genes or transcripts and edges represent how the expression of one relates to the expression of the other.
3. In neuroscience, neurons and their interaction are usually encoded by a graph. The central problem in neuroscience focuses on how the neuronal system captures information about the world; a key question is to how to study this problem using the structure of the interaction graph.

As graphs are pervasive representations of biological data and the simplicial complexes that arise in TDA are generalizations of networks, it is perhaps unsurprising that there are interesting TDA approaches to extract and quantify global properties of biological networks.

423

Next, we describe some natural sets of applications for TDA in medical imaging. For example, magnetic resonance imaging (MRI) is a non-invasive technology that is used to test for many diseases and to obtain data on the real-time activity of living organisms. The output is a function from the three physical dimensions of space (and time if dynamic information is being captured) to the real numbers. Filtrations of this function can be studied using TDA methods and one hopes to relate these topological features to biologically or clinically interpretable characteristics.

Finally, we briefly mention some recent applications of TDA in the context of infectious diseases: first, models of networks of infectious disease spread in a population; second, how organisms respond to infectious diseases.

Our goal in this chapter is to provide a very brief overview of examples of TDA techniques applied to a variety of biological problems beyond the main scope of the book. Our treatment is of necessity superficial, and in particular by no means should be viewed as comprehensive. (We apologize now for work which is omitted; our choices here are not intended to reflect a judgement about the most interesting work.)

9.1 Topological Study of Series Analysis

Time series analysis is an old discipline aiming at extracting patterns and summaries from data arising from weather measurements, financial markets, signal processing, and many other systems. Biological processes are not an exception. In many biological problems data is naturally ordered along a well-defined physical or biological dimension. For example, the position of genes along a chromosome specifies an ordering. Another set of examples come from the time course of a biological process (Figure 9.1).

The first applications we will describe here are time series analysis of expression data. There are a large variety of biological systems that display interesting time dependent expression profiles. For instance, periodicity is observed in circadian regulation, the cell cycle, and the life cycle of malaria [6], among many other examples (Figure 9.2). Genes are regulated according to different temporally orchestrated transcriptional programs, and discovering information about this time dependence can inform theories of how these programs are organized. Several techniques have been used to study time series expression and recently some implementations using ideas from topological data anlysis have appeared. Cohen-Steiner and colleagues [118, 140] proposed a measure of similarity of expression profiles of different genes based on comparing the persistence diagrams arising from level-set persistence (recall Example 2.3.4) applied to a function from time to expression

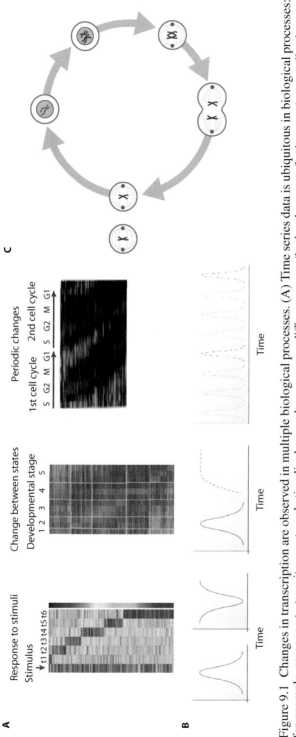

Figure 9.1 Changes in transcription are observed in multiple biological processes. (A) Time series data is ubiquitous in biological processes: for example, response to transitory external stimuli, changes between two different states (in development, for instance), or cyclic changes as observed in the cell cycle or circadian rhythm. There are many interesting biologically interpretable patterns of potentially infinite types. (B) Qualitative phenomena include pulses (a single spike associated to stimuli), sustained changes, periodic changes, among many other examples. (C) An example of a fundamental cyclic process: the cell cycle, where DNA replicates and a mother cell divides into two daughter cells. Source: [35]. Reprinted by permission from Springer Nature: Springer Nature, *Nature Reviews Genetics*, Studying and modelling dynamic biological processes using time-series gene expression data, Ziv Bar-Joseph, Anthony Gitter, and Itamar Simon, 13.8 (2012): 552–564. © 2012.

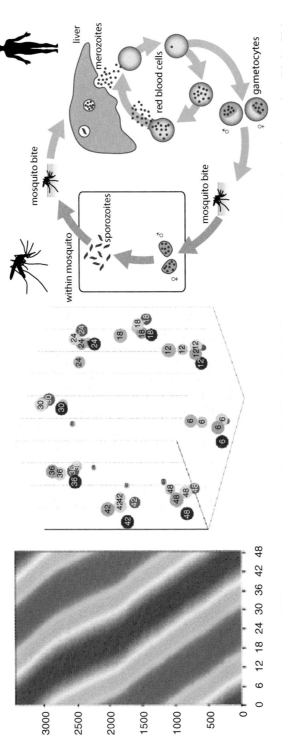

Figure 9.2 Expression of different genes in the malaria parasite life cycle. Left: Genes can be ordered by the time of expression. Right: Using PCA (see Section 4.2 for a brief overview) one can observe a cycle in gene expression reflecting the parasite (*Plasmodium falciparum*) life cycle. Time series data (expression of *P. falciparum* genes at different time points) was analyzed using the fast Fourier transform (FFT) and enrichment of different cell processes. The higher PCA components replicate the cycle of *P. falciparum* replication. Source: [6].

levels. Perea and Harer [404] proposed a method based on a common strategy in time series analysis, applying a sliding window. As we explain below, they regard the sliding window as a map from the time series data to point cloud data, and then explain how to use topological properties of this point cloud to study the periodicity of the original time series data. (There has also been interesting recent work by Khasawneh and Munch [291] on stochastic delay differential equations.)

Finally, we turn to an application to data that uses the natural ordering in genomics coming from the position along a chromosome. As we saw in Chapter 6, chromosomal aberrations (deletions, amplifications and translocations) are very common events in most tumors. The number of copies of particular chromosomal regions is a function of the ordering of genes. Arsuaga and colleagues have also proposed a sliding window approach to study copy number aberrations in cancer [19, 21]. The sliding window, moving across the chromosome, provides a map from the copy number data to point cloud data; hierarchical cluster structure in this point cloud data, as measured by the zeroth Betti number, reflects changes in copy number structure. This approach can be used to identify copy number changes in tumors and to compare the profile of these changes across different tumors.

9.1.1 Time Series Analysis of Gene Expression Data

Biological processes are dynamic, changing at different time scales. For instance, we could be interested in tracking the expression of a particular gene that is involved in the cell cycle. As we saw in Chapter 6, the cell cycle is one of the fundamental processes altered in cancers, and looking at how proteins differentially altered the cell cycle in cancer cells could provide therapeutic opportunities. Another example is the circadian rhythm, which regulates fundamental biological processes in a daily cycle. These biological processes can be studied by measurements $x(t_i)$ taken at different times $\{t_i\}$; this is popularly called a time series. There is an extensive literature on general methods for time series analysis [102, 226], but of particular interest in this chapter is the identification of periodic signals in gene expression data.

Time series expression data presents some distinctive features that make it different from finance or weather time series: it is usually collected for a few cycles, the sampling can be sparse and uneven, there is not usually a characteristic shape (such as a sinusoidal curve), there is significant biological variability and, last but not least, there is a potentially significant amount of experimental noise. All these features make time sampling of genomic data particularly interesting and challenging.

Since the first high-throughput expression experiments at the beginning of the century, there has been a plethora of methods applied to extract signals from time series expression data. These methods usually rely on standard techniques used and developed in other fields, for example Lomb-Scargle periodograms developed for astrophysics [202, 333]. We will briefly summarize some of the most common techniques used for time series analysis to study biological genomic/transcriptomic data.

- **Spectral methods.** Spectral methods express signals in terms of the frequency domain (e.g., via Fourier analysis). A basic and widely used method is direct application of the fast Fourier transform (FFT) algorithm, which transforms discrete data into Fourier components. A particularly useful technique when working with uniformly spaced time-sampled data is to approximate the spectrum by the periodogram:

$$s(\omega) = \frac{\Delta t}{N}\left|\sum_{n=0}^{N-1} x(t_n)e^{2\pi in\omega/N}\right|^2$$

where Δt is the time interval between two observations and N is the total number of observations. Periodic signals can be identified as sharp peaks in the periodogram.

Fourier analysis has been widely applied to study cell cycle genes from expression data (e.g., [537]). Fourier analysis in connection with permutation tests was implemented in [136] and applied to yeast cell cycle data and other species [272].

However, the FFT is suboptimal for sparse and non-uniformly sampled data. The Lomb-Scargle periodogram [333, 448] is a Fourier type of analysis able to infer spectral properties from sparse and irregular sampling at times t_k [202]. For a fixed frequency ω, a time delay τ is defined as a solution to the equations

$$\tan 2\omega\tau = \frac{\left(\sum_k \sin 2\omega t_k\right)}{\left(\sum_k \cos 2\omega t_k\right)}.$$

Then the periodogram at the frequency ω is equal to:

$$s(\omega) = \frac{1}{2}\left(\frac{\left|\sum_{k=0}^{N-1} x_n \cos \omega(t_k - \tau)\right|^2}{\sum_{k=0}^{N-1}(\cos \omega(t_k - \tau))^2} + \frac{\left|\sum_{k=0}^{N-1} x_n \sin \omega(t_k - \tau)\right|^2}{\sum_{k=0}^{N-1}(\sin \omega(t_k - \tau))^2}\right).$$

The Lomb-Scargle periodogram has been used in several biological applications with incomplete data or time series sampled at different time points [422, 441], including the cycle of malaria [202], circadian rhythms in plants [271], and phenotypic behavior in animals [286], among many others. For instance, in reference [202] it was used to study the expression of *Plasmodium falciparum* (the agent causing malaria) genes, during infection. The Lomb-Scargle periodogram analysis showed higher sensitivity than the Fourier transform in the identification of periodic signals, mostly in day and two day cycles.

- **Wavelets**. The Fourier transform decomposes the temporal data as a sum of orthogonal sinusoidal representations; this is typically used for identifying periodic patterns in data. Wavelets provide an alternative decomposition in terms of an orthogonal basis of multiresolution functions (wavelets) $\psi_{j,k}(t)$ localized around a time-frequency region parametrized by the k and j indices; the basis elements are scaled and shifted versions of a generating "mother wavelet." The wavelet basis allows us to express any function as follows:

$$x(t) = \sum\nolimits_{j,k} c_{j,k}\psi_{j,k}(t)$$

where $c_{j,k}$ are the coefficients. (There are many different choices of wavelet bases, including Haar and Daubechies wavelets.)

Wavelets have been used to study clusters in gene expression along the genome [509], regulatory networks from time-varying expression data [189, 295, 475], and functional MRI data [446].

- **Reference curve comparison**. Fourier analysis decomposes data into sinusoidal signals, but the data that we are interested in could have different shapes. For instance, we can be interested in identifying narrow peaks indicating the expression or activity of a particular gene in a narrow time window. If we have a particular time dependence in mind, regression analyses could provide useful insight. Partial least squares regression (PLS) has been used to identify genes with periodic expression along the cell cycle in *Saccharomyces cerevisiae* [275]. In that work, the authors were interested in the identification of periodic signals with a common period (cell cycle) but where different genes obtained the highest expression at different times of the cell cycle. That was modeled by a function $A \sin(\omega t + \phi)$, where $\omega = 2\pi/T$ is the frequency associated to the cell cycle. Every gene received an amplitude and a phase, representing the variability along the cell cycle and the cell cycle phase. The same procedure can be used for a non-sinusoidal family of curves, such as the ones shown in panel B of Figure 9.1, using PLS to find the parameters corresponding to each family member.

The Jonckheere-Terpstra trend test [278] is a non-parametric statistical test for comparing two alternative hypotheses regarding the medians of populations; the null hypothesis is that the medians are the same, and the alternative hypothesis is that the ordered populations have increasing medians. This is closely related to the Kendall τ statistic for analyzing rank correlation. JTK CYCLE is an algorithm based on these statistics that compares data to a set of hypothesized user-defined group orderings [258]. The algorithm has been extensively applied to study different aspects in different systems of circadian rhythms including expression profiles, chromatin changes, metabolic changes and changes in microbiota among many others [132, 258, 303, 496]. An advantage of these methods is that since they are rank based, they are robust invariants and hence resistant to corruption by outliers (recall the discussion from Chapter 3).

- **Stochastic processes and correlograms**. A stochastic process is a set of ordered random variables $x(t_i)$. There is an extensive literature on the study of time series analysis using stochastic processes [102]. Here, we will briefly mention autoregressive models (AR) of order p as a common type of stochastic model used in time series analysis. The main idea behind these models is that the value of the observation at time t_i depends on a linear combination of the previous p-values $\{x_{i-k}\}$ for $k = 1, \ldots, p$, in other words:

$$x(t_i) = \sum_{k=1}^{p} c_k x_{i-k} + \epsilon_{t_i}$$

where c_k are some real coefficients and ϵ_{t_i} is an error term, typically assumed to be Gaussian distributed. These are a generalization of Markov processes; in fact, AR(1) models are precisely Markov processes.

A generalization of AR is given by the moving average process (MA), where the value of the observation depends on a linear combination of q independent random processes ϵ_i with zero mean and equal (finite) variance:

$$x(t_i) = \sum_{k=1}^{q} c_q \epsilon_{i-q}$$

The most general models contain a sum of p terms from an AR model and q terms from an MA model; these are usually called (p, q) ARMA models.

One common assumption in stochastic processes is that the observations $x(t_i)$ are derived (possibly after subtraction of global trends) from a *stationary* stochastic process. A stochastic process is stationary if the joint distribution of every set of variables $x(t_a)$ is the same as $x(t_a + \tau)$ for all τ. In other words, the process does not present any systematic change in mean, variance and higher moments at different time points. Stationarity is a very strong assumption, and it is usually only applicable after all trends (changes in mean, variances, periodic components) are removed from the original data.

A useful function for studying stationary processes is the autocorrelation $\gamma(\tau)$, defined as the covariance between $x(t_i)$ and $x(t_i+\tau)$ divided by the value at $\tau = 0$. Using the data of a stationary process one can define [102]

$$r(k) = \frac{\sum_{i=1}^{N-k} (x_i - \bar{x})(x_{i+k} - \bar{x})}{\sum_{i=1}^{N} (x_i - \bar{x})^2}$$

where N is the total number of observations. The function $r(k)$ is called the correlogram and carries interesting information on the coefficients and memory of stationary stochastic processes. These models have been proposed for gene clustering from time series of expression data [188, 418, 525].

- **Compressibility**. Most patterns in biological data do not have a predefined form; the use of reference curve comparison could preclude the identification of some relevant biological signal. One idea for capturing general regularity is to study the compressibility of the data. One can define the algorithmic complexity of a string of characters as the size of the shortest program that outputs the string [304, 473]. These ideas were applied to yeast cell cycle expression data in [10]. First, the data were transformed to their rank value, for instance, 2.5, 2.7, −1.2, 23 will be transformed to 2, 3, 1, and 4. This corresponds to a permutation of 1, 2, 3, and 4. Then one looks at a function f from the permutations to the real numbers. For example, such a function could be the length of the longest increasing or decreasing sequence (2 in our case), the number of local maxima (2 in our example), the sum of the absolute values of the difference between consecutive numbers ($|2 - 3| + |3 - 1| + |1 - 4| = 6$), etc. (Notice that some permutations will be assigned the same value.) One way of describing the permutation p of interest (e.g. 2, 3, 1, and 4) is to count the number of permutations with the same image under f; denote this quantity by M_f. A bound on the compressibility can be obtained by $k(f) = \log M_t - \log N - \log M_f$, where M_t is the total number of permutations and N is the number of different values the function f can take. The main idea of the method of [10] is that simple functions can be used to identify interesting patterns as corresponding to highly compressible permutations. Note that these patterns are not periodic.

- **Biologically based time dependent models**. In some cases, there are actual models intended to describe the evolution of the biological system. For example, transcription is a classical problem where different models have been proposed that relate the activity of some genes (transcription factors) in the regulation of other genes. Such models are usually represented in the form of a network (e.g., Boolean or Bayesian networks or sets of first order differential equations) [34, 296].

All these methods have tradeoffs [137]. If we are interested in looking for periodic signals using a large collection of longitudinal data, we might be tempted to try Fourier approaches. On the other hand, if we are studying a narrow localized signal, we could try some of the simple wavelet methods. Stochastic methods are useful if we have reason to believe that the hypotheses of the main methods (e.g., AR, MA, ARMA, etc.) hold in data; for instance, when the noise can be modeled with a known distribution and there is a memory of a few previous values. In other cases, we might have a suspicion of what kind of signal we should expect and we can try to fit the expected curve directly to the data.

Of course, in practice, especially when performing exploratory data analysis, we do not expect to have a clear idea of what kind of information we are looking for. The topological approaches that we will describe in the following section provide a more general framework for identification of periodic patterns. At the moment it is an open question which methods will be most informative when working with the time series data arising in genomics. As transcriptomic data becomes more reliable and abundant, we expect to have the opportunity to evaluate the performance of these algorithms.

9.1.2 Time Series Analysis Using Topological Data Analysis

A simple mathematical model of the expression values of a particular gene i evolving in time is simply a function $e_i \colon X \to \mathbb{R}$, where X could be an interval $[a, b]$ (when considering a fixed time interval) or S^1 when looking at periodic systems. Of course, in practice we expect to have access to the values of these functions at a finite collection of times (i.e., points of the domain).

A first natural question is how to compare the expression patterns of two different genes i and j; in this model, we are comparing the functions $e_i(t)$ and $e_j(t)$. One way to answer this question is to consider the sublevel set filtration of $e_i(t)$ and $e_j(t)$ (recall Example 2.3.4); we consider the filtration of spaces induced by considering the collection of sublevel sets $f^{-1}((-\infty, a])$, which are equipped with evident inclusions

$$f^{-1}((-\infty, a]) \to f^{-1}((-\infty, a'])$$

for $a < a'$.

The two functions $e_i(t)$ and $e_j(t)$ can then be compared by measuring the bottleneck or Wasserstein distances between the persistence diagrams arising from this filtration. The stability theorems for persistent homology in this context (recall Theorem 2.4.12) now imply that these measures are fairly robust in the face of sampling variation or noise in the data (Figure 9.3).

Sublevel set persistence and bottleneck distances were used in [118, 140] to study clustering of genes by expression level over different developmental stages in microarray data from the arabidopsis plant. Specifically, the data is structured as vectors of expression levels for each gene, with entries corresponding to developmental stages. In the same papers [118, 140], the authors used topological methods for identification of periodic signals using expression values of 7500 genes across 17 time points within a single period of the formation of somites in mouse embryo. In combination with a variety of other methods (e.g., Lomb-Scargle periodogram

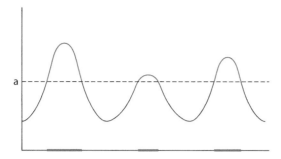

Figure 9.3 Persistence homology can be applied to filtrations from different sublevel sets induced by a function.

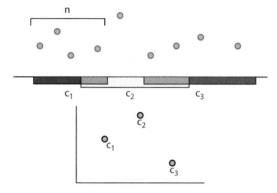

Figure 9.4 The sliding window approach is a very common strategy used in many genomic analysis applications. Data corresponding to a window of constant size n can be represented as a point in n dimensions. Sliding the window, one can generate a point cloud data representing the series. TDA techniques can be then applied to the cloud data to learn properties of the series.

and the cyclohedron method), this study found a new cyclic gene that regulates the segmentation clock. Comparative analysis found the topological methods to be competitive with other methods, although not obviously superior to the best alternatives.

9.1.3 Topological Data Analysis of Sliding Windows

A different strategy of studying time series data is to study the point cloud data generated by sliding windows (Figure 9.4). An illuminating analysis and development of this method for periodicity detection was carried out by Perea and Harer [404].

The basic idea is simple. Suppose we have a series of points $\{x_1, x_2, \ldots, x_n\} \subset \mathbb{R}$, where the subscript indicates the time label of the point. We think of these as the image of a function $f: \mathbb{R} \to \mathbb{R}$ on a collection of values $\{t_1, \ldots, t_n\}$. We define a *window* of size w starting at interval i as the collection of points

$$\{f(t_i), f(t_{i+1}), \ldots, f(t_{i+w-1})\} = \{x_i, x_{i+1}, \ldots x_{i+w-1}\}.$$

An idealized case is where $t_i = t_1 + (i-1)d$, for some shift value d.

As an abstraction, imagine that we simply have a function $f: \mathbb{R} \to \mathbb{R}$, and when fixing a window starting at value x we parametrize by M and τ and instead consider the collection of points

$$\{f(x), f(x+\tau), f(x+2\tau), \ldots, f(x+M\tau)\}.$$

Fixing M and τ, we can regard the window as specifying a curve

$$W(f)_{\tau,M}: \mathbb{R} \to \mathbb{R}^M.$$

To understand what this looks like for a periodic signal, it is interesting to focus on $f(\theta) = \cos(L\theta)$ for some choice of period $L \in \mathbb{N}$. An easy analysis shows that the resulting closed curve traces out an ellipse and that the length of the minor axis of the ellipse is maximized when the window size is close to the period!

This suggests the approach of computing the length of the longest barcode in the persistence diagram for H_1 as an estimate of the periodicity; we can in fact recover the period exactly in this case. More generally, any suitably bounded function can be expressed via the Fourier transform as a linear combination of periodic functions; the stability theorem for persistent homology can then be used to show that in the case where this signal is periodic, the longest barcode still recovers useful periodicity information.

Based on this analysis, Perea and Harer propose the algorithm SW1PerS and demonstrate applications to finding periodicity in gene expression from yeast metabolic and cell cycles [405]. In simple tests, the algorithm compares well to existing tests for periodicity (notably Lomb-Scargle). In particular, SW1PerS has extremely good performance in the face of noise, performing better than Lomb-Scargle in high noise regimes on signals where the magnitude decays over time and where there are separated peaks.

9.1.4 Identification of Copy Number Alterations

Time series are not the only interesting ordered biological data sets. Chromosomes present a natural one-dimensional ordering of genes. In cancer, chromosomal regions are often amplified, deleted, and translocated. Regions that are recurrently amplified could contain oncogenes, regions that are deleted could contain tumor suppressors, and regions with translocations could give rise to gene fusions. For instance, many tumors contain deletions in the 9p21.3 region containing a known tumor suppressor gene *CDKN2A*, or the region 17p13.1 containing the *TP53* gene. A common approach for the identification of genes that could be implicated in cancer is to assess recurrence of alterations across many different patients. Most methods that have been proposed using copy number alterations in cross-sectional samples propose a measure of recurrence and a statistic associated to it [52, 508].

Arsuaga and colleagues have proposed a method for the analysis of copy number data using persistent homology [19, 142]. The idea is based on a sliding window approach similar to the method used for studying periodic signals in time series [404], but instead of using time as the ordering dimension they use the chromosomal position. A sliding window defines a map from the copy number data to a w-dimensional space. In this case, the authors chose a three-dimensional window, so the data can be viewed as point cloud data in three dimensions. If there are no copy number alterations, one should expect that the data should fluctuate around the expected number (two in the case of autosomes), so this will correspond to fluctuations near the point $(2, 2, 2)$ in the three-dimensional point cloud. However, a deletion will change this number to one (heterozygous) or zero (homozygous), changing the concentration point for the point cloud to $(1, 1, 1)$ or $(0, 0, 0)$. In the same fashion, amplifications could be identified as changes in the point cloud data to concentrate around other diagonal values.

The authors use the zero dimensional persistent homology (i.e., the dendrogram representing single-linkage clustering) for the identification of regions containing copy number alterations. To derive a statistical test, the resulting PH_0 barcode was compared to one generated by a non-tumor control. One should expect that controls will become a single cluster at small values of the filtration value whereas the tumor samples containing copy number alterations will cluster in several groups at low filtration value and become a single connected component at larger filtration value. To assess the statistical significance, Arsuaga and colleagues propose a statistic $s = \Sigma_\epsilon (t_\epsilon - c_\epsilon)^2$, where t_ϵ and c_ϵ denote the average number of connected components in the test and control data sets, respectively. An associated null hypothesis (and p-value) was generated by random permutations of the data. When applied to breast tumors from an independent study [249], the authors were able to recapitulate known recurrent alterations and to identify some unreported alterations.

The same approach could be used for the identification of regions of differential gene expression. Using chromosomal position as ordering dimension, and expression of genes as a function, Arsuaga and colleagues [21] generated point cloud data using a sliding window approach. Notice that expression of a gene is correlated to copy number information, i.e., highly amplified genes tend to be more expressed, and deleted genes less expressed. The number of clusters as measured by

H_0 is associated to consistent changes in expression profiles across a chromosomal region. Applying this approach to 251 breast cancer expression profiles from [353] identified specific clustering profiles associated to different expression subtypes.

9.2 Topological Data Analysis in Networks and Neuroscience

Networks have become a common representation of many biological systems, representing different scales of knowledge. For instance, protein-protein interactions are summarized as networks where we can represent proteins in a living system as nodes and their interactions as edges. Neurons in the brain and their interactions provide another example of a biological system where some properties could be loosely captured by networks. The architecture of the brain as captured by the interconnections between different regions provides another example. Researchers are currently exploring the use of topological techniques to characterize the molecular, neuronal, and architectural properties of the brain [98, 130, 201, 407, 424, 463]. The hope is that topological techniques will provide ways to summarize properties of complex networks that generalize and extend the standard invariants based on global statistical properties of local information (e.g., degree distribution, measures of centrality of a vertex, or number of components).

9.2.1 Cellular Scales: Neuronal Activity

One of the central problems in neuroscience is how the ensemble of neurons can efficiently capture and faithfully represent information about the world. Neurons in physical proximity can exchange information across synaptic gaps, reflected in their neuronal activity. The relationship between neuronal connectivity and activity is highly nonlinear. Linear techniques do not suffice to correctly capture its structure. The visual cortex is a classical system in which to study the coding of external physical stimuli into neurons. Singh and colleagues [463] used persistent homology to study the population activity in the primary visual cortex. The invariants derived from persistent homology in natural image stimulation were similar to a spontaneously active cortex. Giusti and colleagues [201] proposed a method based on TDA to extract nonlinear but monotonic relationships. The hippocampus has been found to encode information about the physical environment through pyramidal neurons. Different neurons in the hippocampus respond selectively to different physical locations [390]. In [201], the method is shown to be able to recover geometric information encoded in neuronal correlations (without using the external stimuli); that is, they can recover place cell activity without hypotheses about the stimuli or the receptive fields of the cells.

Reimann et al. [424] explored the relation between neuronal architecture and information processing by constructing directed graphs capturing the direction of synaptic transmission. In particular, they summarized data as a directed graph with nodes representing individual neurons and directed edges representing pre- to postsynaptic neuronal connections. The response to external stimuli can be modeled as time series data in the directed graph. Different aspects of the structure of these graphs can be quantified by identifying different objects at different scales, from local (as indicated by the presence of cliques of neurons) to global (as indicated by the existence of larger topological structures). (See Figure 9.5.) Applying TDA techniques to computational reconstructions of neocortical circuits in the brain of a rat, the authors were able to quantify the presence of these different structures, including large numbers of high-dimensional cliques and "holes." This quantification of structural properties of neuronal networks hopefully provides a first step for understanding the association between brain architecture and function.

9.2.2 Mesoscopic Scales: Brain Functional Networks

Cognitive processes usually involve the coordinated activity of different areas of the brain. The relation between the activity of brain regions can be represented by networks, and statistical properties of these networks can provide information on the functional architecture. Functional imaging can provide information on the activity of the brain at mesoscopic scales of thousands or millions of cells. Petri et al. [407] proposed to use TDA techniques to study the statistical properties of homological cycles in these networks. To test these ideas, they compared the resting state of 15 healthy volunteers receiving placebo or psilocybin, a psychoactive drug. A significant difference was observed in the homological features between the two groups, suggesting that these rough descriptors capture relevant structural properties of brain architecture. The brain architecture of structural connectomes was also explored using topological techniques by Sizemore et al. [464]; in particular, they sought to identify densely connected groups of active regions. Further, they proposed that these cliques were related to local fast processing. Experimentally, they verified that these regions were consistent across a group of eight individuals. Cassidy et al. [98] has applied TDA to compare the activity of the brain using functional MRI (fMRI) in a variety of conditions. This approach improves over standard correlation comparison methods, in the sense that when applied to real data it produces much more sensible functional connectivity predictions. It involves a method to analyze network architecture using persistent homology, accounting for potential artifacts due to spurious spatial and temporal correlations.

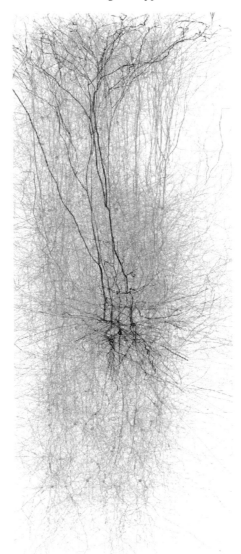

Figure 9.5 Representation from a slice of in silico reconstructed neuronal tissue. In red, a clique formed by five pyramidal cells. Source: [424].

9.3 Topological Approaches to Biomedical Imaging

Imaging is one of the main non-invasive modalities for diagnosing and evaluating the progression of many diseases, including cancers. Solid tumors appear as masses in various imaging technologies, notably including MRI. Tumors have a shape and volume; these can be modeled as the topological and geometric properties of a three-dimensional object. Rough metrics on these masses (e.g., changes in volume) are used as a standard for prognosis and to evaluate therapeutic efficacy. However, it has been found that other geometric invariants can provide interesting clinical information. For instance, glioblastomas, the most common type of brain tumors in adults, come in two types: one single mass or multiple masses (multifocal/multicentric glioblastomas). That is, a glioblastoma is classified by whether it has multiple path components. In [320], it was shown that these two types are associated to different genetics: multifocal/multicentric tumors are strongly enriched in point mutations in *PIK3CA*, a major oncogene, and are genetically highly heterogeneous with different lesions associated to different masses in the tumor (Figure 9.6). This observation has important clinical implications, as multifocal/multicentric glioblastomas have a worse prognosis and drug responses to different masses are extremely heterogeneous.

The observation that simple geometric and topological properties of images (volume or number of path-connected components) can inform prognosis and drug responses prompts the question of how image data relates to genetic and clinical data. Ideally, one would like to systematically explore the map between genetic and phenotypic data (as expressed in the image and in other clinical sources).

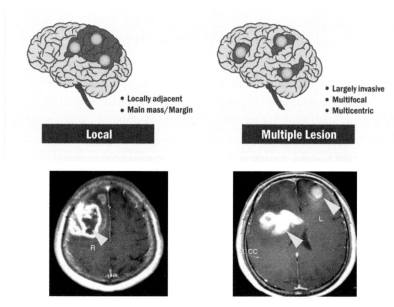

Figure 9.6 Glioblastomas can appear in a single mass (left), or several masses (multifocal/multicentric glioblastomas) (right). This simple topological difference is associated to specific mutations (*PIK3*) and worse prognosis. Source: [320]. Reprinted with permission of Springer-Nature: Lee, Jin-Ku, et al. "Spatiotemporal genomic architecture informs precision oncology in glioblastoma." Nature Genetics 49.4 (2017): 594–599.

Crawford et al. [128] proposed to use the smooth Euler characteristic transform (recall Section 3.8) to decompose tumor image data into a set of topological features amenable to machine learning. The procedure starts by segmenting the tumor image (i.e., identifying from the image data the tumor and reconstructing a three-dimensional manifold [112]), then sectioning it using the smooth Euler characteristic transform to extract a function of different directions that can be used for subsequent functional machine learning analysis. Crawford et al. showed that topological information provides complementary information to other types of genomic, transcriptomic, and volumetric data to predict overall survival in glioblastomas. This work suggests an interesting approach of combining "omic" data with imaging to better characterize the mechanisms of initiation and progression of tumor growth; topological analysis appears naturally as a way of quantifying imaging features.

Another interesting complex network studied using TDA is the blood vessel system. In [489], Szymczak and colleagues proposed reconstructing the vascular trees from three-dimensional images using persistent homology. The method was applied to reconstruct coronary trees from computed tomography (CT) scan data of the heart.

9.4 Spreading of Infectious Diseases

Networks have also been used to capture the spread of infectious agents, where nodes represent infected individuals and direct infection is represented by edges. Understanding the structure of these networks is one of the main objects of study in epidemiology; such analysis provides information about the main routes of transmission (aerosol, food, through other species, etc.), spread via local contact versus long range contact (e.g., airline influence), how effective is the transmission of infection, potential sensitive and resistant populations, and the efficacy of potential ways of curtailing the spread. Taylor and colleagues [491] used TDA techniques to study the mathematical structure of these graphs, their intrinsic dimensionality, and their topological and geometric properties; they connected these invariants to epidemiologically significant quantities.

Another application is the study of immune responses to infectious agents [501]. It is well understood that different hosts will have very different responses when exposed to the same infectious agents: some are resilient, others present mild symptoms, and others could die. Torres et al. propose a phenotypic space, the "disease space," that captures the potential states of an infected host (Figure 9.7). Measuring physiological data (weight, temperature, different cell counts, among others) at different states of infection, one can trace the trajectories of

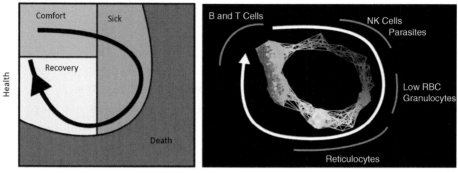

Figure 9.7 Left: Conjectured "disease space" capturing the potential phenotypic states of a host infected with a pathogen. The normal state of an uninfected individual is on the left. When it is exposed it is thrown out of this state. As the host recovers, it goes back to the healthy state through a trajectory that does not track back the previous states. Right: Mapper applied to physiological data from mice exposed to *Plasmodium chabaudi*. Source: [501].

different hosts after exposure. It was observed that resilient hosts do not significantly change states, while less resilient hosts are associated to large "loops" in the disease space. These hypotheses were evaluated in mice exposed to *Plasmodium chabaudi* (a murine analogue of the cause of human malaria) and in malaria patients.

9.5 Summary

There are many exciting directions in the application of TDA methods to biological data beyond genomics, and we expect more to be discovered in the coming years. This work is clearly in its infancy, but already there have been interesting and suggestive results.

- There is a great deal of biological data which comes as an ordered sequence; time series data is a notable example, but not the only one.
- Topological data analysis methods provide a way to detect periodicity in ordered signals, via a sliding window approach, that is competitive with the best standard methods (and has different properties).
- TDA methods for analyzing graph structures have been profitably applied to problems in neuroscience (studying neuronal connectivity and its correlation with neural activity), epidemiology, and analysis of coronary artery structure.

9.6 Suggestions for Further Reading

- There are very good introductory books on time series analysis. A very pedagogical introduction is the book by J. Brockwell and R. Davis, *Time Series: Theory and Methods* [74]. The book reviews spectral methods, autoregressive and moving average processes, state-space models and forecasting.
- A review on time series analysis applications to transcriptomic and epigenetic data is given by Z. Bar-Joseph et al. [35], with a summary of recent interesting problems and standard bioinformatic tools for analysis.
- Regarding the TDA study of the sliding window approach to series data, we recommend reading the work of J. Perea and J. Harer [404].
- For recent applications of TDA techniques to neuroscience, we recommend the review by C. Curto [130].

10
Conclusions

This is an extremely exciting time in biology, where new discoveries happen on a daily basis and there is a vast amount of data available to researchers. In this book, we have made the argument that scientific progress in biology requires the application of more sophisticated methods for representing and analyzing the shape of data. To carry out this argument, we have walked the reader through some basic applications of algebraic topology to biological problems.

The attentive reader has no doubt observed that our efforts serve mostly to highlight the extensive work that needs to be done. On the mathematical side, there are many urgent foundational problems that need to be addressed, including further development of statistical methods attuned to biological applications, finding better ways of applying multiple filtration parameters, defining new simplicial complexes that can capture biological relationships in a natural way, and more. On the biological side, the applications we have described constitute only the tip of the iceberg of potential applications. Developing biologically meaningful metrics and Morse-type functions and accurate and meaningful topological condensed representations remains an art that needs to be systematized.

Our hope is that this book will help inspire the next wave of work in the area.

Appendix A

Algorithms in Topological Data Analysis

A.1 Computing Persistent Homology

To explicitly compute the homology of a simplicial complex, one needs to choose bases for $C_k(X)$ and $C_{k-1}(X)$ and then find the image and kernel of the boundary map

$$\partial_k \colon C_k(X) \to C_{k-1}(X).$$

This can be done via representing the boundary map as a matrix with respect to the chosen bases and putting this matrix in Smith normal form. In order to compute persistent homology, one has to choose compatible bases for each chain group simultaneously; this is what is behind the standard algorithms for computing the persistent homology (recall Section 2.7 or see for example [551]).

  These algorithms are $O(n^3)$ in the number of simplices in the complex, and while they are often linear in practice, it is not hard to construct explicit filtrations that achieve the cubic bound [361]. As discussed in Section 2.7, if the feature scale is close to the diameter of the data, the Vietoris-Rips filtration will have exponentially many simplices. There have been some efforts to achieve better performance by simplifying the complex; [455] describes how to build a hierarchical collection of approximations to suitable finite metric spaces such that for any given accuracy the computation time is linear in the number of points X, and [549] uses simplicial collapses to reduce the complex without changing its homotopy type.

  However, the best performance to date has been achieved by work that uses a series of optimizations of the basic algorithm, most notably the use of persistent cohomology in Ripser [39], which is currently the fastest and most memory-efficient available implementation (for Vietoris-Rips complexes).

A.2 Software for Persistent Homology and Mapper

We begin by describing the software available for persistent homology. There are essentially two categories of efficiency: the first group does not work for large complexes but in general has more functionality, and the second group provides the state of the art performance. A very nice summary of the situation (with detailed performance comparisons) is given in [413]. All of the packages we discuss are under active development at the time of writing.

1. **Javaplex:** handles zigzag persistence and construction of the witness complex [490].
2. **Dionysus:** handles vineyards, zigzag persistence, and persistent cohomology [360].
3. **Perseus:** handles cubical complexes as input [376].

For large data sets, there has been a recent revolution; most notably, Ripser and Eirene work on remarkably large complexes (billions of simplices).

1. **Gudhi:** handles the witness complex, subsampling, persistent cohomology, cubical complexes [341].
2. **Dipha:** distributes computation across many parallel nodes [40, 41].
3. **Ripser:** fastest performance available, only compatible with Vietoris-Rips so far [39].
4. **Eirene:** very good performance (comparable to Ripser), supports identification and rendering of specific cycles that represent homology [238, 239].

Software for multidimensional persistence has been considerably more limited, until the recent development of the excellent Rivet tool.

1. **Rivet:** supports novel simplification techniques and rendering options [324, 325].

A number of the statistical techniques for topological data analysis, most notably persistence landscapes, are now available in free software packages.

1. **Hera:** supports fast computation of bottleneck and Wasserstein distances [289, 290].
2. **TDA statistics R package:** supports confidence sets, bootstrapping, persistence landscapes, and distance to a measure. Incorporates Gudhi, Dionysus, and PHAT [171]. (PHAT is a library for persistence computation [43, 44].)
3. **Persistence Landscape Toolbox:** supports various statistical operations in the context of persistence landscapes [77, 78].

The situation for Mapper is somewhat more limited amongst free software; far and away the best implementation is the Ayasdi version, which requires a license and contacting the company in order to download the software.

1. **Python Mapper:** an experimental package that provides both a GUI interface and a python package [365].
2. **TDA Mapper:** an R package exposing Mapper functionality [402].
3. **KeplerMapper:** an experimental package that provides a python library [447].
4. **Ayasdi:** commercial software for Mapper, supporting a wide range of filters, a well-developed GUI, and elaborate output renderings [267].

Appendix B

Introduction to Population Genetics

B.1 Population Genetics

Population genetics studies the distribution of frequency of alleles in populations, subject to mutation, selection, changes in population size, migration, and many other factors. It is mostly a model-based framework, where estimators of parameters of the model could be derived under certain assumptions. For instance, a common problem is to describe the likelihood of the rise of a certain mutation to become the dominant allele in the population after a few generations. To assess the impact of this mutation, one needs to postulate a model that captures some of the key features of the population. For instance, the simplest assumption is a constant size N and no selective advantage in the mutation. Under these simple assumptions, it is possible to derive the probability for a random mutation to become dominant and compute how many generations this would take. One can also derive probability distributions, for instance, the probability $p(t \mid N)$ that under the hypotheses above, a mutation would become dominant after t generations.

Until the development of genomic techniques and the availability of population data sets, the development of population genetics was largely theoretical. The founding principles of population genetics were set in the 1920s and 1930s by Ronald Fisher, Sewall Wright, and J. B. S. Haldane. Their work quantitatively integrated Mendelian genetics with natural selection. In the 1960s the work of Motoo Kimura connected the probability distributions derived from population genetics models to diffusion theory and Kolmogorov's treatment of stochastic Markov processes [299]. Kimura proposed the neutral theory of evolution, suggesting that, at the molecular level, most of the variation in populations is not the effect of natural selection but of stochastic drift in allele frequencies [300]. Until the 1980s most models in population genetics were evaluating the changes in distributions of alleles, starting from some initial conditions and evolving the population forward in time. In 1982 Kingman proposed a simple framework to study the distribution of potential histories given a sample of n individuals in the population. This approach, known as coalescent theory, has become a standard in the field. We now turn to briefly review these different models.

B.2 Wright-Fisher Model

Ronald Fisher and Sewall Wright, at the beginning of the 1930s, introduced one of the first models in population genetics. The model studies a diploid (each individual has two chromosomal copies) constant population of N individuals, so there are $2N$ copies of each allele in the population. The basic Wright-Fisher model ignores mutations and recombinations to study distributions of alleles in a population. The results are identical for a generation of $2N$ haploid individuals. In the Wright-Fisher model generations are discrete and do not overlap, and copies of an allele are drawn at random from the alleles in the previous generation. This model also does not consider that there can be structure in the parental population, that there are males and females, that the number of individuals in the population can change, and that there can be geographic limitations in mating (Figure B.1).

At generation $t = n + 1$, an allele is picked up from a parent allele at random from generation $t = n$. The number of descendants m of a particular allele follows a binomial distribution with probability $1/2N$:

$$p(m) = \binom{2N}{m}\left(\frac{1}{2N}\right)^m \left(1 - \frac{1}{2N}\right)^{2N-m}.$$

For large populations, the probability that a gene has no descendants in the next generation is $p(0) \sim e^{-1} \sim 0.37$. At each generation, that probability that a particular gene has no descendants increases monotonically. After many generations, only one gene will be present in the population, and we say that it will become fixed. The probability that a gene will become fixed in the population at very large time scales will then be $1/2N$, as in this model all genes have the same probability of becoming fixed in the population. If a particular allele is present in the population at frequency x, the frequency in the next generation will have variance $x(1 - x)/2N$. The effect of random drift depends on the size of the population. Note that this loss of variation has nothing to do with selection but is just random drift. The Wright-Fisher model estimates the average number of generations that have passed until a gene becomes fixed in the population to be $4N$.

One can introduce mutations in this model at a particular rate μ per gene per generation. The expected number of mutations in a generation is $2N\mu$. However, as we have seen, most of the alleles where the mutations occur disappear from the population after a long time ($\geq 4N$).

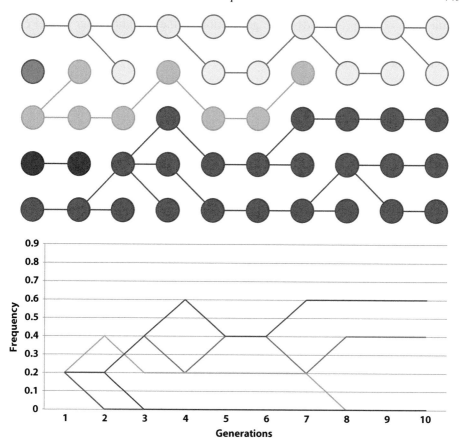

Figure B.1 Wright-Fisher model in population genetics. Assume a population of constant size $2N$ and generations that occur at discrete intervals t_i. A parent gene can randomly replicate into m descendants following a binomial distribution of mean $m = 1$. In some cases, the parent leaves no descendants and the lineage becomes extinct, and only one allele becomes dominant at longer times (becomes fixed). In the top figure we represent an example of how different alleles are related by ancestry. The individuals in the initial population on the left replicate and descendants are colored according to the original parent. After some generations one color dominates. In the bottom figure, we represent the frequency of each of the original parents.

B.3 Moran Model

The Moran model is a variation of the Wright-Fisher model that assumes overlapping generations [358] in a population of $2N$ haploid individuals. Like the Wright-Fisher model, the Moran model does not consider structure in the population or selection of particular alleles. In the Moran model, each step corresponds to (1) selection of a random individual and (2) a decision to reproduce (two copies) or die with equal probability. One generation of the Wright-Fisher model corresponds to $2N$ steps. As expected, the Moran model gives the same results as the Wright-Fisher model.

B.4 Kimura Diffusion Model

One of the main objects of study in the previous models is the probability distribution $f(x', t' \mid x, t)$ of finding a particular allele at a frequency x' and a time t' given that it was observed at a frequency x and time t. The change in allele frequencies in random drift models is then a stochastic stationary Markov process. When the number of individuals or genes in the population is large, the changes in frequency are almost continuous. Kimura in 1964 observed [299] that $f(x', t' \mid x, t)$ obeys an equation that he called the Fokker-Planck equation (or Kolmogorov forward equation).

Consider two alleles A and B in a gene with frequency x and $1 - x$ respectively at time t. Let us consider also the general case where the allele A can mutate to the allele B with a probability of μ_1 per gene per generation, and B can mutate to A at a rate of μ_2 per gene per generation. Kimura found that $f(x', t' \mid x, t)$ follows the equation:

$$\frac{\partial}{\partial t'} f(x', t' \mid x, t) = \frac{1}{4N} \frac{\partial^2}{\partial x'^2} \left(x'(1 - x') f(x', t' \mid x, t) \right)$$
$$+ \frac{\partial}{\partial x'} \left(((\mu_1 + \mu_2)x' - \mu_2) f(x', t' \mid x, t) \right)$$

with initial conditions $f(x', t \mid x, t) = \delta(x' - x)$, where $\delta(x)$ is the Dirac delta. This is a heat or diffusion equation with a drift term (last term on the right). If the mutation rates are very low, the dominant term is the first term on the right, which is a diffusion term that depends on N, the size of the population. If the population size is large, the diffusion term is small, and drift due to mutation dominates. The effect of selection in one of the alleles can also be included as part of the drift term [299].

The stationary solutions of this equation of the two allele model with mutation rates at $t' \to \infty$ are beta distributions, already discovered by Wright in 1931 [541]:

$$f(x) = \frac{\Gamma(4N(\mu_1 + \mu_2))}{\Gamma(4N\mu_1)\Gamma(4N\mu_2)} x^{4N\mu_1 - 1} (1 - x)^{4N\mu_2 - 1}.$$

B.5 Coalescence

Previous models studied the distribution of gene and allele frequencies and how these distributions change as we move forward in time. Coalescence takes a

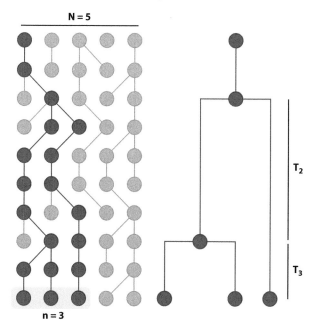

Figure B.2 The coalescence process traces back the ancestry of a set of n lineages/individuals in a pool of $2N$ lineages/individuals. In the simplest model without recombination, two of the n lineages/individuals coalesce, the $n - 1$ coalesce again, etc. generating a tree structure. The times to the branching events $T(n)$ can be easily computed.

different approach [301]. Starting from a sample of n genes, we reconstruct the potential histories. In the case of no recombination, the stories are trees that reflect a common ancestor to different individuals (Figure B.2). Moving back in time, two of the n will be found to be derived from a common ancestor, reducing the problem to $n - 1$ lineages, then to $n - 2$, etc. until all n coalesce in a single ancestor. By studying the distribution of potential histories we can estimate different parameters, for instance, the time to the most recent common ancestor (TMRCA), the number of generations it took for all n alleles to coalesce into a single ancestor.

The simple coalescence model, like the previous forward models, is also based in a homogenous population of constant size N diploid individuals, with no selection, no recombination, and no population structure. In that case results can be obtained analytically. For instance, we can compute the time t when two lineages coalesced. The probability that by random chance two of $2N$ lineages coalesce in one generation is $1/2N$ and the probability that they do not coalesce is $1 - 1/2N$. As these probabilities are the same in every generation (population size is constant), the probability that they coalesce at generation t follows a geometric distribution:

$$p(t) = \frac{1}{2N}\left(1 - \frac{1}{2N}\right)^{t-1}.$$

If we now have m lineages, we can follow the same reasoning and estimate the probability that two of these coalesce. The probability that none of them coalesces in the first generation is $q_m = (1 - 1/2N)(1 - 2/2N)\ldots(1 - (m-1)/2N)$. In the limit of $m \ll 2N$ the probability $q_m \sim \frac{m(m-1)}{4N}$, so the probability that they coalesce in t generations is a geometric distribution:

$$p_m(t) = (1 - q_m)q_m^{t-1}.$$

The expected number of generations is the mean value of the distribution $T_m = 1/q_m \sim \frac{4}{m(m-1)}$. The time to the most recent common ancestor, the expected time to the coalescence of all m lineages, is the sum of T_k for $k = 2, \ldots, m$:

$$T_{MRCA} = \sum_{k=2}^{m} T_k \sim 4N\left(\frac{1}{1\cdot 2} + \frac{1}{2\cdot 3} + \ldots + \frac{1}{(m-1)m}\right) = 4N\left(1 - \frac{1}{m}\right).$$

If there is a mutation rate μ per generation, one can estimate how many mutations S have been accumulated in the whole process by estimating the size of the whole tree. As there are k lineages per time T_k this is simply:

$$S = \mu \sum_{k=2}^{m} kT_k \sim 4N\mu\left(\frac{1}{1} + \frac{1}{2} + \ldots + \frac{1}{(m-1)}\right) \sim 4N\mu \log m.$$

Counting the number of polymorphic sites S by sampling m lineages in a population of $2N$ provides an estimate of $\theta = 4N\mu$, the number of mutations per generation in a population, equal to:

$$\hat{\theta} = \frac{S}{\sum_{k=1}^{m-1}\frac{1}{k}}.$$

This is known as the Watterson estimator [528], one of the most frequently used estimators of θ.

The basic coalescence framework we have just described could be easily extended to recombination, fluctuating population sizes (exponential growth, contracting populations, bottlenecks), population divided into different subpopulations (finite island, population subdivision models), splitting of populations, migrations, and effect of selection, among others. In these cases, there are fewer analytic results but coalescence simulations have been implemented in a variety of contexts.

B.6 Suggestions for Further Reading

There are excellent books and reviews on phylogenetics for the reader who wants to dive into this topic.

- *Principles of Population Genetics* by Daniel Hartl and Andrew G. Clark [230] is a very nice introduction to population genetics, with careful explanations and many examples. It does not require previous knowledge and it is accessible to a wide audience. A beautiful first read on the topic of population genetics.
- *Population Genetics: A Concise Guide* by John H. Gillespie [198], is a concise introduction to population genetics.
- *Gene genealogies, Variation and Evolution: a Primer in Coalescent Theory* by Jotun Hein, and Mikkel Schierup, and Carsten Wiuf [237], provides an introduction to population genetics with a coalescence angle.
- *Coalescent Theory: An Introduction*, by John Wakely [518], like the previous book, focuses on coalescence with a wide target audience, including biologists, concentrating on the mathematical structure of coalescence.
- *ReCombinatorics: a Comprehensive Introduction to Inference of Ancestral Recombination Graphs* by Dan Gusfield [220].
- *Mathematical Population Genetics 1: Theoretical Introduction* by Warren Ewens [166] is a nice book with a mathematical angle. Recommended as an advanced treatment for the more mathematically inclined.

B.7 Data and Software

There are many software packages for population genetics simulations and inference. Here is a very incomplete list of some of the most commonly used.

- GeneTree: by R. C. Griffiths, reconstructs trees describing the mutation history of samples of DNA sequences and estimates maximum likelihood parameters such as mutation rates, changes in population (migration and growth), and distribution of times to most recent common ancestors. Software can be found at `www.stats.ox.ac.uk/griff/software.html`.
- MS: by Hudson [252], can simulate variable population size, migration and admixture, recombination, and gene conversion.
- MSMS: by Gregory Ewing and Joachim Hermisson [167], is a coalescent simulation software that can include different factors such as demographic structure and selection.
- MaCS: by Chen, Marjoram and Wall [114], is a fast algorithm that can simulate different demographic models including admixtures, different population sizes and recombination hotspots.

Appendix C
Molecular Phylogenetics

C.1 Introduction

The main goal in molecular phylogenetics is to reconstruct a phylogenetic tree from a set of protein or genomic sequences. The tree will represent the history of a set of organisms that share a common ancestor. For this purpose, sequences need to be homologous; that is, they should have evolved from a common ancestral sequence. Alignment of homologous sequences is the first step to any phylogenetic inference. For instance, let us try to align two small sequences that we suspect are homologous: $S_1 = ACTGCGAA$ and $S_2 = CCGTCTAA$. An alignment is a map between a set of strings of the same length that contain the same nucleotides and have the same order as in the original data, and could include gaps, represented by $-$. For instance, a potential alignment can be of the form:

$$S'_1 = ACTG - CGAA$$
$$S'_2 = -CCGTCTAA,$$

consisting of two strings of length 9, that differ in position 1 ($A \rightarrow -$), position 3 ($T \rightarrow C$), position 5 ($- \rightarrow T$), and position 7 ($G \rightarrow T$). Differences between two nucleotides represent mutations between the two bases, and a gap indicates a deletion or insertion of a particular nucleotide.

There are many potential alignments between a set of sequences, and we need criteria to determine which are optimal. A common approach is to consider a score for each alignment. For instance, one could count the number of differences. In the case of the alignment $\{S'_1, S'_2\}$, this would be 4. A simple score is just the number of similarities minus the number of differences, in this case $5 - 4 = 1$. But of course, not every change is necessarily weighted equally, indeed, there are some changes that are more likely to occur than others. The genomic material in all known organisms is polymers of nucleotides, each composed of a sugar, a nitrogenous base, and a phosphate group. The five-carbon sugar (ribose or deoxyribose) defines the type

454

of molecule, RNA or DNA. Bases can be divided into two types based on the chemical structure, purines (adenine and guanine) and pyrimidines (cytosine, uracil and thymine). A substitution, changing a base into another, could happen as a result of different chemical processes. For instance, one of the most common chemical processes is a spontaneous deamination (removing an amine group) of a cytosine, changing it into uracil. If not correctly repaired, that will lead to a $C \to T$ mutation in DNA. It turns out that changes within purines and within pyrimidines (transitions) are much more common than changes between purines and pyrimidines (transversions). The probabilities of these changes can be inferred experimentally, by considering data across different homologous sequences and estimating likelihoods. Similarly, one can evaluate the effect of small insertions and deletions by working with protein sequences.

A popular type of score assigns to every alignment a linear score that adds the weights for every substitution and adds a gap penalty for indels. There are several classic dynamic programming algorithms, like Needleman-Wunsch [379] and Smith-Waterman [468], for optimal alignment between a pair of sequences. When dealing with multiple sequences one could consider adding the pairwise scores, but this problem has been shown to be NP-complete [521]. In practice, heuristics are used.

Once the sequences are aligned one can start inferring trees (see Figure C.1). There are several kinds of trees to consider. In some cases, trees can be rooted, e.g., if they have a node that represents the common ancestor to all the analyzed taxa, which gives information about the temporal order of nodes in the tree. Alternatively, unrooted trees display the evolutionary relationships among taxa, without any ancestral root. The root is usually, but not always, determined by using an outgroup taxon that falls outside the group of taxa of interest. In general, when this information is not available or not used, one can construct an unrooted tree. For m sequences, an unrooted tree has m external nodes (or leaves) each of which is labeled with a different sequence. Internal nodes can be labeled by inferred sequences that represent the genomic information of common ancestors of two of the contiguous nodes. With m labeled leaves it is easy to see that the number of tree topologies is $(2m - 5)!! = (2m - 5)(2m - 7)(2m - 9) \cdots 1$. That is, for $m = 3$ branches, there is only one labeled unrooted tree, for $m = 4$ there are 3, for $m = 5$ there are 15, etc.

Edges could be weighted by a positive number that is associated to the number of changes between nodes (for instance, it could be the number of changes, or a weighted version of this, assigning each change a different weight). The space of potential trees is enormous and some criterion is needed to find the optimal tree. There are many methods that have been proposed. Here we explain some of the most popular ones. Many methods work directly with the aligned sequences,

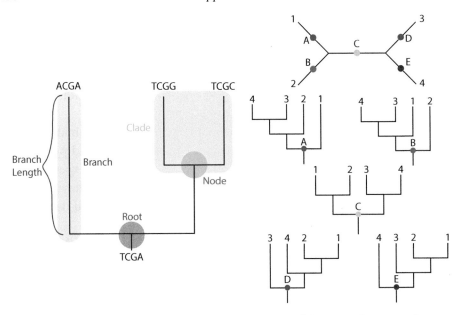

Figure C.1 Notation on a tree. There are $2m - 3$ possible ways of constructing a rooted tree from an unrooted tree of m leaves.

referred to as alignment-based methods. In all these methods the strategy is to minimize a criterion (a likelihood, for instance) by exploring a large number of trees. As previously described, the number of trees increases as $m!$, making it unfeasible to explore all possible tree topologies. Different heuristics are used to explore a reasonable set of topologies.

A second type of approach computes a distance metric from the data, and works directly using the distance data. These methods have the advantage that they scale polynomially in the number of sequences and genome length, and so one can easily work with thousands of sequences. On the other hand, the results are sometimes less biologically plausible than likelihood-based methods.

C.2 Sequence Based Methods

C.2.1 Parsimony

The parsimony principle is the preference for the simplest explanation of some facts. In the case of phylogenetic reconstruction, parsimony selects the tree with the minimum number of changes required to explain an alignment. Given a tree T, and a set of sequences S from a multiple alignment attached to the leaves (external vertices), we can assign hypothetical sequences H to internal nodes. We can compute for each edge a distance using the Hamming metric or a weighted version of it. Adding the results for all edges, we obtain the parsimony score $P(T, H | S)$. The (large) maximum parsimony tree is the tree T and hypothetical internal sequences assignment H that minimizes $P(T, H | S)$. The task of computing the best H, given a particular tree topology T, from some external data S, is called the small parsimony problem and can be computed in polynomial time, using for example, some classical algorithms from phylogenetics such as the Fitch algorithm [176]. The large parsimony problem requires going through all possible topologies and for each one computing the optimal H. The output is the topology that minimizes the parsimony score. This problem has been shown to be NP-complete [180].

There are, however, some heuristic methods to explore possible solutions, without, of course, any guarantee that they will be the optimal solution. Branch and bound methods start with a subset of three sequences S_3 from the original data S. In this case, there is a unique tree and a maximum parsimony solution can easily be found. Now, we can select a sequence from S that is not in S_3 and attach a new leaf to any of the three leaves. There are three different possibilities to consider. Now we can select another sequence from S that was not previously considered, and repeat the procedure, but now there are five potential edges. In this way, one can construct iteratively all possible trees in a hierarchical fashion (a tree of trees). Now in each of these iterations we can compute the parsimony score, which will always increase when considering more branches. In the branch and bound method, one proceeds iteratively and selects the best tree in each iteration and only considers subsequent iterations along those particular branches in the tree of trees (see Figure C.2).

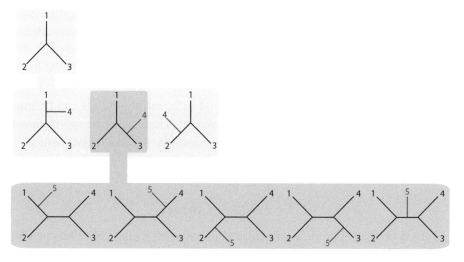

Figure C.2 Branch and bound.

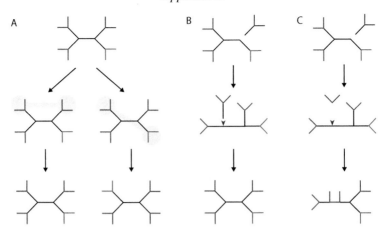

Figure C.3 As the number of taxa increases, the number of potential trees becomes extremely large. There are techniques to explore the space of potential trees by branch swapping strategies: (A) nearest neighbor interchange (NNI), (B) pruning and regrafting (SPR), and (C) tree bisection and reconnection (TBR).

A second type of strategy is based on the idea of swapping branches. Here we describe three major strategies, nearest neighbor interchange (NNI), pruning and regrafting (SPR), and tree bisection and reconnection (TBR). Any internal edge on a bifurcating tree has four neighbor subtrees, two attached to one vertex and two to the other. NNI is an operation that exchanges a tree of one of the neighbor vertices with another one (see Figure C.3). There are several implementations of these methods, but in the simplest version, a NNI procedure is accepted if it reduces the parsimony score. Pruning and regrafting is an idea along the same lines where a subtree is cut and regrafted in one of the edges, creating a new node (see Figure C.3). Tree bisection and reconnection (TBR) selects an edge and removes it completely from the larger tree, generating two smaller subtrees. Then one edge from each subtree is selected and two new nodes are introduced in each of the edges and finally joined by a new edge (see Figure C.3). NNI, SPR and TBR are operations in the space of trees, and different algorithms can be implemented to make sure that local minima are avoided. These heuristic approaches are commonly used in other phylogenetic techniques, like the likelihood methods we discuss next.

C.2.2 Likelihood Methods

Likelihood methods optimize a likelihood function $\Psi(S \mid T, L, M)$, that calculates the probability of obtaining the observed sequence data S given a tree T with branch lengths L and a model M that determines the probability of a particular mutation to occur. The advantage of probabilistic models is that they incorporate realistic assumptions based on empirical data and they can be used to estimate parameters. For instance, it is easy to incorporate the probability of back mutations, different rates for transitions and transversions, and so forth, and estimate the likelihood for those parameters. For instance, one can compute the probability of a base to change after some particular time l (associated to a branch of length l in the tree) and at a constant mutation rate μ to be $p(l) = \frac{3}{4}(1 - e^{-\frac{4}{3}\mu l})$. The formula can be easily adapted to allow different mutation rates across different bases, and even different rates in different branches and genomic positions.

Most likelihood methods assume that the likelihood for a sequence is the product of likelihoods for all positions (independence among sites): $\Psi(S \mid T, L, M) = \prod \Psi(s_i \mid T, L, M)$, where s_i is the alignment data for genomic position i. For each edge on the tree and site i, one can associate characters and compute the probability for change ($p(l)$) or staying the same ($1 - p(l)$), where l is the length associated to the edge. For a given tree T with edge length L, the likelihood for the observed data S can be computed using the Felsenstein algorithm [173, 174].

However, the full solution to the likelihood problem requires that all different tree topologies and branch lengths are explored, and as such finding the maximum likelihood tree is NP-hard [115]. There are, however, good approximations that adjust tree topology and branch lengths simultaneously. For instance, in [216] a hill-climbing algorithm is proposed, that starts from a fast distance-based method and modifies this tree to improve its likelihood at each iteration.

C.2.3 Bayesian Methods

Bayesian methods are based on a similar idea to likelihood methods, but instead of estimating the probability of the observed data S given a tree, they estimate the posterior probability $P(T, L \mid S)$ of a weighted tree (T, L) given the observed data S [257]. The basic object here is a distribution on the space of all potential trees. The whole distribution cannot be estimated analytically, but it is possible to sample the distribution. Most of the implementations are based on variations of Markov chain Monte Carlo (MCMC) approaches. The main idea is simple: one can take a tree T_i, modify it to obtain a new tree T', and compute the ratio between the posterior probabilities:

$$R = \frac{P(T', L' \mid S)}{P(T_i, L_i \mid S)}.$$

Using Bayes theorem, this can be shown to be equivalent to:

$$R = \frac{P(S \mid T', L')P(T', L')}{P(S \mid T_i, L_i)P(T, L)},$$

where $P(T, L)$ is the prior probability of observing a weighted tree (T, L).

Based on the posterior ratio R, we can decide whether or not to accept the modification T'. If so, we take the accepted tree as new starting point and we continue the operation. This procedure generates a random walk on the space of trees sampling a distribution that approximates $P(T, L|S)$. One of the most popular implementations is the Metropolis-Hastings algorithm [234]: we accept the new tree with a probability $\min(1, R)$. If so, we define $(T_{i+1}, L_{i+1}) = (T', L')$, and now we iterate. Trees with higher posterior probability will tend to be sampled more frequently. In the end, the result is a set of trees with high posterior probabilities; this allows us to account for uncertainty. This set of trees can then be summarized in a consensus tree if necessary.

Different implementations have been carried out, involving different perturbations of the trees (for instance, the ones discussed above, NNI, SPR, TRD), different evolutionary models (constant and non-constant rates, different rates for different mutation types), variations on initial location in tree space, and many others.

In some simple cases, Bayesian techniques have been reported to be more accurate (the output tree topology displays evolutionary relationships closer to reality) than parsimony or distance based methods, especially when analyzing highly divergent taxa [256].

C.3 Distance Based Methods

Distance based methods reduce the complexity of the inference tree dramatically by considering only the distances between the sequences S; the problem is then to reconstruct a weighted tree. As not all the information regarding particular positions is used, there is no attempt to reconstruct sequences attached to internal nodes (ancestral states).

Several distance functions can be constructed from a set of sequences. The simplest one is the Hamming distance d_H, which is the fraction of bases that differ between two sequences. The Hamming distance considers all substitutions equally likely and it does not consider the probability that for long times there could be mutations in already mutated positions. A natural way to assign a distance that takes into account the possibility of back mutations is estimating μt by the fraction of bases changed after a time t if the mutation rate is μ, which we can compute by inverting $p(l) = \frac{3}{4}(1 - e^{-\frac{4}{3}\mu l})$. This defines the Jukes-Cantor distance:

$$d_{JC} = -\frac{3}{4} \log(1 - \frac{4}{3} d_H).$$

The Jukes-Cantor distance is just a transformation of the unit interval into the positive numbers. When $d_H \to 0$, $d_{JC} \sim d_H$, i.e. when they are near zero both distances are similar. But when two random sequences with four bases are aligned and there is an equal number of the four bases, only one quarter of the bases will be the same. Then $d_H \to 3/4$ and $d_{JC} \to \infty$. More complicated models incorporate different rates of transitions and transversions, and different frequencies of nucleotides. For instance the K80 model [298] considers that all bases are equally frequent but that there are different rates for transitions and transversions. If p is the fraction of transitions (like the Hamming distance but only counting transitions) and q the fraction of transversions, the K80 distance is defined as:

$$d_{K80} = -\frac{1}{2}\log(1 - 2p - q) - \frac{1}{4}\log(1 - 2q).$$

If the frequency of the four nucleotides is different from 25%, Jukes-Cantor can be modified to the Tajima-Nei distance:

$$d_{TN84} = -\beta \log(1 - \frac{d_H}{\beta})$$

where $\beta = \sum_i f_i^2$, and f_i is the frequency of the nucleotide i. Further generalizations include different rates for each mutation and different frequencies per nucleotide.

Now assume the whole data is reduced to a distance metric. How can we infer a weighted tree from this metric? One of the oldest methods is the least squares method, proposed in 1967 by Fitch and Margoliash [177]. The basic idea is to find the weighted tree that minimizes the sum of the squares of differences between the distances between two sequences and the sequence in the tree (sum of branches connecting the two leaves d_{ij}^T, also called patristic distance):

$$s = \sum_{i,j}(d_{ij} - d_{ij}^T)^2.$$

The method requires exploration of all topologies; unsurprisingly, the method is NP-complete [135].

Agglomerative or clustering methods are usually much more convenient and faster, generating a solution in polynomial time. The main idea of these methods is to start from the pair of closest sequences that are linked. Then eliminate the columns and rows from these sequences and introduce a new one where distances are computed using a particular rule. Now, the distance matrix contains one fewer column and row. By iterations, one quickly arrives at a single element. The most popular algorithm of this type is neighbor-joining (NJ) [442], which proceeds as follows.

1. First calculate the matrix $T_{ij} = (m - 2)d_{ij} - \sum_k d_{ik} - \sum_k d_{jk}$, where m is the number of sequences.
2. Find the leaves with lowest T_{ij}, i and j.
3. Define a new leaf k, and join i and j with k.
4. Compute distances to the new node from the leaves being joined:

$$d_{ik} = \frac{1}{2}d_{ij} + \frac{1}{2(m-2)}\left(\sum_s (d_{is} - d_{js})\right).$$

5. Compute distances to the other leaves from the new node k,

$$d_{ks} = \frac{(d_{is} + d_{js} - d_{ij})}{2}$$

6. Replace the joined neighbors with a new node, using the recomputed distance. And restart the algorithm.

The NJ algorithm generates a tree in polynomial time, and generates the right tree if the distance matrix satisfies the four point condition. However, in more general cases, it could lead to strange results such as negative branch lengths.

C.4 Phylogenetic Networks

As we have seen in Chapter 5, phylogenetic trees fail to capture reticulate events including recombinations and reassortments in viruses, horizontal gene transfer in bacteria, and meiotic recombination and species hybridization in eukaryotes. Phylogenetic networks aim to represent these events as a generalization of a tree with external nodes representing the observed data and a graph, with cycles representing incompatibilities. Like phylogenetic trees, phylogenetic networks can be constructed from sequences or distances. We will briefly mention a few methods that we have discussed in this book.

A common approach to capture reticulate events is to use split networks. Split networks represent incompatible splits, but the interpretation in terms of biological processes that could generate these splits is obscure. For example, it not easy to tell if an incompatible tree was generated by recombination, back mutations, or a horizontal gene transfer event. Nor can one determine how many events generated the incompatibility, whether just one reticulate event is enough to generate an incompatibility, how the number of incompatibilities scale with recombination rates, etc. The lack of interpretability of split representations constitutes a serious obstacle to a wider adoption of these representations for the biological community.

Clearer biological interpretations arise from a reticulate network, where each loop is supposed to represent a reticulate event (recombination, gene transfer, reassortment, etc.). Ancestral recombinant graphs (ARGs), for instance, introduce nodes that correspond to potential recombination.

C.4.1 Split Networks

Let X be a set. Then a split $S_1 \mid S_2$ is any partition of X into two non-empty sets:

- $S_1 \neq \emptyset$ and $S_2 \neq \emptyset$,
- $S_1 \cup S_2 = S$,
- $S_1 \cap S_2 = \emptyset$.

A weighted set of splits (S, L) is a collection of splits $\{S_i\}$ together with a set of weights $\{l_i \geq 0\}$. In a tree, each edge provides a split: if we cut the edge, the data splits into two non-overlapping subsets. More interestingly, trees provide a set of splits $\{S_i\}$ that satisfy an extra condition – they are compatible. Two splits $S_i = X_i \mid Y_i$ and $S_j = X_j \mid Y_j$ are compatible if and only if one of the intersections $X_i \cap X_j$, $X_i \cap Y_j$, $Y_i \cap X_j$, or $Y_i \cap Y_j$ is empty. A set of splits S is compatible if all possible pairs are compatible.

Trees generate compatible sets of splits, and compatible sets of splits can be represented by trees. A weighted tree (T, L) is in this way equivalent to a weighted set of compatible splits (S, L). But splits that are not compatible generalize the notion of a tree. A representation of incompatible splits is through a split network. A split network represents a set (S, L) where each element of S labels a single node, and each edge is labeled by splits, in such a way that removing the edges corresponding to a split partitions the graph into two, where labeled nodes are split correspondingly. In a split network one can use one or more edges to represent a split, in such a way that the deletion of such edges generates the two elements of the split (see Figure C.4).

C.4.2 Sequence Based Methods

We will briefly mention two sequence based network techniques that we described in Section 5.10: median networks and ancestral recombinant graphs (ARGs).

Median networks. Median networks take as input a set of aligned sequences S, that we will assume have letters 0 and 1. First a simplification of S is performed by taking a condensed representation that discards positions that are the same in all sequences, and taking only one representative position for every set of positions that displays the same pattern. Each of the representatives is assigned a weight β_i corresponding to the number of positions that are in the same class. Let us call the set of representatives with weights the condensed representation, S'. The median operation takes any three binary sequences and computes another sequence that for each character takes the median (the most common character in that position). For

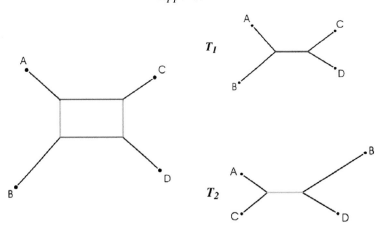

Figure C.4 A split network captures incompatible splits. Each taxon or sequence is associated to a vertex. Each edge on the network is labeled by a a split, in such a way that cutting the edges cuts the network into two, separating the labeled vertices corresponding to the split.

instance, the median of 0000, 1100, and 0111 is 0100. Adding the median sequence to the set and iterating the procedure until no new sequences are generated gives the median closure. A network, the median network or Buneman graph, can be constructed by taking as many nodes as binary sequences in the condensed representation S' and edges connecting them if they differ by only one character. The median network has nice properties as it is made of cubes of different dimensions and it contains all trees with minimal parsimony scores. But typically the number of nodes generated by the median operation is extremely large, and the biological interpretation is extremely obscure. The reduced median (RM) network algorithm and median-joining algorithm [31, 33] selects a subnetwork in the median network, reducing significantly the complexity of the network. Generalizations to more than two states sequences are called quasi-median networks.

Ancestral recombinant graphs (ARGs). Ancestral recombinant graphs constitute the most interpretable of all phylogenetic networks. An ARG provides a potential reconstruction of the history that gave rise to the data S through a series of mutations and recombinations. For a full explanation of ARGs and extensions using topological data analysis we refer the reader to Section 5.10.

C.4.3 Distance Based Methods

In distance based methods, we compute a distance between a set of sequences S, and we work exclusively with the distance matrix. These methods, as with phylogenetic trees, are fast and easily implementable, with the caveat that the interpretation of cycles in the network is obscure.

Split decomposition. In split decomposition, one takes advantage of the unique decomposition of a finite metric space into a set of splits using the Bandelt and Dress theorem [30]. The main idea of the Bandelt-Dress decomposition is that a finite metric space can be decomposed into a sum of independent metrics associated to weighted splits plus a remnant. The weight of each split $S_1|S_2$, called the isolation index, can be computed as follows:

$$\alpha_S = \frac{1}{2} \min_{\substack{i_1,j_1 \in S_1 \\ i_2,j_2 \in S_2}} (\max(d_{i_1,j_1} + d_{i_2,j_2}, d_{i_1,i_2} + d_{j_1,j_2}, d_{i_1,j_2} + d_{i_2,j_1}) - d_{i_1,j_1} - d_{i_2,j_2}).$$

For every split S one can define a split metric d_S to be 0 if two elements are in the same split and 1 if not. The Bandelt-Dress result decomposes the original metric:

$$d = \sum_S \alpha_S d_S + r.$$

The simplest example of a residue metric corresponds to 5 points with distances derived from a complete bipartite graph $K_{2,3}$. For s sequences, the approach provides at most $\binom{s}{2}$ non-zero weight splits. Remember that a tree is a set of compatible $2s - 3$ splits: if the set (S, L) is compatible then there is a single tree with edges corresponding to compatible splits and edge weights corresponding to split weights. In particular, finite metric spaces that satisfy the four point condition correspond to totally decomposable metrics and compatible splits, corresponding to the underlying tree. Focusing on the non-remnant part, this construction allows finite metric spaces to be mapped to weighted splits that can be represented by a split network. But in the case of non-compatible splits the split decomposition generalizes this result.

Neighbor-net (NN). Developed in 2004 by David Bryant and Vincent Moulton [75], this is an agglomerative distance based method that generalizes the neighbor-joining algorithm we discussed before. Given a finite metric space, NN constructs a collection of weighted splits and then represents the results using a split graph. Like NJ, this is an agglomerative method that starts by selecting pairs of nodes, but instead of replacing them immediately by a new node, it waits until it is paired a second time. Then the three linked nodes become two nodes and the distance matrix is reduced (see Figure C.5). The procedure continues until the number of nodes is reduced to two or three. Being an agglomerative distance based method, the speed and throughput is very high, similar to NJ. Like other split networks, the main problem is the interpretability of the results. A nice mathematical description of the NN algorithm in terms of some discrete metric spaces, called circular decomposable metrics, can be found in [329].

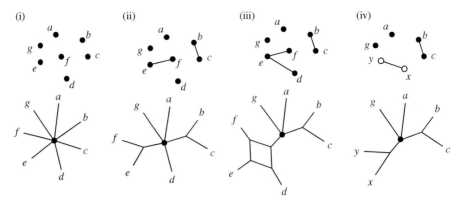

Figure C.5 This figure illustrates how the neighbor-net algorithm works. It is an agglomerative method that takes distance matrices as input. In (i) each node represents a single sequence. One first looks for closest neighbors; in (ii) the closest to e is f and the closest to b is c. Other neighbors are identified in (iii); e has as neighbors f and d. Two incompatible splits $ef|abcdg$ and $de|acdfg$ are represented and d, e, and f are substituted by new nodes x and y. Source: [75]. Bryant, David, and Vincent Moulton. "Neighbor-net: an agglomerative method for the construction of phylogenetic networks." Molecular Biology and Evolution, 2004, 21.2: 255–265, by permission of Oxford University Press.

C.5 Suggestions for Further Reading

There are excellent books and reviews on phylogenetics for the reader who wants to dive into this topic.

- *Inferring Phylogenies*, by J. Felsenstein [174] is a complete and clear exposition on different approaches to phylogenetic trees which is highly recommended.
- *Molecular Evolution and Phylogenetics*, by Masatoshi Nei and Sudhir Kumar [380] is a nice didactical overview on phylogenetic methods.
- *The Phylogenetic Handbook: A Practical Approach to Phylogenetic Analyses and Hypothesis Testing*, by Philippe Lemey, Marco Salemi and A. M. Vandamme [443].
- *ReCombinatorics: The Algorithmics of Ancestral Recombination Graphs and Explicit Phylogenetic Networks*, by Dan Gusfield [220] is highly recommended to learn more about biologically interpretable phylogenetic networks.
- *Phylogenetic Networks*, by Daniel Huson, Regula Rupp and Celine Scornavacca [262] is a nice clear survey on methods for inference of phylogenetic networks.
- *Basic Phylogenetic Combinatorics*, by Andreas Dress and colleagues [151] is a nice mathematical introduction to the relationship between finite metric spaces, split systems, and systems of quartets.

C.6 Data and Software

There is a large number of software programs for phylogenetic inference. Here is a very incomplete list of some of the most commonly used.

- PHYLIP: phylogenetic tree inference package by Felsenstein incorporating now classical methods, such as maximum parsimony, distance based algorithms, and maximum likelihood. It can be found at `http://evolution.genetics.washington.edu/phylip/general.html`.
- PhyML [217] and RaxML [482]: two of the most commonly used likelihood algorithms.
- Bayesian Evolutionary Analysis Sampling Trees (BEAST) [153]: a very commonly used Bayesian method for tree inference and parameter estimation.
- MrBayes: Bayesian posterior probability estimation for phylogenetic trees [436].
- SplitsTree [260]: a very wide platform that provides a wide range of phylogenetic tree and network inference methods, including median networks, parsimony splits, spectral analysis, split decomposition, and neighbor-net.
- Dendroscope [264]: provides a platform for visualizing trees and networks.

References

[1] Eddie Aamari et al. "Estimating the reach of a manifold." 2017. arXiv: 1705.04565.

[2] Francesco Abate et al. "Distinct viral and mutational spectrum of endemic Burkitt lymphoma." *PLoS Pathog.* 11.10 (2015), e1005158.

[3] Henry Adams and Joshua Mirth. Metric thickenings of euclidean submanifolds. 2017. arXiv: 1709.02492 [math.AT].

[4] Henry Adams et al. "Persistence images: a stable vector representation of persistent homology." *J. Mach. Learn. Res.* 18.1 (2017), 218–252.

[5] Aaron Adcock, Erik Carlsson, and Gunnar Carlsson. "The ring of algebraic functions on persistence bar codes." *Homology Homotopy Appl.* 18.1 (2016), 381–402.

[6] Sophie H. Adjalley et al. "Genome-wide transcriptome profiling reveals functional networks involving the *Plasmodium falciparum* drug resistance transporters PfCRT and PfMDR1." *BMC Genomics* 16.1 (2015), 1090.

[7] R. J. Adler, O. Bobrowski, and S. Weinberger. "Crackle: the homology of noise." *Discrete Comput. Geom.* 52.4 (2014), 680–704.

[8] R. J. Adler and J. E. Taylor. *Random Fields and Geometry*. Springer Monographs in Mathematics. Springer Verlag, 2007.

[9] R. Adler et al. "Persistent homology for random fields and complexes." 2010. arXiv: 1003.1001 [math.PR].

[10] Sebastian E. Ahnert et al. "Unbiased pattern detection in microarray data series." *Bioinformatics* 22.12 (2006), 1471–1476.

[11] A. D. Alexandrov. "Über eine Verallgemeinerung der Riemannschen Geometrie." *Schr. Orschungsinst. Math. Berlin* 1 (1957), 33–84.

[12] Ludmil B. Alexandrov et al. "Signatures of mutational processes in human cancer." *Nature* 500.7463 (2013), 415–421.

[13] El-ad David Amir et al. "viSNE enables visualization of high dimensional single-cell data and reveals phenotypic heterogeneity of leukemia." *Nature Biotechnol.* 31.6 (2013), 545–552.

[14] David Anick. "The computation of rational homotopy groups is #P-hard." *Lecture Notes in Pure and Applied Mathematics*, volume 114, pp. 1–56. Springer, 1989.

[15] Luis Aparicio et al. "Quasi-universality in single-cell sequencing data." 2018. biorXiv: www.biorxiv.org/content/early/2018/10/05/426239.

[16] Peter Armitage and Richard Doll. "The age distribution of cancer and a multi-stage theory of carcinogenesis." *Br. J. Cancer* 8.1 (1954), 1.

[17] Luis Arnes et al. "Comprehensive characterisation of compartment-specific long non-coding RNAs associated with pancreatic ductal adenocarcinoma." *Gut* (2018). doi: 10.1136/gutjnl-2017-314353.

[18] Norman Arnheim, Peter Calabrese, and Irene Tiemann-Boege. "Mammalian meiotic recombination hot spots." *Annu. Rev. Genet.* 41 (2007), 369–399.

[19] Javier Arsuaga et al. "Identification of copy number aberrations in breast cancer subtypes using persistence topology." *Microarrays* 4.3 (2015), 339–369.

[20] Javier Arsuaga et al. "Knotting probability of DNA molecules confined in restricted volumes: DNA knotting in phage capsids." *Proc. Natl. Acad. Sci. USA* 99.8 (2002), 5373–5377.

[21] Javier Arsuaga et al. "Topological analysis of gene expression arrays identifies high risk molecular subtypes in breast cancer." *Applic. Alg. Eng. Commun. Comput.* 23.1–2 (2012), pp. 3–15.

[22] Michael Artin. *Algebra.* Prentice Hall, Englewood Cliffs, NJ, 1991.

[23] Kevin Atteson. "The performance of neighbor-joining algorithms of phylogeny reconstruction." *Computing and Combinatorics: Third Annual International Conference, COCOON '97 Shanghai, China, August 20–22, 1997*, pp. 101–110. Ed. by Tao Jiang and D. T. Lee. Springer, Berlin, 1997.

[24] Sheldon Jay Axler. *Linear Algebra Done Right.* Third edition. Undergraduate Texts in Mathematics. Springer International Publishing, 2015.

[25] Ferhat Ay and William S Noble. "Analysis methods for studying the 3D architecture of the genome." *Genome Biol.* 16.1 (2015), 1306.

[26] Gayane Badalian-Very et al. "Recurrent BRAF mutations in Langerhans cell histiocytosis." *Blood* 116.11 (2010), 1919–1923.

[27] David J. Balding, Richard A. Nichols, and David M. Hunt. "Detecting gene conversion: primate visual pigment genes." *Proc. R. Soc. London, Ser. B* 249.1326 (1992), 275–280.

[28] David Baltimore. "Expression of animal virus genomes." *Bacteriol. Rev.* 35.3 (1971), 235.

[29] David Baltimore. "RNA-dependent DNA polymerase in virions of RNA tumour viruses." *A Century of Nature: Twenty-One Discoveries that Changed Science and the World*, p. 173. Ed. by Laura Garwin and Tim Lincoln. University of Chicago Press, Chicago, IL, 2003.

[30] Hans-Jürgen Bandelt and Andreas W. M. Dress. "Split decomposition: a new and useful approach to phylogenetic analysis of distance data." *Mol. Phylogenet. Evol.* 1.3 (1992), 242–252.

[31] Hans-Jürgen Bandelt, Peter Forster, and Arne Röhl. "Median-joining networks for inferring intraspecific phylogenies." *Mol. Biol. Evol.* 16.1 (1999), 37–48.

[32] Hans-Jürgen Bandelt, Vincent Macaulay, and Martin Richards. "Median networks: speedy construction and greedy reduction, one simulation, and two case studies from human mtDNA." *Mol. Phylogenet. Evol.* 16.1 (2000), 8–28.

[33] Hans-J. Bandelt et al. "Mitochondrial portraits of human populations using median networks." *Genetics* 141.2 (1995), 743–753.

[34] Ziv Bar-Joseph. "Analyzing time series gene expression data." *Bioinformatics* 20.16 (2004), 2493–2503.

[35] Ziv Bar-Joseph, Anthony Gitter, and Itamar Simon. "Studying and modelling dynamic biological processes using time-series gene expression data." *Nature Rev. Genet.* 13.8 (2012), 552–564.

[36] F. Barre-Sinoussi et al. "Isolation of a T-lymphotropic retrovirus from a patient at risk for acquired immune deficiency syndrome (AIDS)." *Science* 220.45995 (1983), 868–871.

[37] Frédéric Baudat, Yukiko Imai, and Bernard de Massy. "Meiotic recombination in mammals: localization and regulation." *Nature Rev. Genet.* 14.11 (2013), 794–806.

[38] F. Baudat et al. "PRDM9 is a major determinant of meiotic recombination hotspots in humans and mice." *Science* 327.5967 (2010), 836–840.

[39] Ulrich Bauer. Ripser. 2016. https://github.com/Ripser/ripser.

[40] Ulrich Bauer, Michael Kerber, and Jan Reininghaus. DIPHA. 2014. https://github.com/DIPHA/dipha.

[41] Ulrich Bauer, Michael Kerber, and Jan Reininghaus. "Distributed computation of persistent homology." *Proceedings of the Sixteenth Workshop on Algorithm Engineering and Experiments, ALENEX 2014, Portland, OR, USA, January 5, 2014*, pp. 31–38. doi.org/10.1137/1.9781611973198.4.

[42] Ulrich Bauer and Michael Lesnick. "Induced matchings of barcodes and the algebraic stability of persistence." *Proceedings of the Thirtieth Annual Symposium on Computational Geometry. SOCG'14, Kyoto, Japan*, pp. 355:355–355:364. ACM, 2014. doi: 10.1145/2582112.2582168.

[43] Ulrich Bauer et al. PHAT: Persistent homology algorithms toolbox. 2013. https://bitbucket.org/phat-code/phat.

[44] Ulrich Bauer et al. "Phat – Persistent Homology Algorithms Toolbox." *J. Symbol. Comput.* 78.Supplement C (2017), 76–90.

[45] Mikhail Belkin and Partha Niyogi. "Laplacian eigenmaps for dimensionality reduction and data representation." *Neural Comput.* 15.6 (2003), 1373–1396. doi: 10.1162/089976603321780317.

[46] Mikhail Belkin and Partha Niyogi. "Towards a theoretical foundation for Laplacian-based manifold methods." *J. Comput. Syst. Sci.* 74.8 (2008), 1289–1308. doi: 10.1016/j.jcss.2007.08.006.

[47] Sean C. Bendall et al. "Single-cell trajectory detection uncovers progression and regulatory coordination in human B cell development." *Cell* 157.3 (2014), 714–725.

[48] Paul Bendich, Bei Wang, and Sayan Mukherjee. "Local homology transfer and stratification learning." *Proceedings of the Twenty-Third Annual ACM-SIAM Symposium on Discrete Algorithms*, pp. 1355–1370.

[49] P. Bendich et al. "Topological and statistical behavior classifiers for tracking applications." *IEEE Trans. Aero. Electron. Syst.* 52.6 (2016), 2644–2661.

[50] Yoshua Bengio, Aaron Courville, and Pascal Vincent. "Representation learning: a review and new perspectives." *IEEE Trans. Pattern Anal. Mach. Intell.* 35.8 (2013), 1798–1828. doi: 10.1109/TPAMI.2013.50.

[51] Mira Bernstein et al. *Graph Approximations to Geodesics on Embedded Manifolds.* Stanford University technical report. 2000.

[52] Rameen Beroukhim et al. "Assessing the significance of chromosomal aberrations in cancer: methodology and application to glioma." *Proc. Natl. Acad. Sci. USA* 104.50 (2007), 20007–20012.

[53] Eva Bianconi et al. "An estimation of the number of cells in the human body." *Ann. Human Biol.* 40.6 (2013), 463–471.

[54] Peter J. Bickel and David A. Freedman. "Some asymptotic theory for the bootstrap." *Ann. Stat.* 9.6 (1981), 1196–1217. doi: 10.1214/aos/1176345637.

[55] Louis J. Billera, Susan P. Holmes, and Karen Vogtmann. "Geometry of the space of phylogenetic trees." *Adv. Appl. Math.* 27.4 (2001), 733–767. doi: 10.1006/aama.2001.0759.

[56] P. Billingsley. *Convergence of Probability Measures.* Wiley, New York, 1968.

[57] Patrick Billingsley. *Probability and Measure.* Third edition. John Wiley and Sons, 1995.

[58] Andrew J. Blumberg and Michael Lesnick. "Universality of the homotopy interleaving distance." 2017. arXiv: 1705.01690 [math.AT].

[59] Andrew J. Blumberg and Michael A. Mandell. "Quantiative homotopy theory in topological data analysis." *Found. Comput. Math.* 13.6 (2013), 885–911.

[60] Andrew J. Blumberg et al. "Robust statistics, hypothesis testing, and confidence intervals for persistent homology on metric measure spaces." *Found. Comput. Math.* 14.4 (2014), 745–789.

[61] O. Bobrowski and M. Kahle. "Topology of random geometric complexes: a survey." *Topology in Statistical Inference.* Proceedings of Symposia in Applied Mathematics. 2014.

[62] Omer Bobrowski and Sayan Mukherjee. "The topology of probability distributions on manifolds." *Prob. Theor. Relat. Fields* 161.3 (2015), 651–686.

[63] O. Bobrowski, M. Kahle, and P. Skraba. "Maximally persistent cycles in random geometric complexes." *Ann. Appl. Prob.* (2017). arXiv: 1509.04347 [math.PR].

[64] B. Bollobas. *Modern Graph Theory.* Graduate Texts in Mathematics, volume 184. Springer Verlag, 1998.

[65] Magnus Bordewich and Charles Semple. "Computing the minimum number of hybridization events for a consistent evolutionary history." *Discrete Appl. Math.* 155.8 (2007), 914–928.

[66] Magnus Bordewich and Charles Semple. "On the computational complexity of the rooted subtree prune and regraft distance." *Ann. Combinatorics* 8.4 (2005), 409–423.

[67] Magnus Bakke Botnan and Michael Lesnick. "Algebraic stability of zigzag persistence modules." 2016. arxiv.org/abs/1604.00655.

[68] Theodor (Biologe) Boveri and Henry Harris. *Concerning the Origin of Malignant Tumours.* Cold Spring Harbor Laboratory Press, 2008.

[69] Matthew Brand. "Charting a manifold." *Adv. Neur. Inf. Process. Syst.* 15 (2002), 961–968.

[70] J. R. Bray and J. T. Curtis. "An ordination of the upland forest communities of Southern Wisconsin." *Ecol. Monogr.* 27 (1957), 325–349.

[71] Cameron W. Brennan et al. "The somatic genomic landscape of glioblastoma." *Cell* 155.2 (2013), 462–477.

[72] Bryn A. Bridges. "Microbial genetics: hypermutation under stress." *Nature* 387.6633 (1997), 557–558.

[73] M. R. Bridson and A. Häfliger. *Metric Spaces of Non-Positive Curvature.* Grundlehren der Mathematischen Wissenschaften. Springer, Berlin, 2011.

[74] Peter J. Brockwell and Richard A. Davis. *Time Series: Theory and Methods.* Springer Science & Business Media, 2013.

[75] David Bryant and Vincent Moulton. "Neighbor-net: an agglomerative method for the construction of phylogenetic networks." *Mol. Biol. Evol.* 21.2 (2004), 255–265.

[76] Peter Bubenik. "Statistical topological data analysis using persistence landscapes." *J. Mach. Learn. Res.* 16.1 (2015), 77–102.

[77] Peter Bubenik and Pawel Dlotko. A persistence landscapes toolbox for topological statistics. 2015. www.math.upenn.edu/~dlotko/persistenceLandscape.html.

[78] Peter Bubenik and Pawel Dlotko. "A persistence landscapes toolbox for topological statistics." *J. Symb. Comput.* 78 (2017), 91–114. doi: 10.1016/j.jsc.2016.03.009.

[79] Mickaël Buchet et al. "Efficient and robust persistent homology for measures." *Comput. Geom.* 58 (2016), 70–96.

[80] Dorothy Buck. "DNA topology." Applications of Knot Theory. *Proc. Symp. Appl. Math.,* 66 (2009), 47–79.

[81] O. Peter Buneman. "The recovery of trees from measures of dissimilarity." *Math. Archaeol. Hist. Sci.* (1971), 387–395.

[82] Peter Buneman. "A note on the metric properties of trees." *J. Combinatorics Theor., Ser. B* 17.1 (1974), 48–50.

[83] Dmitri Burago, Yuri Burago, and Sergei Ivanov. *A Course in Metric Geometry.* Graduate Studies in Mathematics, volume 33. American Mathematical Society, Providence, RI, 2001.

[84] Denis Burkitt. "A sarcoma involving the jaws in African children." *Hematology: Landmark Papers of the Twentieth Century* 1.8 (2000), 353.

[85] Martin Cadek et al. "Polynomial-time computation of homotopy groups and Postnikov systems in fixed dimension." *SIAM J. Comput.* 43.5 (2014), 1728–1780. doi: 10.1137/120899029.

[86] John Cairns, Julie Overbaugh, and Stephan Miller. "The origin of mutants." *Nature* 335.6186 (1988), 142–145.

[87] Pablo G. Cámara. "Topological methods for genomics: present and future directions." *Curr. Opin. Syst. Biol.* 1 (2017), 95–101.

[88] Pablo G. Cámara, Arnold J. Levine, and Raúl Rabadán. "Inference of ancestral recombination graphs through topological data analysis." *PLoS Comput. Biol.* 12.8 (2016), e1005071.

[89] Emmanuel J. Candès et al. "Robust principal component analysis?" *J. ACM* 58.3 (2011), 11:1–11:37. doi: 10.1145/1970392.1970395.

[90] Gunnar Carlsson. "Topological pattern recognition for point cloud data." *Acta Num.* 23 (2014), 289–368.

[91] Gunnar E. Carlsson, Vin de Silva, and Dmitriy Morozov. "Zigzag persistent homology and realvalued functions." *Proceedings of the 25th ACM Symposium on Computational Geometry, Aarhus, Denmark, June 8-10, 2009*, pp. 247–256.

[92] Gunnar E. Carlsson and Afra Zomorodian. "The theory of multidimensional persistence." *Discrete Comput. Geom.* 42.1 (2009), pp. 71–93. doi: 10.1007/s00454-009-9176-0.

[93] Gunnar Carlsson and Vin De Silva. "Zigzag persistence." *Found. Comput. Math.* 10.4 (2010), 367–405.

[94] Gunnar Carlsson and Sara Kališnik Verovšek. "Symmetric and r-symmetric tropical polynomials and rational functions." *J. Pure Appl. Algebra* 220.11 (2016), 3610–3627. doi: 10.1016/j.jpaa.2016.05.002.

[95] Gunnar Carlsson et al. "On the local behavior of spaces of natural images." *Int. J. Comput. Vision* 76.1 (2008), 1–12.

[96] Manfredo Perdigão do Carmo. *Riemannian Geometry. Mathematics: Theory & Applications.* Translated from the second Portuguese edition by Francis Flaherty. Birkhauser Boston, MA, 1992. doi: 10.1007/978-1-4757-2201-7.

[97] Mathieu Carrière, Steve Y. Oudot, and Maks Ovsjanikov. "Stable topological signatures for points on 3D shapes." *Proceedings of the Eurographics Symposium on Geometry Processing. SGP '15. Graz, Austria*, pp. 1–12. Eurographics Association, 2015. doi: 10.1111/cgf.12692.

[98] Ben Cassidy, Caroline Rae, and Victor Solo. "Brain activity: conditional dissimilarity and persistent homology." *Biomedical Imaging (ISBI), Proceedongs of the IEEE 12th International Symposium on Biomedical Imaging*, pp. 1356–1359. IEEE, 2015

[99] Andrea Cerri et al. "Betti numbers in multidimensional persistent homology are stable functions." *Math. Methods Appl. Sci.* 36.12 (2013), 1543–1557.

[100] Joseph Minhow Chan, Gunnar Carlsson, and Raúl Rabadán. "Topology of viral evolution." *Proc. Natl. Acad. Sci. USA* 110.46 (2013), 18566–18571.

[101] Yuan Chang et al. "Identification of herpesvirus-like DNA sequences in AIDS-associated Kaposi's sarcoma." *Science* 266.5192 (1994), 1865.

[102] Chris Chatfield. *The Analysis of Time Series: An Introduction.* CRC Press, 2013.

[103] Frédéric Chazal. "Higher-dimensional topological data analysis." To appear in *Handbook of Discrete and Computational Geometry.* 2017.

[104] Frédéric Chazal, David Cohen-Steiner, and Quentin Mérigot. "Geometric inference for probability measures." *Found. Comput. Math.* 11.6 (2011), 733–751. doi: 10.1007/s10208-011-9098-0.

[105] Frédéric Chazal, Vin de Silva, and Steve Oudot. "Persistence stability for geometric complexes." *Geometriae Dedicata* 173.1 (2014), 193–214.

[106] Frédéric Chazal and Jian Sun. "Gromov-Hausdorff approximation of filament structure using Reeb-type graph." *Proceedings of the Thirtieth Annual Symposium on Computational Geometry, SOCG'14. Kyoto, Japan, June 08–11, 2014*, p. 491. ACM, 2014. doi: 10.1145/2582112.2582129.

[107] Frédéric Chazal et al. "Proximity of persistence modules and their diagrams." *Proceedings of the Twenty-fifth Annual Symposium on Computational Geometry, SOCG '09. Aarhus, Denmark*, pp. 237–246. ACM, 2009. doi: 10.1145/1542362.1542407.

[108] Frédéric Chazal et al. "Robust topological inference: distance to a measure and kernel distance." *J. Mach. Learn. Res.* 18.159 (2018), 1–40. arXiv: 1412.7197 [math.ST].

[109] Frédéric Chazal et al. "Stochastic convergence of persistence landscapes and silhouettes." *Proceedings of the Thirtieth Annual Symposium on Computational Geometry, SOCG'14. Kyoto, Japan*, pp. 474:474–474:483. ACM, 2014. doi: 10.1145/2582112.2582128.

[110] Frédéric Chazal et al. "Subsampling methods for persistent homology." 2014. arXiv: 1406.1901 [math.AT].

[111] Frédéric Chazal et al. *The Structure and Stability of Persistence Modules.* Springer Briefs in Mathematics. Springer, 2016. doi: 10.1007/978-3-319-42545-0.

[112] Andrew X. Chen and Raúl Rabadán. "A fast semi-automatic segmentation tool for processing brain tumor images." *Towards Integrative Machine Learning and Knowledge Extraction*, pp. 170–181. Lecture Notes in Computer Science, volume 10344. Springer, 2017.

[113] Chao Chen and Daniel Freedman. "Hardness results for homology localization." *Discrete Comput. Geom.* 45.3 (2011), 425–448.

[114] Gary K. Chen, Paul Marjoram, and Jeffrey D. Wall. "Fast and flexible simulation of DNA sequence data." *Genome Res.* 19.1 (2009), 136–142.

[115] Benny Chor and Tamir Tuller. "Maximum likelihood of evolutionary trees: hardness and approximation." *Bioinformatics* 21.suppl 1 (2005), i97–i106.

[116] Fan R. K. Chung. *Spectral Graph Theory.* American Mathematical Society, 1997.

[117] David Cohen-Steiner, Herbert Edelsbrunner, and John Harer. "Stability of persistence diagrams." *Discrete Comput. Geom.* 37.1 (2007), 103–120.

[118] David Cohen-Steiner et al. "Lipschitz functions have L p-stable persistence." *Found. Comput. Math.* 10.2 (2010), 127–139.

[119] Ronald R. Coifman and Mauro Maggioni. "Diffusion wavelets." *Appl. Comput. Harmonic Anal.* 21.1 (2006), 53–94. doi: 10.1016.

[120] Rodney Colina et al. "Evidence of intratypic recombination in natural populations of hepatitis C virus." *J. Gen. Virol.* 85.1 (2004), 31–37.

[121] Michael Collins and Nigel Du_y. "Convolution kernels for natural language." *Proceedings of the 14th International Conference on Neural Information Processing Systems: Natural and Synthetic. NIPS'01. Vancouver, British Columbia, Canada*, pp. 625–632. MIT Press, 2001.

[122] 1000 Genomes Project Consortium et al. "An integrated map of genetic variation from 1,092 human genomes." *Nature* 491.7422 (2012), 56–65.

[123] ENCODE Project Consortium et al. "An integrated encyclopedia of DNA elements in the human genome." *Nature* 489.7414 (2012), 57–74.

[124] J. E. Cooper and Edward J. Feil. "The phylogeny of *Staphylococcus aureus* – which genes make the best intra-species markers?" *Microbiology* 152.5 (2006), 1297–1305.

[125] Thomas H. Cormen et al. *Introduction to Algorithms*. Third edition. MIT Press, 2009.

[126] Nicholas R. Cozzarelli and James C. Wang. *DNA Topology and its Biological Effects*. Cold Spring Harbor Laboratory Press, 1990.

[127] Lorin Crawford et al. "Functional data analysis using a topological summary statistic: the smooth Euler characteristic transform." 2016. arXiv: 1611.06818.

[128] Lorin Crawford et al. "Topological summaries of tumor images improve prediction of disease free survival in glioblastoma multiforme." 2016. arXiv: 1611.06818.

[129] José M. Cuevas et al. "Extremely high mutation rate of HIV-1 in vivo." *PLoS Biol.* 13.9 (2015), e1002251.

[130] Carina Curto. "What can topology tell us about the neural code?" *Bull. Am. Math. Soc.* 54.1 (2017), 63–78.

[131] Riccardo Dalla-Favera et al. "Human c-myc oncogene is located on the region of chromosome 8 that is translocated in Burkitt lymphoma cells." *Proc. Natl. Acad. Sci. USA* 79.24 (1982), 7824–7827.

[132] Robert Dallmann et al. "The human circadian metabolome." *Proc. Natl. Acad. Sci. USA* 109.7 (2012), 2625–2629.

[133] Charles Darwin. *On the Origin of Species*. Murray, London, 1859.

[134] Julian Davies and Dorothy Davies. "Origins and evolution of antibiotic resistance." *Microbiol. Mol. Biol. Rev.* 74.3 (2010), 417–433.

[135] William H. E. Day. "Computational complexity of inferring phylogenies from dissimilarity matrices." *Bull. Math. Biol.* 49.4 (1987), 461–467.

[136] Ulrik De Lichtenberg et al. "Comparison of computational methods for the identification of cell cycle-regulated genes." *Bioinformatics* 21.7 (2005), 1164–1171.

[137] Anastasia Deckard et al. "Design and analysis of large-scale biological rhythm studies: a comparison of algorithms for detecting periodic signals in biological data." *Bioinformatics* 29.24 (2013), 3174–3180.

[138] Job Dekker, Marc A. Marti-Renom, and Leonid A. Mirny. "Exploring the three-dimensional organization of genomes: interpreting chromatin interaction data." *Nature Rev. Genet.* 14 (2013), 390–403.

[139] Albert B. DeLeo et al. "Detection of a transformation-related antigen in chemically induced sarcomas and other transformed cells of the mouse." *Proc. Natl. Acad. Sci. USA* 76.5 (1979), 2420–2424.

[140] Mary-Lee Dequeant et al. "Comparison of pattern detection methods in microarray time series of the segmentation clock." *PLoS One* 3.8 (2008), e2856.

[141] Satyan L. Devadoss, Daoji Huang, and Dominic Spadacene. "Polyhedral covers of tree space." *SIAM J. Discrete Math.* 28.3 (2014), 1508–1514. doi: 10.1137/130947532.

[142] D. DeWoskin et al. "Applications of computational homology to the analysis of treatment response in breast cancer patients." *Topol. Appl.* 157.1 (2010), 157–164.

[143] Tamal K. Dey, Facundo Mémoli, and Yusu Wang. "Topological analysis of nerves, Reeb spaces, mappers, and multiscale mappers." *Proceedings of the 33rd International Symposium on Computational Geometry, SoCG 2017. July 4–7, 2017, Brisbane, Australia*, pp. 36:1–36:16. 2017.

[144] Persi Diaconis, Susan Holmes, and Mehrdad Shahshahani. "Sampling from a manifold." *Advances in Modern Statistical Theory and Applications: A Festschrift in honor of Morris L. Eaton*. Contemporary Mathematics, volume 453, pp. 102–125. Institute of Mathematical Statistics, 2013.

[145] Persi Diaconis and Bernd Sturmfels. "Algebraic algorithms for sampling from conditional distributions." *Ann. Stat.* 26.1 (1998), 363–397. doi: 10.1214/aos/1030563990.

[146] David L. Donoho and Carrie Grimes. "Hessian eigenmaps: locally linear embedding techniques for high-dimensional data." *Proc. Natl. Acad. Sci. USA* 100.10 (2003), 5591–5596.

[147] W. Ford Doolittle. "Phylogenetic classification and the universal tree." *Science* 284.5423 (1999), 2124–2128. doi: 10.1126/science.284.5423.2124.

[148] Boryana Doyle et al. "Chromatin loops as allosteric modulators of enhancer-promoter interactions." *PLoS Comput. Biol.* 10.10 (2014), e1003867.

[149] John W. Drake. "A constant rate of spontaneous mutation in DNA-based microbes." *Proc. Natl. Acad. Sci. USA* 88.16 (1991), 7160–7164.

[150] John W. Drake et al. "Rates of spontaneous mutation." *Genetics* 148.4 (1998), 1667–1686.

[151] Andreas Dress, Katharina T. Huber, and Jacobus Koolen. *Basic Phylogenetic Combinatorics*. Cambridge University Press, 2012.

[152] Brian J. Druker. "Perspectives on the development of imatinib and the future of cancer research." *Nature Med.* 15.10 (2009), 1149–1152.

[153] Alexei J. Drummond and Andrew Rambaut. "BEAST: Bayesian evolutionary analysis by sampling trees." *BMC Evol. Biol.* 7.1 (2007), 1.

[154] H. Edelsbrunner, D. Letscher, and A. Zomorodian. "Topological persistence and simplification." *Proceedings of the 41st Annual Symposium on Foundations of Computer Science, FOCS '00*, pp. 454–463. IEEE Computer Society, Washington, DC, 2000.

[155] Herbert Edelsbrunner and John Harer. *Computational Topology – an Introduction*. American Mathematical Society, 2010.

[156] Herbert Edelsbrunner and John Harer. "Persistent homology – a survey." *Survey on Discrete and Computational Geometry. Twenty Years Later*. Contemporary Mathematics, volume 453. Ed. by J. E. Goodman, J. Pach, and R. Pollback. American Mathematical Society, 2008.

[157] Herbert Edelsbrunner and Dmitriy Morozov. "Persistent homology." To appear in *Handbook of Discrete and Computational Geometry*. 2017.

[158] Editorial. "Microbiology by numbers." *Nature Rev. Microbiol.* 9.9 (2011), 628.

[159] Peter Eirew et al. "Dynamics of genomic clones in breast cancer patient xenografts at single-cell resolution." *Nature* 518.7539 (2014), 422–426.

[160] Gábor Elek. "Betti numbers are testable." *Fete of Combinatorics and Computer Science*, pp. 139–149. Ed. by Gyula O. H. Katona et al. Springer, Berlin, 2010,.

[161] Kevin J. Emmett and Raúl Rabadán. "Characterizing scales of genetic recombination and antibiotic resistance in pathogenic bacteria using topological data analysis." *Brain Informatics and Health*, pp. 540–551. Springer, 2014.

[162] Kevin Emmett and Raúl Rabadán. "Quantifying reticulation in phylogenetic complexes using homology." *Proceedings of the 9th EAI International Conference on Bio-inspired Information and Communications Technologies (formerly BIONETICS)*, pp. 193–196. ICST (Institute for Computer Sciences, Social-Informatics and Telecommunications Engineering), 2016.

[163] Kevin Emmett, Benjamin Schweinhart, and Raúl Rabadán. "Multiscale topology of chromatin folding." *Proceedings of the 9th EAI International Conference on Bio-inspired Information and Communications Technologies (formerly BIONETICS)*, pp. 177–180. ICST (Institute for Computer Sciences, Social-Informatics and Telecommunications Engineering), 2016.

[164] Kevin Emmett et al. "Parametric inference using persistence diagrams: a case study in population genetics." *ICML Workshop on Topological Methods in Machine Learning*. 2014. arXiv: 1406.4582.

[165] Michael Anthony Epstein, Bert G. Achong, and Yvonne M. Barr. "Virus particles in cultured lymphoblasts from Burkitt's lymphoma." *The Lancet* 283.7335 (1964), 702–703.

[166] Warren J. Ewens. *Mathematical Population Genetics 1: Theoretical Introduction*. Springer Science & Business Media, 2012.

[167] Gregory Ewing and Joachim Hermisson. "MSMS: a coalescent simulation program including recombination, demographic structure and selection at a single locus." *Bioinformatics* 26.16 (2010), 2064–2065.

[168] Jittat Fakcharoenphol, Satish Rao, and Kunal Talwar. "A tight bound on approximating arbitrary metrics by tree metrics." *Proceedings of the Thirty-Fifth Annual ACM Symposium on Theory of Computing. STOC '03. San Diego, CA, USA*, pp. 448–455. ACM, 2003. doi: 10.1145/780542.780608.

[169] Nuno R. Faria et al. "The early spread and epidemic ignition of HIV-1 in human populations." *Science* 346.6205 (2014), 56–61.

[170] Brittany T. Fasy et al. "Confidence sets for persistence diagrams." *Ann. Stat.* 42.6 (2014), 2301–2339. doi: 10.1214/14-AOS1252.

[171] Brittany T. Fasy et al. TDA: Statistical tools for topological data analysis. 2017. https://cran.r-project.org/web/packages/TDA/index.html.

[172] J. Felsenstein. *Inferring Phylogenies*. Sinauer Associates, Sunderland, MA, 2003.

[173] Joseph Felsenstein. "Evolutionary trees from DNA sequences: a maximum likelihood approach." *J. Mol. Evol.* 17.6 (1981), 368–376.

[174] Joseph Felsenstein. *Inferring Phylogenies*, volume 2. Sinauer Associates Sunderland, MA, 2004.

[175] Huichen Feng et al. "Clonal integration of a polyomavirus in human Merkel cell carcinoma." *Science* 319.5866 (2008), 1096–1100.

[176] Walter M. Fitch. "Toward defining the course of evolution: minimum change for a specific tree topology." *Systematic Biol.* 20.4 (1971), 406–416.

[177] Walter M. Fitch, Emanuel Margoliash et al. "Construction of phylogenetic trees." *Science* 155.3760 (1967), 279–284.

[178] Erica Flapan. *When Topology meets Chemistry: A Topological Look at Molecular Chirality*. Cambridge University Press, 2000.

[179] S. Jane Flint et al. *Principles of Virology*. Third edition. ASM Press, 2009.

[180] Leslie R. Foulds and Ronald L. Graham. "The Steiner problem in phylogeny is NP-complete." *Adv. Appl. Math.* 3.1 (1982), 43–49.

[181] Kyle R. Fowler et al. "Evolutionarily diverse determinants of meiotic DNA break and recombination landscapes across the genome." *Genome Res.* 24.10 (2014), 1650–1664.

[182] Christophe Fraser, William P. Hanage, and Brian G. Spratt. "Recombination and the nature of bacterial speciation." *Science* 315.5811 (2007), 476–480.

[183] Veronique Frattini et al. "The integrated landscape of driver genomic alterations in glioblastoma." *Nature Genet.* 45.10 (2013), 1141–1149.

[184] David A. Freedman. *Statistical Models: Theory and Practice.* Second edition. Cambridge University Press, 2009.

[185] A. E. Friedman-Kien et al. "Kaposi's sarcoma and *Pneumocystis pneumonia* among homosexual men – New York City and California." *MMWR. Morbidity and Mortality Weekly Report* 30.25 (1981), 305–308.

[186] S. S. Froland et al. "HIV-1 infection in a Norwegian family before 1970." *The Lancet* 331.8598 (1988), 1344–1345.

[187] P. Frosini. "Discrete computation of size functions." *J. Combinatorics Inf. Syst. Sci.* 17.3–4 (1992), 232–250.

[188] André Fujita et al. "Modeling gene expression regulatory networks with the sparse vector autoregressive model." *BMC Syst. Biol.* 1.1 (2007), 39.

[189] André Fujita et al. "Time-varying modeling of gene expression regulatory networks using the wavelet dynamic vector autoregressive method." *Bioinformatics* 23.13 (2007), 1623–1630.

[190] Robert C. Gallo et al. "Frequent detection and isolation of cytopathic retroviruses (HTLV-III) from patients with AIDS and at risk for AIDS." *Science* 224.4648 (1984), 500–503.

[191] Rongbao Gao et al. "Human infection with a novel avian-origin influenza A (H7N9) virus." *N. Engl. J. Med.* 368.20 (2013), 1888–1897. doi: 10.1056/NEJMoa13 04459.

[192] Charles Gawad, Winston Koh, and Stephen R. Quake. "Single-cell genome sequencing: current state of the science." *Nature Rev. Genet.* 17.3 (2016), 175–188.

[193] Robert Ghrist. "Barcodes: the persistent topology of data." *Bull. Am. Math. Soc.* 45 (2008), 61–75.

[194] Robert Ghrist. *Elementary Applied Topology.* Createspace, 2014.

[195] Alison L. Gibbs and Francis Edward Su. "On choosing and bounding probability metrics." *Int. Stat. Rev.* 70.3 (2002), 419–435. doi: 10.1111/j.1751-5823.2002 .tb00178.x.

[196] Anna C. Gilbert and Lalit Jain. "If it ain't broke, don't fix it: sparse metric repair." *Allerton Conf. on Communication, Control, and Computing.* 2017.

[197] Brian J. Gill et al. "MRI-localized biopsies reveal subtype-specific differences in molecular and cellular composition at the margins of glioblastoma." *Proc. Natl. Acad. Sci. USA* 111.34 (2014), 12550–12555.

[198] John H. Gillespie. *Population Genetics: A Concise Guide.* JHU Press, 2010.

[199] Janet R. Gilsdorf, Carl F. Marrs, and Betsy Foxman. "Haemophilus influenzae: genetic variability and natural selection to identify virulence factors." *Infect. Immun.* 72.5 (2004), 2457–2461.

[200] M. Ginsberg et al. "Swine influenza A (H1N1) infection in two children – Southern California, March–April 2009." *MMWR. Morbidity and Mortality Weekly Report* 58.15 (2009), 400–402.

[201] Chad Giusti et al. "Clique topology reveals intrinsic geometric structure in neural correlations." *Proc. Natl. Acad. Sci. USA* 112.44 (2015), 13455–13460.

[202] Earl F. Glynn, Jie Chen, and Arcady R. Mushegian. "Detecting periodic patterns in unevenly spaced gene expression time series using Lomb–Scargle periodograms." *Bioinformatics* 22.3 (2005), 310–316.

[203] Mitchell Goldfarb et al. "Isolation and preliminary characterization of a human transforming gene from T 24 bladder carcinoma cells." *Nature* 296.5856 (1982), 404–409.

[204] Oded Goldreich and Dana Ron. "Property testing in bounded degree graphs." *Proceedings of the Twenty-ninth Annual ACM Symposium on Theory of Computing. STOC '97. El Paso, Texas, USA*, pp. 406–415. ACM, 1997.

[205] P. Grassberger and I. Procaccia. "Measuring the strangeness of strange attractors." *Physica D* 9 (1983), 170–189.

[206] Benjamin D. Greenbaum et al. "Viral reassortment as an information exchange between viral segments." *Proc. Natl. Acad. Sci. USA* 109.9 (2012), 3341–3346.

[207] Andreas Greven, Peter Pfaffelhuber, and Anita Winter. "Convergence in distribution of random metric measure spaces (Λ-coalescent measure trees)." *Prob. Theor. Relat. Fields* 145.1 (2009), 285–322.

[208] Robert C. Griffths and Paul Marjoram. "An ancestral recombination graph." *Inst. Math. Appl.* 87 (1997), 257.

[209] Robert C. Griffths and Paul Marjoram. "Ancestral inference from samples of DNA sequences with recombination." *J. Comput. Biol.* 3.4 (1996), 479–502.

[210] N. R. Grist. "Pandemic influenza 1918." *Br. Med. J.* 2.6205 (1979), 1632–1633.

[211] Ann Griswold. "Genome packaging in prokaryotes: the circular chromosome of *E. coli*." *Nature Educ.* 1.1 (2008), 57.

[212] Mikhail Gromov. *Metric Structures for Riemannian and Non-Riemannian Spaces*. Ed. by J. LaFontaine and P. Pansu. Modern Birkhauser Classics. Birkhauser, Boston, MA, 2007.

[213] Mikhail Gromov. "Hyperbolic groups." *Essays in Group Theory*, pp. 75–263. Ed. by S. M. Gersten. Mathematical Sciences Research Institute Publications, volume 8. Springer, New York, 1987. doi: 10.1007/978-1-4613-9586-7_3.

[214] Dominic Grün et al. "Single-cell messenger RNA sequencing reveals rare intestinal cell types." *Nature* 525.7568 (2015), 251–255.

[215] Leonidas J. Guibas and Steve Oudot. "Reconstruction using witness complexes." *Discrete Comput. Geom.* 40.3 (2008), 325–356. doi: 10.1007/s00454-008-9094-6.

[216] Stéphane Guindon and Olivier Gascuel. "A simple, fast, and accurate algorithm to estimate large phylogenies by maximum likelihood." *Systematic Biol.* 52.5 (2003), 696–704.

[217] Stephane Guindon et al. "PHYML Online – a web server for fast maximum likelihood-based phylogenetic inference." *Nucleic Acids Res.* 33.suppl 2 (2005), W557–W559.

[218] Vishal Gupta, David J. Earl, and Michael W. Deem. "Quantifying influenza vaccine efficacy and antigenic distance." *Vaccine* 24.18 (2006), 3881–3888.

[219] Dan Gusfield. "Optimal, efficient reconstruction of root-unknown phylogenetic networks with constrained and structured recombination." *J. Comput. Syst. Sci.* 70.3 (2005), 381–398.

[220] Dan Gusfield. *ReCombinatorics: The Algorithmics of Ancestral Recombination Graphs and Explicit Phylogenetic Networks*. MIT Press, 2014.

[221] Dan Gusfield, Satish Eddhu, and Charles Langley. "Optimal, efficient reconstruction of phylogenetic networks with constrained recombination." *J. Bioinformatics Comput. Biol.* 2.01 (2004), 173–213.

[222] Daniel A. Haber and Jeff Settleman. "Cancer: drivers and passengers." *Nature* 446.7132 (2007), 145–146.

[223] Laleh Haghverdi, Florian Buettner, and Fabian J. Theis. "Diffusion maps for high-dimensional single-cell analysis of differentiation data." *Bioinformatics* 31.18 (2015), 2989–2998.

[224] Paul R. Halmos. *Naive Set Theory*. Reprint of the 1960 edition. Undergraduate Texts in Mathematics. Springer-Verlag, New York, 1974.

[225] Jihun Ham et al. "A kernel view of the dimensionality reduction of manifolds." *Proceedings of the Twenty-First International Conference on Machine Learning. ICML '04. Banff, Alberta, Canada*, p. 47. ACM, 2004. doi: 10.1145/1015330.1015417.

[226] James Douglas Hamilton. *Time Series Analysis*, volume 2. Princeton University Press, Princeton, NJ, 1994.

[227] Douglas Hanahan and Robert A. Weinberg. "The hallmarks of cancer." *Cell* 100.1 (2000), 57–70.

[228] Shaun Harker et al. "Discrete Morse theoretic algorithms for computing homology of complexes and maps." *Found. Comput. Math.* 14.1 (2014), 151–184.

[229] John A. Hartigan. *Clustering Algorithms*. John Wiiley & Sons, New York, 1975.

[230] Daniel L. Hartl and Andrew G Clark. *Principles of Population Genetics*. Sinauer Associates, Sunderland, MA, 1997.

[231] Masami Hasegawa, Hirohisa Kishino, and Taka-aki Yano. "Dating of the human-ape splitting by a molecular clock of mitochondrial DNA." *J. Mol. Evol.* 22.2 (1985), 160–174.

[232] Tamar Hashimshony et al. "CEL-Seq: single-cell RNA-Seq by multiplexed linear amplification." *Cell Rep.* 2.3 (2012), 666–673.

[233] Trevor Hastie, Robert Tibshirani, and Jerome Friedman. *The Elements of Statistical Learning: Data Mining, Inference and Prediction*. Second edition. Springer, 2009.

[234] W. Keith Hastings. "Monte Carlo sampling methods using Markov chains and their applications." *Biometrika* 57.1 (1970), 97–109.

[235] Allen Hatcher. *Algebraic Topology*. Cambridge University Press, Cambridge, 2002.

[236] Ingrid Hedenfalk et al. "Gene-expression profiles in hereditary breast cancer." *N Engl. J. Med.* 344.8 (2001), 539–548.

[237] Jotun Hein, Mikkel Schierup, and Carsten Wiuf. *Gene Genealogies, Variation and Evolution: A Primer in Coalescent Theory*. Oxford University Press, 2004.

[238] Gregory Henselman. Eirene. http://gregoryhenselman.org/eirene/. 2016.

[239] Gregory Henselman and Robert Ghrist. "Matroid filtrations and computational persistent homology." 2016. arXiv: 1606.00199.

[240] Sander Herfst et al. "Airborne transmission of influenza A/H5N1 virus between ferrets." *Science* 336.6088 (2012), 1534–1541. doi: 10.1126/science.1213362.

[241] Carlos Xavier Hernández et al. "Understanding the origins of a pandemic virus." 2011. arXiv: 1104.4568.

[242] Geoffrey Hinton and Sam Roweis. "Stochastic neighbor embedding." *Proceedings of the 15th International Conference on Neural Information Processing Systems. NIPS'02. Cambridge, MA, USA*, pp. 857–864. MIT Press, 2002.

[243] Yorio Hinuma et al. "Adult T-cell leukemia: antigen in an ATL cell line and detection of antibodies to the antigen in human sera." *Proc. Natl. Acad. Sci. USA* 78.10 (1981), 6476–6480.

[244] The Joint United Nations Programme on HIV/AIDS (UNAIDS). *Global AIDS Update 2016*. World Health Organization, 2016.

[245] E. Hodis et al. "A landscape of driver mutations in melanoma." *Cell* 150.2 (2012), 251–63. doi: 10.1016/j.cell.2012.06.024.

[246] Edward C. Holmes. *The Evolution and Emergence of RNA Viruses*. Oxford University Press, 2009.

[247] Susan Holmes. "Bootstrapping phylogenetic trees: theory and methods." *Stat. Sci.* (2003), 241–255.

[248] Susan Holmes. "Statistical approach to tests involving phylogenies." *Mathematics of Evolution and Phylogeny*, pp. 91–120. Ed. by Olivier Gascuel. Oxford University Press, Oxford, 2005.

[249] Hugo M. Horlings et al. "Integration of DNA copy number alterations and prognostic gene expression signatures in breast cancer patients." *Clin. Cancer Res.* 16.2 (2010), 651–663.

[250] Katharina T. Huber et al. "Pruned median networks: a technique for reducing the complexity of median networks." *Mol. Phylogenet. Evol.* 19.2 (2001), 302–310.

[251] Peter J. Huber and Elvezio M. Ronchetti. *Robust Statistics*. Second edition. Wiley Series in Probability and Statistics. Wiley, Hoboken, NJ, 2009.

[252] Richard R. Hudson. "Generating samples under a Wright–Fisher neutral model of genetic variation." *Bioinformatics* 18.2 (2002), 337–338.

[253] Richard R. Hudson. "Properties of a neutral allele model with intragenic recombination." *Theor. Population Biol.* 23.2 (1983), 183–201.

[254] Richard R. Hudson and Norman L. Kaplan. "Statistical properties of the number of recombination events in the history of a sample of DNA sequences." *Genetics* 111.1 (1985), 147–164.

[255] R. R. Hudson, D. Dr Boos, and N. L. Kaplan. "A statistical test for detecting geographic subdivision." *Mol. Biol. Evol.* 9.1 (1992), 138–151.

[256] John P. Huelsenbeck and David M. Hillis. "Success of phylogenetic methods in the four-taxon case." *Systematic Biol.* 42.3 (1993), 247–264.

[257] John P. Huelsenbeck, Fredrik Ronquist et al. "MRBAYES: Bayesian inference of phylogenetic trees." *Bioinformatics* 17.8 (2001), 754–755.

[258] Michael E. Hughes, John B. Hogenesch, and Karl Kornacker. "JTK CYCLE: an efficient nonparametric algorithm for detecting rhythmic components in genome-scale data sets." *J. Biol. Rhythms* 25.5 (2010), 372–380.

[259] Devon P. Humphreys et al. "Fast estimation of recombination rates using topological data analysis." *Genetics* 211.4 (2019), 1191–1204.

[260] Daniel H. Huson. "SplitsTree: analyzing and visualizing evolutionary data." *Bioinformatics* 14.1 (1998), 68–73.

[261] Daniel H. Huson and David Bryant. "Application of phylogenetic networks in evolutionary studies." *Mol. Biol. Evol.* 23.2 (2006), 254–267.

[262] Daniel H. Huson, Regula Rupp, and Celine Scornavacca. *Phylogenetic Networks: Concepts, Algorithms and Applications*. Cambridge University Press, 2010.

[263] Daniel H. Huson and Celine Scornavacca. "A survey of combinatorial methods for phylogenetic networks." *Genome Biol. Evol.* 3 (2011), 23–35.

[264] Daniel H. Huson and Celine Scornavacca. "Dendroscope 3: an interactive tool for rooted phylogenetic trees and networks." *Systematic Biol.* 61.6 (2012), 1061–1067.

[265] Francisco J. Iborra et al. "Active RNA polymerases are localized within discrete transcription 'factories' in human nuclei." *J. Cell Sci.* 109.6 (1996), 1427–1436.

[266] Masaki Imai et al. "Experimental adaptation of an influenza H5 HA confers respiratory droplet transmission to a reassortant H5 HA/H1N1 virus in ferrets." *Nature* 486.7403 (2012), 420–428.

[267] Ayasdi Inc. Ayasdi Iris. www.ayasdi.com.

[268] Ayasdi Inc. Iris. 2015. www.ayasdi.com.

[269] Toshihiro Ito et al. "Molecular basis for the generation in pigs of influenza A viruses with pandemic potential." *J. Virol.* 72.9 (1998), 7367–7373.

[270] Aniek Janssen et al. "Chromosome segregation errors as a cause of DNA damage and structural chromosome aberrations." *Science* 333.6051 (2011), 1895–1898.

[271] Jose A. Jarillo et al. "An Arabidopsis circadian clock component interacts with both CRY1 and phyB." *Nature* 410.6827 (2001), 487–490.

[272] Lars Juhl Jensen et al. "Co-evolution of transcriptional and post-translational cell-cycle regulation." *Nature* 443.7111 (2006), 594–597.

[273] Slade O. Jensen and Bruce R. Lyon. "Genetics of antimicrobial resistance in *Staphylococcus aureus.*" *Future Microbiol.* 4.5 (2009), 565–582.

[274] Fulai Jin et al. "A high-resolution map of the three-dimensional chromatin interactome in human cells." *Nature* 503 (2013), 290–294.

[275] Daniel Johansson, Petter Lindgren, and Anders Berglund. "A multivariate approach applied to microarray data for identification of genes with cell cycle-coupled transcription." *Bioinformatics* 19.4 (2003), 467–473.

[276] Niall P. A. S. Johnson and Juergen Mueller. "Updating the accounts: global mortality of the 1918–1920 'Spanish' influenza pandemic." *Bull. Hist. Med.* 76.1 (2002), 105–115.

[277] Keith A. Jolley and Martin C. J. Maiden. "BIGSdb: Scalable analysis of bacterial genome variation at the population level." *BMC Bioinformatics* 11.1 (2010), 595.

[278] Aimable R. Jonckheere. "A distribution-free k-sample test against ordered alternatives." *Biometrika* 41.1/2 (1954), 133–145.

[279] Thomas H. Jukes, Charles R. Cantor et al. "Evolution of protein molecules." *Mammal. Protein Metabol.* 3.21 (1969), 132.

[280] Tomasz Kaczynski, Konstantin Mischaikow, and Marian Mrozek. *Computational Homology.* Applied Mathematical Sciences, volume 157. Springer, 2004.

[281] Matt Kahle. "Random simplicial complexes." To appear in *Handbook of Discrete and Computational Geometry.* 2017.

[282] Matthew Kahle and Elizabeth Meckes. "Limit the theorems for Betti numbers of random simplicial complexes." *Homology Homotopy Appl.* 15.1 (2013), 343–374.

[283] Sara Kališnik Verovšek. "Tropical coordinates on the space of persistence barcodes." *Found. Comput. Math.* (2018), 1–29.

[284] Irving Kaplansky. *Set Theory and Metric Spaces.* Second edition. Chelsea Publishing Co., New York, 1977.

[285] H. Karcher. "Riemannian center of mass and mollifier smoothing." *Commun. Pure Appl. Math.* 30.5 (1977), 509–541. doi: 10.1002/cpa.3160300502.

[286] T. Kasahara et al. "Mice with neuron-specific accumulation of mitochondrial DNA mutations show mood disorder-like phenotypes." *Mol. Psychiatry* 11.6 (2006), 577–593.

[287] Marcus Kaul. "HIV-1 associated dementia: update on pathological mechanisms and therapeutic approaches." *Curr. Opin. Neurol.* 22.3 (2009), 315.

[288] Liisa Kauppi, Alec J. Jeffreys, and Scott Keeney. "Where the crossovers are: recombination distributions in mammals." *Nature Rev. Genet.* 5.6 (2004), 413–424.

[289] Michael Kerber, Dmitriy Morozov, and Arnur Nigmetov. "Geometry helps to compare persistence diagrams." *Proceedings of the Eighteenth Workshop on Algorithm Engineering and Experiments (ALENEX),* pp. 103–112. 2016.

[290] Michael Kerber, Dmitriy Morozov, and Arnur Nigmetov. Hera. https://bitbucket.org/grey_narn/hera. 2015.

[291] Firas A. Khasawneh and Elizabeth Munch. "Utilizing topological data analysis for studying signals of time-delay systems." *Time Delay Systems: Theory, Numerics, Applications, and Experiments*, pp. 93–106. Ed. by Tamás Insperger, Tulga Ersal, and Gábor Orosz. Springer International Publishing, 2017.

[292] Hossein Khiabanian, Vladimir Trifonov, and Raúl Rabadán. "Reassortment patterns in swine influenza viruses." *PloS One* 4.10 (2009), e7366.

[293] Hossein Khiabanian et al. "Viral diversity and clonal evolution from unphased genomic data." *BMC Genomics* 15.Suppl 6 (2014), S17.

[294] Pavel P. Khil et al. "Sensitive mapping of recombination hotspots using sequencing-based detection of ssDNA." *Genome Res.* 22.5 (2012), 957–965.

[295] Bong-Rae Kim et al. "Wavelet-based functional clustering for patterns of high-dimensional dynamic gene expression." *J. Comput. Biol.* 17.8 (2010), 1067–1080.

[296] Sunyong Kim, Seiya Imoto, and Satoru Miyano. "Dynamic Bayesian network and nonparametric regression for nonlinear modeling of gene networks from time series gene expression data." *Biosystems* 75.1 (2004), 57–65.

[297] Edna T. Kimura et al. "High prevalence of BRAF mutations in thyroid cancer genetic evidence for constitutive activation of the RET/PTC-RAS-BRAF signaling pathway in papillary thyroid carcinoma." *Cancer Res.* 63.7 (2003), 1454–1457.

[298] Motoo Kimura. "A simple method for estimating evolutionary rates of base substitutions through comparative studies of nucleotide sequences." *J. Mol. Evol.* 16.2 (1980), 111–120.

[299] Motoo Kimura. "Diffusion models in population genetics." *J. Appl. Probab.* 1.2 (1964), 177–232.

[300] Motoo Kimura. *The Neutral Theory of Molecular Evolution.* Cambridge University Press, 1984.

[301] John Frank Charles Kingman. "The coalescent." *Stochastic Process. Appl.* 13.3 (1982), 235–248.

[302] Allon M. Klein et al. "Droplet barcoding for single-cell transcriptomics applied to embryonic stem cells." *Cell* 161.5 (2015), 1187–1201.

[303] Nobuya Koike et al. "Transcriptional architecture and chromatin landscape of the core circadian clock in mammals." *Science* 338.6105 (2012), 349–354.

[304] Andrei Nikolaevich Kolmogorov. "Three approaches to the definition of the concept 'quantity of information'." *Probl. Peredachi Inf.* 1.1 (1965), 3–11.

[305] Eugene V. Koonin, Kira S. Makarova, and L. Aravind. "Horizontal gene transfer in prokaryotes: quantification and classification." *Annu. Rev. Microbiol.* 55 (2001), 709–742.

[306] Arthur Kornberg, Tania A. Baker et al. *DNA Replication.* W. Freeman, San Francisco, CA, 1980.

[307] Dmitry N. Kozlov. *Combinatorial Algebraic Topology.* Algorithms and Computation in Mathematics, volume 21. Springer, 2008. doi: 10.1007/978-3-540-71962-5.

[308] Mark A. Krasnow et al. "Determination of the absolute handedness of knots and catenanes of DNA." *Nature* 304.5926 (1983), 559–560.

[309] Martin Kreitman. "Nucleotide polymorphism at the alcohol dehydrogenase locus of *Drosophila melanogaster.*" *Nature* 304, (1983), 412–417.

[310] Susanna L. Lamers et al. "Extensive HIV-1 intra-host recombination is common in tissues with abnormal histopathology." *PLoS One* 4.3 (2009), e5065.

[311] Susanna L. Lamers et al. "Human immunodeficiency virus-1 evolutionary patterns associated with pathogenic processes in the brain." *J. Neurovirol.* 16.3 (2010), 230–241.

[312] Sangeet Lamichhaney et al. "Evolution of Darwin's finches and their beaks revealed by genome sequencing." *Nature* 518.7539 (2015), 371–375.

[313] D. P. Lane and L. V. Crawford. "T antigen is bound to a host protein in SY40-transformed cells." *Nature* 278 (1979), 261–263.

[314] Serge Lang. *Undergraduate Algebra*. Undergraduate Texts in Mathematics. Springer, 2005.

[315] J. Latschev. "Vietoris-Rips complexes of metric spaces near a closed Riemannian manifold." *Arch. Math.* 77.6 (2001), 522–528.

[316] Michael S. Lawrence et al. "Mutational heterogeneity in cancer and the search for new cancer associated genes." *Nature* 499.7457 (2013), 214–218.

[317] Tung B. K. Le et al. "High-resolution mapping of the spatial organization of a bacterial chromosome." *Science* 342.6159 (2013), 731–734.

[318] Joshua Lederberg and Edward L. Tatum. "Gene recombination in *Escherichia coli*." *Nature* 158 (1946), 558.

[319] Daniel D. Lee and H. Sebastian Seung. "Algorithms for non-negative matrix factorization." *Advances in Neural Information Processing Systems*, volume 13, pp. 556–562. Ed. by T. K. Leen, T. G. Dietterich, and V. Tresp. MIT Press, 2001.

[320] Jin-Ku Lee et al. "Spatiotemporal genomic architecture informs precision oncology in glioblastoma." *Nature Genet.* 49.4 (2017), 594–599.

[321] Jin-Ku Lee et al. "Pharmacogenomic landscape of patient-derived tumor cells informs precision oncology therapy." *Nature Genet.* 50.10 (2018), 1399–1411.

[322] Christina Leslie et al. "Mismatch string kernels for SVM protein classification." *Proceedings of the 15th International Conference on Neural Information Processing Systems. NIPS'02. Cambridge, MA, USA*, pp. 1441–1448. MIT Press, 2002.

[323] Michael Lesnick, Raúl Rabadán, and Daniel Rosenbloom. "Quantifying genetic innovation: mathematical foundations for the topological study of reticulate evolution." 2018. arXiv: 1804.01398 [math.AT].

[324] Michael Lesnick and Mathew Wright. Rivet. http://rivet.online/. 2015.

[325] Michael Lesnick and Matthew Wright. "Interactive visualization of 2-D persistence modules." 2015. arXiv: 1512.00180.

[326] Mike Lesnick. *Studying the Shape of Data Using Topology*. IAS Letter, 2013.

[327] Elizaveta Levina and Peter J. Bickel. "Maximum likelihood estimation of intrinsic dimension." *Advances in Neural Information Processing Systems*, volume 17, pp. 777–784. Ed. by L. K. Saul, Y. Weiss, and L. Bottou. MIT Press, 2005.

[328] Arnold J. Levine. *Viruses*. Scientific American Library, New York, 1992.

[329] Dan Levy and Lior Pachter. "The neighbor-net algorithm." *Adv. Appl. Math.* 47.2 (2011), 240–258.

[330] E. Lieberman-Aiden et al. "Comprehensive mapping of long-range interactions reveals folding principles of the human genome." *Science* 326.5950 (2009), 289–293.

[331] Daniel I. H. Linzer and Arnold J. Levine. "Characterization of a 54K dalton cellular SV40 tumor antigen present in SV40-transformed cells and uninfected embryonal carcinoma cells." *Cell* 17.1 (1979), 43–52.

[332] A. V. Little, M. Maggioni, and L. Rosasco. Multiscale geometric methods for data sets I: multiscale SVD, noise, and curvature. MIT-CSAIL-TR-2012-029. 2012.

[333] Nicholas R. Lomb. "Least-squares frequency analysis of unequally spaced data." *Astrophys. Space Sci.* 39.2 (1976), 447–462.

[334] P. Y. Lum et al. "Extracting insights from the shape of complex data using topology." *Sci. Rep.* 3 (2013). doi: 10.1038/srep01236.

[335] Laurens J. P. van der Maaten, Eric O. Postma, and H. Jaap van den Herik. "Dimensionality reduction: a comparative review." *J. Mach. Learn. Res.* 10.1-41 (2009), 66–71.

[336] L. J. P. van der Maaten and G. E. Hinton. "Visualizing high-dimensional data using *t*-SNE." *J. Mach. Learn. Res.* 9 (2008), 2579–2605.

[337] Saunders MacLane. *Categories for the Working Mathematician*. Graduate Texts in Mathematics, volume 5. Springer-Verlag, New York, 1971.

[338] Evan Z. Macosko et al. "Highly parallel genome-wide expression profiling of individual cells using nanoliter droplets." *Cell* 161.5 (2015), 1202–1214.

[339] Robert MacPherson and Benjamin Schweinhart. "Measuring shape with topology." *J. Math. Phys.* 53.7 (2012), 073516.

[340] Eugenio Marco et al. "Bifurcation analysis of single-cell gene expression data reveals epigenetic landscape." *Proc. Natl. Acad. Sci. USA* 111.52 (2014), E5643–E5650.

[341] Clément Maria et al. "The Gudhi Library: simplicial complexes and persistent homology." *Mathematical Software – ICMS 2014: Proceedings of the Fourth International Congress, Seoul, South Korea, August 5–9, 2014*, pp. 167–174. Ed. by Hoon Hong and Chee Yap. Springer, Berlin, 2014.

[342] J. P. May. *A Concise Course in Algebraic Topology*. Chicago Lectures in Mathematics. University of Chicago Press, Chicago, IL, 1999.

[343] Roger McLendon et al. "Comprehensive genomic characterization defines human glioblastoma genes and core pathways." *Nature* 455.7216 (2008), 1061–1068.

[344] Gilean A. T. McVean et al. "The fine-scale structure of recombination rate variation in the human genome." *Science* 304.5670 (2004), 581–584.

[345] Emma Meats et al. "Characterization of encapsulated and noncapsulated Haemophilus influenzae and determination of phylogenetic relationships by multilocus sequence typing." *J. Clin. Microbiol.* 41.4 (2003), 1623–1636.

[346] Facundo Mémoli. "A spectral notion of Gromov-Wasserstein distances and related methods." Appl. Comput. Math. 30 (2011), 363–401. doi: 10.1016.

[347] Facundo Mémoli. "Gromov–Wasserstein distances and the metric approach to object matching." *Found. Comput. Math.* (2011), 1–71. doi: 10.1007/s10208-011-9093-5.

[348] Ignacio Mena et al. "Origins of the 2009 H1N1 influenza pandemic in swine in Mexico." *eLife* (2016), e16777.

[349] Carl D. Meyer (editor). *Matrix Analysis and Applied Linear Algebra*. Society for Industrial and Applied Mathematics, Philadelphia, PA, 2000.

[350] Folker Meyer, Ross Overbeek, and Alex Rodriguez. "FIGfams: yet another set of protein families." *Nucleic Acids Res.* 37.20 (2009), 6643–6654.

[351] Radu Mihaescu, Dan Levy, and Lior Pachter. "Why neighbor-joining works." *Algorithmica* 54.1 (2009), 1–24.

[352] Yuriy Mileyko, Sayan Mukherjee, and John Harer. "Probability measures on the space of persistence diagrams." *Inverse Probl.* 27.12 (2011), 124007.

[353] Lance D. Miller et al. "An expression signature for p53 status in human breast cancer predicts mutation status, transcriptional effects, and patient survival." *Proc. Natl. Acad. Sci. USA* 102.38 (2005), 13550–13555.

[354] J. Milnor. *Morse Theory. Based on lecture notes by M. Spivak and R. Wells*. Annals of Mathematics Studies, No. 51. Princeton University Press, Princeton, NJ, 1963.

[355] John W. Milnor. *Topology from the Differentiable Viewpoint. Based on notes by David W. Weaver*. Princeton Landmarks in Mathematics. Revised reprint of the 1965 original. Princeton University Press, Princeton, NJ, 1997.

[356] Mark J. Minichiello and Richard Durbin. "Mapping trait loci by use of inferred ancestral recombination graphs." *Am. J. Human Genet.* 79.5 (2006), 910–922.

[357] Anthea Monod et al. "Tropical sufficient statistics for persistent homology." 2017. arXiv: 1709.02647.

[358] Patrick Alfred Pierce Moran. "A general theory of the distribution of gene frequencies. I. Overlapping generations." *Proc. R. Soc. London, Ser. B* 149.934 (1958), 102–112.

[359] Thomas Hunt Morgan. *A Critique of the Theory of Evolution.* Princeton University Press, 1916.

[360] Dmitriy Morozov. Dionysus. 2017. www.mrzv.org/software/dionysus2/.

[361] Dmitriy Morozov. *Persistence algorithm takes cubic time in the worst case.* 2005.

[362] David A. Morrison. *Introduction to Phylogenetic Networks.* RJR Productions, 2011.

[363] Siddhartha Mukherjee. *The Emperor of all Maladies: A Biography of Cancer.* Simon and Schuster, 2011.

[364] Hermann J. Muller. "The production of mutations by X-rays." *Proc. Natl. Acad. Sci. USA* 14.9 (1928), 714.

[365] Daniel Müllner and Aravindakshan Babu. Python mapper. 2016. http://danifold.net/mapper/.

[366] Bei Munch and Elizabeth Wang. "Convergence between categorical representations of Reeb space and Mapper." *32nd International Symposium on Computational Geometry, SoCG 2016*, pp. 53:1–53:16. Ed. by Sándor Fekete and Anna Lubiw. Leibniz International Proceedings in Informatics (LIPIcs), volume 51. Leibniz-Zentrum fuer Informatik, Dagstuhl, 2016. doi: 10.4230/LIPIcs.SoCG.2016.53.

[367] Elizabeth Munch et al. "Probabilistic Fréchet means and statistics on vineyards." 2013. arXiv: 1307.6530.

[368] James R. Munkres. *Elements of Algebraic Topology.* Addison-Wesley, Menlo Park, CA, 1984.

[369] James R. Munkres. *Topology: A First Course.* Prentice-Hall, Englewood Cliffs, NJ, 1975.

[370] Simon R. Myers and Robert C. Griffiths. "Bounds on the minimum number of recombination events in a sample history." *Genetics* 163.1 (2003), 375–394.

[371] Simon Myers et al. "A fine-scale map of recombination rates and hotspots across the human genome." *Science* 310.5746 (2005), 321–324.

[372] Simon Myers et al. "Drive against hotspot motifs in primates implicates the PRDM9 gene in meiotic recombination." *Science* 327.5967 (2010), 876–879.

[373] Peter Mysling, Soren Hauberg, and Kim S. Pedersen. "An empirical study on the performance of spectral manifold learning techniques." *Artificial Neural Networks and Machine Learning – ICANN 2011*, pp. 347–354. Lecture Notes in Computer Science, volume 6791. Springer, Berlin, 2011.

[374] Takashi Nagano et al. "Single-cell Hi-C for genome-wide detection of chromatin interactions that occur simultaneously in a single cell." *Nature Protocols* 10.12 (2015), 1986–2003.

[375] Takashi Nagano et al. "Single-cell Hi-C reveals cell-to-cell variability in chromosome structure." *Nature* 502.7469 (2013), 59–64.

[376] Vidit Nanda. Perseus. 2017. http://people.maths.ox.ac.uk/nanda/perseus/index.html.

[377] Nicholas E. Navin. "The first five years of single-cell cancer genomics and beyond." *Genome Res.* 25.10 (2015), 1499–1507.

[378] Nicholas Navin et al. "Tumour evolution inferred by single-cell sequencing." *Nature* 472.7341 (2011), 90–94.

[379] Saul B. Needleman and Christian D. Wunsch. "A general method applicable to the search for similarities in the amino acid sequence of two proteins." *J. Mol. Biol.* 48.3 (1970), 443–453.

[380] Masatoshi Nei and Sudhir Kumar. *Molecular Evolution and Phylogenetics.* Oxford University Press, 2000.

[381] Cancer Genome Atlas Research Network et al. "Comprehensive, integrative genomic analysis of diffuse lower-grade gliomas." *N. Engl. J. Med.* 2015.372 (2015), 2481–2498.

[382] Harold C. Neu. "The crisis in antibiotic resistance." *Science* 257.5073 (1992), 1064–1073.

[383] Monica Nicolau, Arnold J. Levine, and Gunnar Carlsson. "Topology based data analysis identifies a subgroup of breast cancers with a unique mutational profile and excellent survival." *Proc. Natl. Acad. Sci. USA* 108.17 (2011), 7265–7270.

[384] Partha Niyogi, Stephen Smale, and Shmuel Weinberger. "Finding the homology of submanifolds with high confidence from random samples." *Discrete Comput. Geom.* 39.1–3 (2008), 419–441.

[385] Takeshi Noda et al. "Architecture of ribonucleoprotein complexes in influenza A virus particles." *Nature* 439.7075 (2006), 490–492.

[386] C. O. Nordling. "A new theory on the cancer-inducing mechanism." *Br. J. Cancer* 7.1 (1953), 68.

[387] Peter C. Nowell. "A minute chromosome in human granulocytic leukemia." *Science* 132 (1960), 1497–1501.

[388] Peter C. Nowell. "The clonal evolution of tumor cell populations." *Science* 194.4260 (1976), 23–28.

[389] Howard Ochman, Jeffrey G. Lawrence, and Eduardo A. Groisman. "Lateral gene transfer and the nature of bacterial innovation." *Nature* 405.6784 (2000), 299–304.

[390] John O'Keefe and Jonathan Dostrovsky. "The hippocampus as a spatial map. Preliminary evidence from unit activity in the freely-moving rat." *Brain Res.* 34.1 (1971), 171–175.

[391] World Health Organization. *Antimicrobial Resistance: global report on surveillance 2014.* Technical Report. World Health Organization, 2014.

[392] Steve Y. Oudot. *Persistence Theory – From Quiver Representations to Data Analysis.* Mathematical Surveys and Monographs, volume 209. American Mathematical Society, 2015.

[393] Steve Y. Oudot and Donald R. Sheehy. "Zigzag zoology: rips zigzags for homology inference." *Proceedings of the 29th Annual Symposium on Computational Geometry*, pp. 387–396. 2013.

[394] Megan Owen and J. Scott Provan. "A fast algorithm for computing geodesic distances in tree space." *IEEE/ACM Trans. Comput. Biol. Bioinformatics* 8.1 (2011), 2–13.

[395] L. Pachter and B. Sturmfels. *Algebraic Statistics for Computational Biology.* Cambridge University Press, New York, 2005.

[396] Kenneth Paigen and Petko Petkov. "Mammalian recombination hot spots: properties, control and evolution." *Nature Rev. Genet.* 11.3 (2010), 221–233.

[397] Jing Pan et al. "A hierarchical combination of factors shapes the genome-wide topography of yeast meiotic recombination initiation." *Cell* 144.5 (2011), 719–731.

[398] Luis F. Parada et al. "Human EJ bladder carcinoma oncogene is homologue of Harvey sarcoma virus ras gene." *Nature* 297.5866 (1982), 474–478.

[399] Donald Maxwell Parkin. "The global health burden of infection-associated cancers in the year 2002." *Int. J. Cancer* 118.12 (2006), 3030–3044.

[400] Emil D. Parvanov, Petko M. Petkov, and Kenneth Paigen. "Prdm9 controls activation of mammalian recombination hotspots." *Science* 327.5967 (2010), 835–835.

[401] Anoop P. Patel et al. "Single-cell RNA-seq highlights intratumoral heterogeneity in primary glioblastoma." *Science* 344.6190 (2014), 1396–1401.

[402] Paul Pearson. TDAmapper. 2017. https://cran.r-project.org/web/packages/TDAmapper/README.html.

[403] M. Penrose. *Random Geometric Graphs*. Oxford Studies in Probability. Oxford University Press, 2003.

[404] Jose A. Perea and John Harer. "Sliding windows and persistence: an application of topological methods to signal analysis." *Found. Comput. Math.* 15.3 (2015), 799–838.

[405] Jose A. Perea et al. "SW1PerS: sliding windows and 1-persistence scoring; discovering periodicity in gene expression time series data." *BMC Bioinformatics* 16.1 (2015), 257.

[406] K. Petren et al. "Comparative landscape genetics and the adaptive radiation of Darwin's finches: the role of peripheral isolation." *Mol. Ecol.* 14.10 (2005), 2943–2957.

[407] Giovanni Petri et al. "Homological scaffolds of brain functional networks." *J. R. Soc. Interface* 11.101 (2014), 20140873.

[408] Sophie Petropoulos et al. "Single-Cell RNA-Seq reveals lineage and X chromosome dynamics in human preimplantation embryos." *Cell* 165.4 (2016), 1012–1026.

[409] Brett E. Pickett and Elliot J. Lefkowitz. "Recombination in West Nile Virus: minimal contribution to genomic diversity." *Virol. J.* 6.1 (2009), 1.

[410] Peter Piot et al. "Acquired immunodeficiency syndrome in a heterosexual population in Zaire." *The Lancet* 324.8394 (1984), 65–69.

[411] Joshua B. Plotkin and Hunter B. Fraser. "Assessing the determinants of evolutionary rates in the presence of noise." *Mol. Biol. Evol.* 24.5 (2007), 1113–1121.

[412] Dimitris N. Politis and Joseph P. Romano. "Large sample confidence regions based on subsamples under minimal assumptions." *Ann. Stat.* 22.4 (1994), 2031–2050. doi: 10.1214/aos/1176325770.

[413] M. Porter et al. "A roadmap for the computation of persistent homology." *EPJ Data Sci.* 6.17 (2017).

[414] David Posada. "Evaluation of methods for detecting recombination from DNA sequences: empirical data." *Mol. Biol. Evol.* 19.5 (2002), 708–717.

[415] Lisa Postow et al. "Topological domain structure of the *Escherichia coli* chromosome." *Genes Dev.* 18.14 (2004), 1766–1779.

[416] Raúl Rabadán, Arnold J. Levine, and Michael Krasnitz. "Non-random reassortment in human influenza A viruses." *Influenza Other Respir. Viruses* 2.1 (2008), 9–22.

[417] Raúl Rabadán and Harlan Robins. "Evolution of the influenza A virus: some new advances." *Evol. Bioinformatics Online* 3 (2007), 299.

[418] Marco F. Ramoni, Paola Sebastiani, and Isaac S. Kohane. "Cluster analysis of gene expression dynamics." *Proc. Natl. Acad. Sci. USA* 99.14 (2002), 9121–9126.

[419] Daniel Ramsköld et al. "Full-length mRNA-Seq from single-cell levels of RNA and individual circulating tumor cells." *Nature Biotechnol.* 30.8 (2012), 777–782.

[420] Suhas S. P. Rao et al. "A 3D map of the human genome at kilobase resolution reveals principles of chromatin looping." *Cell* 159.7 (2014), 1665–1680.

[421] Matthew D. Rasmussen et al. "Genome-wide inference of ancestral recombination graphs." *PLoS Genet.* 10.5 (2014), e1004342.

[422] Roberto Refinetti, Germaine Cornélissen, and Franz Halberg. "Procedures for numerical analysis of circadian rhythms." *Biol. Rhythm Res.* 38.4 (2007), 275–325.

[423] Ann H. Reid et al. "Origin and evolution of the 1918 'Spanish' influenza virus hemagglutinin gene." *Proc. Natl. Acad. Sci. USA* 96.4 (1999), 1651–1656.

[424] Michael W. Reimann et al. "Cliques of neurons bound into cavities provide a missing link between structure and function." *Front. Comput. Neurosci.* 11 (2017), 48.

[425] J. Reininghaus et al. "A stable multi-scale kernel for topological machine learning." *Proceedings of the 2015 IEEE Conference on Computer Vision and Pattern Recognition (CVPR)*, pp. 4741–4748. IEEE, 2015.

[426] Rebeca Rico-Hesse. "Microevolution and virulence of dengue viruses." *Adv. Virus Res.* 59 (2003), 315–341.

[427] Emily Riehl. Categorical Homotopy Theory. New Mathematical Monographs, volume 24. Cambridge University Press, Cambridge, 2014.

[428] Emily Riehl. *Category Theory in Context*. Aurora: Modern Math Originals. Dover Publications, 2016.

[429] Philippe Rigollet and Régis Vert. "Optimal rates for plug-in estimators of density level sets." *Bernoulli* 15.4 (2009), 1154–1178. doi: 10.3150/09-BEJ184.

[430] Christian Rinke et al. "Insights into the phylogeny and coding potential of microbial dark matter." *Nature* 499.7459 (2013), 431–437.

[431] Abbas H. Rizvi et al. "Single-cell topological RNA-seq analysis reveals insights into cellular differentiation and development." *Nature Biotechnol.* 35.6 (2017), 551–560.

[432] David L. Robertson, Beatrice H. Hahn, and Paul M. Sharp. "Recombination in AIDS viruses." *J. Mol. Evol.* 40.3 (1995), 249–259.

[433] Vanessa Robins. "Toward computing homology from finite approximations." *Topology Proc.* 24 (1999), 503–532.

[434] Mark D. Robinson and Alicia Oshlack. "A scaling normalization method for differential expression analysis of RNA-seq data." *Genome Biol.* 11.3 (2010), R25.

[435] Holger Rohde et al. "Open-source genomic analysis of shiga-toxin–producing *E. coli* O104:H4." *N. Engl. J. Med.* 365.8 (2011), 718–724.

[436] Fredrik Ronquist and John P. Huelsenbeck. "MrBayes 3: Bayesian phylogenetic inference under mixed models." *Bioinformatics* 19.12 (2003), 1572–1574.

[437] Peyton Rous. "A sarcoma of the fowl transmissible by an agent separable from the tumor cells." *J. Exp. Med.* 13.4 (1911), 397–411.

[438] Peyton Rous. "A transmissible avian neoplasm (sarcoma of the common fowl)" *J. Exp. Med.* 12.5 (1910), 696–705.

[439] S. Roweis and L. Saul. "Nonlinear dimensionality reduction by locally linear embedding." *Science* 290.5500 (2000), 2323–2326.

[440] Walter Rudin. *Principles of Mathematical Analysis*. Third edition. International Series in Pure and Applied Mathematics. McGraw-Hill, New York, 1976.

[441] T. Ruf. "The Lomb-Scargle periodogram in biological rhythm research: analysis of incomplete and unequally spaced time-series." *Biol. Rhythm Res.* 30.2 (1999), 178–201.

[442] Naruya Saitou and Masatoshi Nei. "The neighbor-joining method: a new method for reconstructing phylogenetic trees." *Mol. Biol. Evol.* 4.4 (1987), 406–425.

[443] Marco Salemi, Philippe Lemey, and Anne-Mieke Vandamme. *The Phylogenetic Handbook: A Practical Approach to Phylogenetic Analysis and Hypothesis Testing.* Cambridge University Press, 2009.

[444] Eugenio Santos et al. "T24 human bladder carcinoma oncogene is an activated form of the normal human homologue of BALB-and Harvey-MSV transforming genes." *Nature* 298 (1982), 343–347.

[445] Rahul Satija et al. "Spatial reconstruction of single-cell gene expression data." *Nature Biotechnol.* 33.5 (2015), 495–502.

[446] Joao Ricardo Sato et al. "A method to produce evolving functional connectivity maps during the course of an fMRI experiment using wavelet-based time-varying Granger causality." *Neuroimage* 31.1 (2006), 187–196.

[447] Nathaniel Saul and Hendrik Jacob van Veen. "MLWave/kepler-mapper: 186f." 2017. doi: 10.5281/zenodo.1054444.

[448] Jeffrey D Scargle. "Studies in astronomical time series analysis. II – Statistical aspects of spectral analysis of unevenly spaced data." *Astrophys. J.* 263 (1982), 835–853.

[449] Benjamin Schweinhart. *Statistical Topology of Embedded Graphs*. PhD thesis. Princeton University Press, 2015.

[450] Lars Seemann, Jason Shulman, and Gemunu H. Gunaratne. "A robust topology-based algorithm for gene expression profiling." *ISRN Bioinformatics* 2012 (2012), 381023.

[451] Stefan Semrau and Alexander van Oudenaarden. "Studying lineage decision-making in vitro: emerging concepts and novel tools." *Annu. Rev. Cell Dev. Biol.* 31 (2015), 317–345.

[452] Manu Setty et al. "Wishbone identifies bifurcating developmental trajectories from single-cell data." *Nature Biotechnol.* 34.6 (2016), 637–645.

[453] Paul M. Sharp and Beatrice H. Hahn. "Origins of HIV and the AIDS pandemic." *Cold Spring Harbor Persp. Med.* 1.1 (2011), a006841.

[454] Ann M. Sheehy, Nathan C. Gaddis, and Michael H. Malim. "The antiretroviral enzyme APOBEC3G is degraded by the proteasome in response to HIV-1 Vif." *Nature Med.* 9.11 (2003), 1404–1407.

[455] Donald R. Sheehy. "Linear-size approximations to the Vietoris-Rips filtration." *Discrete Comput. Geom.* 49.4 (2013), 778–796.

[456] Chiaho Shih and Robert A. Weinberg. "Isolation of a transforming sequence from a human bladder carcinoma cell line." *Cell* 29.1 (1982), 161–169.

[457] Rebecca Siegel et al. "Cancer statistics, 2014." *CA Cancer J. Clin.* 64.1 (2014), 9–29. doi: 10.3322/caac.21208.

[458] Vin de Silva and Gunnar Carlsson. "Topological estimation using witness complexes." *Proceedings of the First Eurographics Conference on Point-Based Graphics* , pp. 157–166. Ed. by Marc Alex et al. Eurographics Association, Zurich, 2004.

[459] Vin de Silva, Elizabeth Munch, and Amit Patel. "Categorified Reeb graphs." *Discrete Comput. Geom.* 55.4 (2016), 854–906. doi: 10.1007/s00454-016-9763-9.

[460] Vin de Silva and Joshua B. Tenenbaum. "Global versus local methods in nonlinear dimensionality reduction." *Advances in Neural Information Processing Systems*, volume 15, pp. 705–712. MIT Press, 2002.

[461] Devendra Singh et al. "Transforming fusions of FGFR and TACC genes in human glioblastoma." *Science* 337.6099 (2012), 1231–1235. doi: 10.1126/science.1220834.

[462] Gurjeet Singh, Facundo Mémoli, and Gunnar Carlsson. "Topological methods for the analysis of high dimensional data sets and 3d object recognition." *Eurographics Symposium on Point-Based Graphics*, pp. 91–100. The Eurographics Association, 2007.

[463] Gurjeet Singh et al. "Topological analysis of population activity in visual cortex." *J. Vision* 8.8 (2008), 11–11.

[464] Ann Sizemore et al. "Closures and cavities in the human connectome." 2016. arXiv: 1608.03520 (2016).

[465] J. Skryzalin and G. Carlsson. "Numeric invariants from multidimensional persistence." *J. Appl. Comput. Topology* 1.1 (2017), 89–119.

[466] Fatima Smagulova et al. "Genome-wide analysis reveals novel molecular features of mouse recombination hotspots." *Nature* 472.7343 (2011), 375–378.

[467] Derek J. Smith et al. "Mapping the antigenic and genetic evolution of influenza virus." *Science* 305.5682 (2004), 371–376.

[468] Temple F. Smith and Michael S. Waterman. "Identification of common molecular subsequences." *J. Mol. Biol.* 147.1 (1981), 195–197.

[469] Gordon K. Smyth. "Limma: linear models for microarray data." *Bioinformatics and Computational Biology Solutions using R and Bioconductor*, pp. 397–420. Springer, 2005.

[470] Peter H. A. Sneath, Michael J. Sackin, and Richard P. Ambler. "Detecting evolutionary incompatibilities from protein sequences." *Systematic Biol.* 24.3 (1975), 311–332.

[471] Alexandra Snyder et al. "Genetic basis for clinical response to CTLA-4 blockade in melanoma." *N. Engl. J. Med.* 371.23 (2014), 2189–2199.

[472] American Cancer Society. *Cancer Facts and Figures 2015*. American Cancer Society, Atlanta, GA, 2015.

[473] Ray J. Solomonoff. "A formal theory of inductive inference. Part II." *Inf. Control* 7.2 (1964), 224–254.

[474] A. Solovyov et al. "Cluster analysis of the origins of the new influenza A (H1N1) virus." *Euro Surveill.* 14.21 (2009), 19224.

[475] Jiuzhou Z Song et al. "The wavelet-based cluster analysis for temporal gene expression data." *EURASIP J. Bioinformatics Syst. Biol.* 2007 (2007), 2–2.

[476] Yun S. Song and Jotun Hein. "Constructing minimal ancestral recombination graphs." J. Comput. Biol. 12.2 (2005), 147–169. doi: 10.1089/cmb.2005.12.147.

[477] Yun S. Song, Yufeng Wu, and Dan Gusfield. "Efficient computation of close lower and upper bounds on the minimum number of recombinations in biological sequence evolution." *Bioinformatics* 21.suppl 1 (2005), i413–i422.

[478] Andrea Sottoriva et al. "Intratumor heterogeneity in human glioblastoma reflects cancer evolutionary dynamics." *Proc. Natl. Acad. Sci. USA* 110.10 (2013), 4009–4014.

[479] David Speyer and Bernd Sturmfels. "The tropical grassmannian." *Adv. Geom.* 4.3 (2004), 389–411.

[480] David I. Spivak. *Category Theory for the Sciences*. MIT Press, 2014.

[481] B. St. Thomas et al. "Learning subspaces of different dimensions." 2014. arXiv: 1404.6841.

[482] Alexandros Stamatakis. "RAxML-VI-HPC: maximum likelihood-based phylogenetic analyses with thousands of taxa and mixed models." *Bioinformatics* 22.21 (2006), 2688–2690.

[483] Oliver Stegle, Sarah A. Teichmann, and John C. Marioni. "Computational and analytical challenges in single-cell transcriptomics." Nature Rev. Genet. 16.3 (2015), 133–145.

[484] Dominique Stehelin et al. "DNA related to the transforming gene(s) of avian sarcoma viruses is present in normal avian DNA." *Nature* 260 (1976), 170–173.

[485] J. Claiborne Stephens. "Statistical methods of DNA sequence analysis: detection of intragenic recombination or gene conversion." *Mol. Biol. Evol.* 2.6 (1985), 539–556.

[486] Tim J. Stevens et al. "3D structures of individual mammalian genomes studied by single-cell Hi-C." *Nature* 544.7648 (2017), 59–64.

[487] Karl-Theodor Sturm. "Probability measures on metric spaces of nonpositive curvature." *Heat Kernels and Analysis on Manifolds, Graphs, and Metric Spaces.* Contemporary Mathematics, volume 338. American Mathematical Society, 2003.

[488] Curtis A. Suttle. "Marine viruses major players in the global ecosystem." *Nature Rev. Microbiol.* 5.10 (2007), 801–812.

[489] Andrzej Szymczak et al. "Coronary vessel trees from 3D imagery: a topological approach." *Med. Image Anal.* 10.4 (2006), 548–559.

[490] Andrew Tausz, Mikael Vejdemo-Johansson, and Henry Adams. "JavaPlex: a research software package for persistent (co)homology." *Proceedings of ICMS 2014*, pp. 129–136. Ed. by Han Hong and Chee Yap. Lecture Notes in Computer Science, volume 8592. Springer, 2014.

[491] Dane Taylor et al. "Topological data analysis of contagion maps for examining spreading processes on networks." *Nature Commun.* 6 (2015), 7723.

[492] H. M. Temin and S. Mizutani. "RNA-dependent DNA polymerase in virions of Rous sarcoma virus." *Nature* 226 (1970), 1211–1213.

[493] Howard M. Temin. "The effects of actinomycin D on growth of Rous sarcoma virus in vitro." *Virology* 20.4 (1963), 577–582.

[494] Alan Templeton. "Out of Africa again and again." *Nature* 416.6876 (2002), 45–51.

[495] J. B. Tenenbaum, V. de Silva, and J. C. Langford. "A global geometric framework for nonlinear dimensionality reduction." *Science* 290.5500 (2000), 2319–2323.

[496] Christoph A. Thaiss et al. "Transkingdom control of microbiota diurnal oscillations promotes metabolic homeostasis." *Cell* 159.3 (2014), 514–529.

[497] Christopher M. Thomas and Kaare M Nielsen. "Mechanisms of, and barriers to, horizontal gene transfer between bacteria." *Nature Rev. Microbiol.* 3.9 (2005), 711–721.

[498] Enrico Tiacci et al. "BRAF mutations in hairy-cell leukemia." *N. Engl. J. Med.* 364.24 (2011), 2305–2315.

[499] Itay Tirosh et al. "Dissecting the multicellular ecosystem of metastatic melanoma by single-cell RNA-seq." *Science* 352.6282 (2016), 189–196.

[500] Suxiang Tong et al. "A distinct lineage of influenza A virus from bats." *Proc. Natl. Acad. Sci. USA* 109.11 (2012), 4269–4274.

[501] Brenda Y. Torres et al. "Tracking resilience to infections by mapping disease space." *PLoS Biol.* 14.4 (2016), e1002436.

[502] Csaba D. Toth, Joseph O'Rourke, and Jacob E. Goodman. *Handbook of Discrete and Computational Geometry.* Third edition. CRC Press, 2017.

[503] Valerio Tozzi et al. "Persistence of neuropsychologic deficits despite long-term highly activeantiretroviral therapy in patients with HIV-related neurocognitive impairment: prevalence and risk factors." *JAIDS J. Acquir. Immune Defic. Syndr.* 45.2 (2007), 174–182.

[504] Cole Trapnell et al. "The dynamics and regulators of cell fate decisions are revealed by pseudotemporal ordering of single cells." *Nature Biotechnol.* 32.4 (2014), 381–386.

[505] Barbara Treutlein et al. "Reconstructing lineage hierarchies of the distal lung epithelium using single-cell RNA-seq." *Nature* 509.7500 (2014), 371–375.

[506] Vladimir Trifonov, Hossein Khiabanian, and Raúl Rabadán. "Geographic dependence, surveillance, and origins of the 2009 influenza A (H1N1) virus." *N. Engl. J. Med.* 361.2 (2009), 115–119.

[507] Vladimir Trifonov, Vincent Racaniello, and Raúl Rabadán. "The contribution of the PB1-F2 protein to the fitness of influenza A viruses and its recent evolution in the 2009 influenza A (H1N1) pandemic virus." *PLOS Curr.* (2009), RRN1006.

[508] Vladimir Trifonov et al. "MutComFocal: an integrative approach to identifying recurrent and focal genomic alterations in tumor samples." *BMC Syst. Biol.* 7.1 (2013), 25.

[509] Federico E. Turkheimer et al. "Chromosomal patterns of gene expression from microarray data: methodology, validation and clinical relevance in gliomas." *BMC Bioinformatics* 7.1 (2006), 1.

[510] Katharine Turner, Sayan Mukherjee, and Douglas Boyer. "Persistent homology transform for modeling shapes and surfaces." *Inform. Inference* 3.4 (2014), 310–344.

[511] Katharine Turner et al. "Fréchet means for distributions of persistence diagrams." *Discrete Comput. Geom.* 52.1 (2014), 44–70.

[512] Marc J. Van De Vijver et al. "A gene-expression signature as a predictor of survival in breast cancer." *N. Engl. J. Med.* 347.25 (2002), 1999–2009.

[513] J. Craig Venter et al. "The sequence of the human genome." *Science* 291.5507 (2001), 1304–1351.

[514] Annelien Verfaillie et al. "Decoding the regulatory landscape of melanoma reveals TEADS as regulators of the invasive cell state." Nature Commun. 6 (2015), 6683. doi: 10.1038/ncomms7683.

[515] Roel G. W. Verhaak et al. "Integrated genomic analysis identifies clinically relevant subtypes of glioblastoma characterized by abnormalities in PDGFRA, IDH1, EGFR, and NF1." *Cancer Cell* 17.1 (2010), 98–110.

[516] Nicole Vidal et al. "Unprecedented degree of human immunodeficiency virus type 1 (HIV-1) group M genetic diversity in the Democratic Republic of Congo suggests that the HIV-1 pandemic originated in Central Africa." *J. Virol.* 74.22 (2000), 10498–10507.

[517] C. Villani. *Optimal Transport – Old and New*. Grundlehren der Mathematischen Wissenschaften, volume 338. Springer, Berlin, 2009.

[518] John Wakeley. *Coalescent Theory: An Introduction*. Roberts and Company, 2008.

[519] Jianzhong Wang. "Local tangent space alignment." *Geometric Structure of High-Dimensional Data and Dimensionality Reduction*, pp. 221–234. Springer, Berlin, 2011.

[520] Jiguang Wang et al. "Clonal evolution of glioblastoma under therapy." *Nature Genet.* 48 (2016), 768–776. doi:10.1038/ng.3590.

[521] Lusheng Wang and Tao Jiang. "On the complexity of multiple sequence alignment." *J. Comput. Biol.* 1.4 (1994), 337–348.

[522] Lusheng Wang, Kaizhong Zhang, and Louxin Zhang. "Perfect phylogenetic networks with recombination." *J. Comput. Biol.* 8.1 (2001), 69–78.

[523] Xindan Wang, Paula Montero Llopis, and David Z Rudner. "Organization and segregation of bacterial chromosomes." *Nature Rev. Genet.* 14.3 (2013), 191–203.

[524] Yong Wang and Nicholas E. Navin. "Advances and applications of single-cell sequencing technologies." *Mol. Cell* 58.4 (2015), 598–609.

[525] Zidong Wang et al. "Stochastic dynamic modeling of short gene expression time-series data." *IEEE Trans. NanoBiosci.* 7.1 (2008), 44–55.

[526] Larry Wasserman. *All of Statistics: A Concise Course in Statistical Inference.* Springer, 2010.

[527] A. R. Wattam et al. "PATRIC, the bacterial bioinformatics database and analysis resource." *Nucleic Acids Res.* 42.D1 (2013), D581–D591.

[528] G. A. Watterson. "On the number of segregating sites in genetical models without recombination." *Theor. Population Biol.* 7.2 (1975), 256–276.

[529] R. G. Webster et al. "Evolution and ecology of influenza A viruses." *Microbiol. Rev.* 56.1 (1992), 152–179.

[530] Georg F. Weiller. "Phylogenetic profiles: a graphical method for detecting genetic recombinations in homologous sequences." *Mol. Biol. Evol.* 15.3 (1998), 326–335.

[531] Robert Weinberg. *The Biology of Cancer.* Garland Science, 2013.

[532] Killan Q. Weinberger and Lawrence K. Saul. "An introduction to nonlinear dimensionality reduction by maximum variance unfolding." *Proceedings of the 21st National Conference on Artificial Intelligence, AAAI'06, Boston, MA*, volume 2, p. 1683. AAAI Press, 2006.

[533] Shmuel Weinberger. "The complexity of some topological inference problems." *Found. Comput. Math.* 14.6 (2014), 1277–1285. doi: 10.1007/s10208-013-9152-1.

[534] Shmuel Weinberger. *The Topological Classification of Stratified Spaces.* Chicago Lectures in Mathematics. University of Chicago Press, Chicago, IL, 1994.

[535] Shmuel Weinberger. "What is ... persistent homology?" *Notices AMS* 58.1 (2011), 36–39.

[536] I. Bernard Weinstein and Andrew Joe. "Oncogene addiction." *Cancer Res.* 68.9 (2008), 3077–3080. doi: 10.1158/0008-5472.can-07-3293.

[537] Michael L. Whitfield et al. "Identification of genes periodically expressed in the human cell cycle and their expression in tumors." *Mol. Biol. Cell* 13.6 (2002), 1977–2000.

[538] Carl R. Woese, Otto Kandler, and Mark L. Wheelis. "Towards a natural system of organisms: proposal for the domains Archaea, Bacteria, and Eucarya." *Proc. Natl. Acad. Sci. USA* 87.12 (1990), 4576–4579.

[539] Michael Worobey, Andrew Rambaut, and Edward C. Holmes. "Widespread intra-serotype recombination in natural populations of dengue virus." *Proc. Natl. Acad. Sci. USA* 96.13 (1999), 7352–7357.

[540] Michael Worobey et al. "Direct evidence of extensive diversity of HIV-1 in Kinshasa by 1960." *Nature* 455.7213 (2008), 661–664.

[541] Sewall Wright. "Evolution in Mendelian populations." *Genetics* 16.2 (1931), 97–159.

[542] Katsusaburo Yamagiwa and Koichi Ichikawa. "Experimental study of the pathogenesis of carcinoma." *J. Cancer Res.* 3.1 (1918), 1–29.

[543] D. Yogeshwaran, Eliran Subag, and Robert J. Adler. "Random geometric complexes in the thermodynamic regime." *Prob. Theor. Relat. Fields* 167.1 (2017), 107–142.

[544] Xianghui Yu et al. "Induction of APOBEC3G ubiquitination and degradation by an HIV-1 Vif-Cul5-SCF complex." *Science* 302.5647 (2003), 1056–1060.

[545] Sakellarios Zairis et al. "Genomic data analysis in tree spaces." 2016. arXiv: 1607.07503 [q-bio.GN].

[546] Sakellarios Zairis et al. "Moduli spaces of phylogenetic trees describing tumor evolutionary patterns." 2014. arXiv: 1410.0980 [q-bio.GN].

[547] Dmitriy Zamarin, Mila B. Ortigoza, and Peter Palese. "Influenza A virus PB1-F2 protein contributes to viral pathogenesis in mice." *J. Virol.* 80.16 (2006), 7976–7983.

[548] H. Zhu et al. "Infectivity, transmission, and pathology of human H7N9 influenza in ferrets and pigs." *Science* 341.6142 (2013), 183–186. doi: 10.1126/science .1239844.

[549] Afra Zomorodian. "The tidy set: a minimal simplicial set for computing homology of clique complexes." *Proceedings of the Twenty-sixth Annual Symposium on Computational Geometry, SoCG '10, Snowbird, UT, USA*, pp. 257–266. ACM, 2010. doi: 10.1145/1810959.1811004

[550] Afra J. Zomorodian. *Topology for Computing*. Cambridge University Press, New York, 2009.

[551] Afra Zomorodian and Gunnar Carlsson. "Computing persistent homology." *Discrete Comput. Geom.* 33.2 (2005), 249–274.

[552] Hui Zou, Trevor Hastie, and Robert Tibshirani. "Sparse principal component analysis." *J. Comput. Graph. Stat.* 15 (2004), 2006.

Index